Production diseases in farm animals

Production diseases in farm animals

12th international conference

edited by:
Nanda P. Joshi
Thomas H. Herdt

Wageningen Academic
P u b l i s h e r s

ISBN-10: 90-76998-57-4
ISBN-13: 978-90-76998-57-2

First published, 2006

Wageningen Academic Publishers
The Netherlands, 2006

International Scientific Board

Jens Fredrick Agger, Denmark
Dave Beede, USA
Jurg W. Blum, Switzerland
Jeanne Burton, USA
Paul Coussens, USA
Ronald J. Erskine, USA
Manford Fürll, Germany
Jesse Goff, USA
Thomas H. Herdt, USA (Chairperson)
Ronald Horst, USA
Knut Hove, Norway
Holger Martens, Germany
Barbara Straw, USA
Theo Wensing, The Netherlands

Chairpersons of Recent ICPD Meetings

Fran Kallfelz, USA, 1989
Jurg Blum, Switzerland, 1992
Holger Martens, Germany, 1995
Theo Wensing, The Netherlands, 1998
Jens Agger, Denmark, 2001

Conference Venue

Kellogg Hotel and Conference Center
Michigan State University
East Lansing, MI48824
United States of America

Local Organizing Committee

Dave Beede, MSU
Jeanne Burton, MSU
Paul Coussens, MSU
Ronald Erskine, MSU
Jesse Goff, USDA NADC-Laboratory
Thomas Herdt, MSU (Chairperson)
Ronald Horst, USDA NADC-Laboratory
Nanda Joshi, MSU
Barbara Straw, MSU

Acknowledgements

The ICPD, since it's inception has been dependent on the generosity of institutions and sponsors for its support. The organizing committee would like to recognize the major contributions made by Michigan State University, Pfizer Animal Health, and the United States Department of Agriculture. In addition, other important sponsors include

ABS Global
Balchem Encapsulates
Chr.Hansen Inc.
DSM
Land O'Lakes-Purina Mills
Merial
West Central Soy

With out these sponsors, this conference would not have been possible.

List of previous ICPDs

I. Urbana, Illinois, USA, 1968
II. Reading University, England, 1972
III. Wageningen, The Netherlands, 1976
IV. Munich, Germany, 1980
V. Uppsala, Sweden, 1983
VI. Belfast, Northern Ireland, 1986
VII. Ithaca, New York, 1989
VIII. Bern, Switzerland, 1992
IX. Berlin Germany, 1995
X. Utrecht, The Netherlands, 1998
XI. Copenhagen, Denmark, 2001

Distributions

Copies of the Proceedings may be obtained from:
Wageningen Academic Publishers, the Netherlands.

Copyright

Introduction

The goal of the International Conference on Production Diseases in Farm Animals has always been to unite scientists from many disciplines and many countries in the study of production disease, and the 12th ICPD has met that goal. The breadth of national representation at the ICPD continues to expand, as do the number of scientific disciplines represented. Disciplines such as ethology and molecular genetics, which were little known at the beginning of this conference series, have become important components of the ICPD.

In the preface to the proceedings of the second of these conferences, Dr. Jack Payne defined production disease as *"a number of metabolic disorders of increasing importance in agriculture"*. Over the years, this definition has expanded to include not only metabolic and nutritional diseases, but also many diseases of an infectious and genetic nature. The common theme of all these diseases is their association with management and selection of animals for efficient agricultural production. We now appreciate that this association goes well beyond the important "input", "output", and "throughput" equations of Dr. Payne's. We must now work also to understand the effects of management and selection on a wide variety of factors such as animal behavior, immunity, and gene expression. The complexity of these interactions makes the interdisciplinary nature of the ICPD even more important than before. The themes of animal welfare, health, and productivity in agricultural systems go well beyond "metabolic disorders", but they remain "of increasing importance in agriculture". Thus, the management and prevention of production diseases will become evermore challenging as agricultural systems advance. These challenges have provided rich and productive discussions for these ICPD proceedings, and should provide continued themes for productive ICPD conferences for many years to come.

On behalf of the organizing committee,

T.H. Herdt
Chairperson, 12th ICPD

Table of contents

Introduction   7

The History and Influence of the ICPD   19
Thomas H. Herdt

Section A. Transition cow biology and management

Advances in transition cow biology: new frontiers in production diseases   24
J. K. Drackley

Research priorities from a producer's point of view   35
Walter M. Guterbock

Metabolic profiling to assess health status of transition dairy cows   41
Robert J. Van Saun

A consideration about the energy supply in peripartum of dairy cows on the basis of
change in plasma free amino acid concentration   42
S. Kawamura., K. Shibaro., R. Hakamada., M. Otsuka, S. Sat and H. Hoshi

Metabolic predictors of displaced abomasum in transition dairy cows   43
S. LeBlanc, T. Duffield and K. Leslie

Evaluation of a rapid test for NEFA in bovine serum   44
L. Gooijer, K. Leslie, S. LeBlanc and T. Duffield

Association of rump fat thickness and plasma NEFA concentration with postpartum
metabolic diseases in Holstein cows   45
N.P. Joshi, T.H. Herdt and L. Neuder

Effect of pre-partum feeding intensity on postpartum energy status of Estonian Holstein
cows   46
H. Jaakson, K. Ling, H. Kaldmae, J. Samarütel, T. Kaart and O. Kärt

Using a pooled sample technique for herd metabolic profile screening   47
Robert J. Van Saun

Section B. Metabolic effects of immune mediators

Metabolic effects of immune mediators   50
Kirk C. Klasing

The effect of bovine respiratory disease on carcass traits   51
R.L. Larson

Does calf health during the feedlot period affect gain and carcass traits?   52
L.R. Corah, W.D. Busby, P. Beedle, D. Strohbehn, J.F. Stika

Effects of dexamethasone on mRNA levels and binding sites of hepatic β-adrenergic
receptors in neonatal calves and dependence on colostrum feeding   53
H.M. Hammon, J. Carron, C. Morel, and J.W. Blum

Pathogenesis of avian growth plate dyschondroplasia   55
N.C. Rath, J.M. Balog, G.R. Huff, and W.E. Huff

Concentrations of haptoglobin and fibrinogen during the first ten days after calving in
dairy cows with acute endometritis   56
M. Drillich, D. Voigt, D. Forderung and W. Heuwieser

Extracellular pH alters innate immunity by decreasing the production of reactive
oxygen and nitrogen species, but enhancing phagocytosis in bovine neutrophils and
monocytes   57
D. Donovan

High nitrite/nitrate status in neonatal calves is associated with increased plasma levels
of S-nitrosoalbumin and other S-nitrosothiols   58
I. Cattin, S. Christen, S.G. Shaw and J.W. Blum

Effects of 0.03% dietary β-glucans on nonspecific/specific immunity, oxidative/
antioxidative status and growth performance in weanling pigs   59
S. Schmitz, S. Hiss and Helga Sauerwein

Influence of organic nutrition and housing on selected immunological and metabolic
properties in fattening pigs   61
S. Millet, E. Cox, M. Van Paemel, B.M. Goddeeris and G.P.J. Janssens

Section C. Animal behaviour and welfare in intensive production systems

Associations of cow comfort indices with total lying time, stall standing time, and lameness 64
K.V. Nordlund, N.B. Cook and T.B. Bennett

Interaction of lameness with sand and mattress surfaces in dairy cow freestalls   66
K.V. Nordlund, N.B. Cook and T.B. Bennett

Electric and magnetic fields (EMF) affect milk production and bevavior of cows   68
D. Hillman, C.L. Goeke, D. Stetzer, K. Mathson, M.H. Graham, H.H. VanHorn, C.J. Wilcox

Effect of roughages added to the milk replacer diet of veal calves on behavior and gastric
development   69
A.M. van Vuuren, N. Stockhofe, J.J. Heeres-van der Tol, L.F.M. Heutink and C.G. van Reenen

Transportation stress of cattle by railway   71
Z. Cui, B. Han, G. Gao and D. Cui

Section D. Infection and infectious diseases associated with production systems

Postweaning Multisystemic Wasting Syndrome (PMWS) in pigs   74
Matti Kiupel

Influence of growth conditions on caprine nasal bacterial flora in the presence of
Mannhemia hameolytica   90
D.R. McWhinney and Kingsley Dunkley

The relationship between exposure to bovine viral diarrhea virus and fertility in a
commercial dairy herd   91
W. Raphael, S. Bolin, W. Guterbock, J. Kaneene, and D. Grooms

Tritrichomonas foetus cysteine protease CP30 induces cell death in host cells   92
B.N. Singh, J.J. Lucas, G.R. Hayes, C.E. Costello, U. Sommer and R.O. Gilbert

Immunological approaches to enhanced production in food animals   93
R.J. Yancey, Jr.

Colostrum management in calves: effects of drenching versus bottle feeding   102
M. Kaske, A. Werner, H.-J. Schubert, H.W. Kehler and J. Rehage

Immune system activation increases the tryptophan requirement in post weaning pigs   103
Nathalie Le Floc'h and Bernard Sève

Periparturient negative energy balance and neutrophil function suppression are
associated with uterine health disorders and fever in Holstein cows   104
D.S. Hammon, I.M. Evjen, J.P. Goff, T.R. Dhiman

Activities of enzymes in peripheral leukocytes reflect metabolic conditions in fattening
steers   106
T. Arai, A. Inoue, A. Takeguchi, S. Urabe, T. Sako, I. Yoshimura and N. Kimura

Effects of 0.3% dietary β-glucan on nonspecific/specific immunity, oxidative/
antioxidative status and growth performance in weanling pigs   107
S. Schmitz, S. Hiss and H. Sauerwein

Evaluation of Liver Abscess Incidence in Feedlot Cattle Fed a Dietary Antioxidant
(AGRADO®) across Four Studies   109
M. Vázquez-Añón, F. Scott, B. Miller and T. Peters

Pathogenic *Escherichia coli* in dairy cows held in farms with organic production (OP)
and with integrated production (IP)   110
P. Kuhnert, C. Dubosson, M. Roesch, E. Homfeld, M.G. Doherr and J.W. Blum

Section E. Reproductive health

Rearing conditions and disease influencing the reproductive performance of swedish dairy heifers 112
J. Hultgren

Factors influencing conception rate after synchronization of ovulation and timed artificial insemination 114
B.-A. Tenhagen, C. Ruebesam and W. Heuwieser

The incidence of endometritis and its effect on reproductive performance of dairy cows 116
R.O. Gilbert, S.T. Shin, M. Frajblat, C.L. Guard, H.N. Erb and H. Roman

Use of **Neospora caninum** vaccine in a dairy herd undergoing an abortion outbreak 117
R.W. Meiring and P.J. Rajala-Schultz

Ovulatory cycles, metabolic profiles, body condition scores and their relation to fertility of multiparous Holstein dairy cows 118
J. Samarütel, K. Ling, A. Waldmann, H. Jaakson, A. Leesmäe and T. Kaart

Field trial on blood metabolites, body condition score (BCS) and their relation to the recurrence of ovarian cyclicity in Estonian Holstein cows 119
Katri Ling, Andres Waldmann, Jaak Samarütel, Hanno Jaakson, Tanel Kaart, Andres Leesmäe

In vitro embryo production: growth performance, feed efficiency, health status, and hematological, metabolic and endocrine traits in veal calves 120
M. Rérat, Y. Zbinden, R. Saner, H. Hammon, and J.W. Blum

Dairy farms with organic production (OP) and with integrated production (IP): comparison of management, feeding, production, reproduction and udder health 121
M. Roesch, E. Homfeld, M.G. Doherr and J.W. Blum

Milk production, nutritional status, and fertility in dairy cows held in organic farms and in farms with integrated production 122
E. Homfeld, M. Roesch, M.G. Doherr and J.W. Blum

Oviductal prolapse led to more than eleven percent of hens dead in a highly inbred line of white leghorn chickens 123
H.M. Zhang, A.R. Pandiri and G.B. Kulkarni

Section F. Epidemiology of production diseases

Epidemiology of subclinical production diseases in dairy cows with an emphasis on ketosis 126
T.F. Duffield

Associations of cow and management factors with culture positive milk samples 136
R. van Dorp and D. Kelton

A mathematical model for the dynamics of digital dermatitis in groups of cattle to study
the efficacy of group-based therapy and prevention strategies   137
D. Dopfer, M.R. van Boven, and M.C.M. de Jong

Diagnostic test evaluation without gold standard using two different bayesian
approaches for detecting verotoxinogenic *Escherichia coli* (VTEC) in cattle   138
D. Dopfer, L. Geue, W. Buist and B. Engel

A relative comparison of diagnostic tests for bovine paratuberculosis (Johne's disease)   140
S.H. Hendrick, T.F. Duffield, D.F. Kelton, K.E. Leslie, K. Lissemore and M. Archambault

Etiology, pathophysiology and prevention of fatty liver in dairy cows   141
R.R. Grummer

Relationships between mild fatty liver and health and reproductive performance in
Holstein cows   154
G. Bobe, B.N. Ametaj, R.A. Nafikov, D.C. Beitz and J.W. Young

Prevention of fatty liver in transition dairy cows by subcutaneous glucagon injections   155
R.A. Nafikov, B.N. Ametaj, G. Bobe, K.J. Koehler, J.W. Young and D.C. Beitz

Clinical relevance and therapy of fatty liver in cows   156
M. Fürll, H. Bekele, D. Röchert, L. Jäckel, U. Delling and Th. Wittek

The inflammation could have a role in the liver lipidosis occurrence in dairy cows   157
G. Bertoni, Trevisi Erminio, Calamari Luigi and Bionaz Massimo

Effect of propylene glycol on fatty liver development and hepatic gluconeogenesis in
periparturient dairy cows   159
T. Rukkwamsuk, Apassara Choothesa, Sunthorn Rungruagn and Theo Wensing

The report of classical swine fever in Romania between april 2001 and april 2003   161
C. Pascu, V. Herman and N. Catana

Epidemiological survey on the downers in Korea   163
H.R. Han, C.W. Lee, H.S. Yoo, B.K. Park, Bo Han and D. Kim

Section G. Alternatives to growth promoting antibiotics

Alternatives to antibiotic feed additives   166
Chad R. Risley and John Lopez

Continuous use of a dry disinfectant/antiseptic to enhance health and well being in
food-animal production facilities   178
Bud G. Harmon

Bovine colostrum as an alternative to feed additive in weaning diet improves gut health
of piglets   179
I. Le Huërou-Luron, A. Huguet and J. Le Dividich

Production responses from antimicrobial and rendered animal protein inclusions in
swine starter diets 180
B.V. Lawrence, S.A. Hansen and D. Overend

Section H. Macromineral metabolism and production diseases

Strategies for controlling hypocalcemia in dairy cows in confinement and pasture settings 182
Jorge M. Sánchez and Jesse P. Goff

Phosphorus digestion and metabolism in ruminants: applications to production disease
and environmental considerations 188
Ernst Pfeffer

Prevalence of subclinical hypocalcemia in U.S. dairy operations 215
R.L. Horst, J.P. Goff and B.J. McCluskey

The effect of low phosphorus intake in early lactation on apparent digestibility of
phosphorus and bone metabolism in dairy cows 216
A. Ekelund, R. Spörndly and K. Holtenius

Bone metabolism of milk goat and sheep during pregnancy and lactation 218
A. Liesegang and Juha Risteli

Influence of starvation on fermentation in bovine rumen fluid (*in vivo*) 220
M. Höltershinken, A. Höhling, N. Holsten and H. Scholz

Serum mineral concentrations and periparturient disease in Holstein dairy cows 221
Robert J. Van Saun, Amy Todd and Gabriella Varga

The relevance of hypophosphataemia in cows 223
M. Fürll, T. Sattler and M. Hoops

Section I. Micromineral nutrition, metabolism and homeostasis

Trace element nutriture and immune function 226
Jerry W. Spears

Expected changes of the selenium content in foods of animal origin at a change from
inorganic to organic selenium compounds for supplementation of the diet of farm animals 236
B.G. Pehrson

Long acting injectable remedies to prevent cobalt and selenium deficiencies in grazing
lambs and calves 238
N.D. Grace and S.O. Knowles

Parenteral and oral selenium supplementation of weaned beef calves 240
W.S. Swecker, Jr., R.K. Shanklin, K.H. Hunter, C.L. Pickworth, G. Scaglia, and J.P. Fontenot

Selenium yeast prevents nutritional muscular degeneration (NMD) in nursing calves 241
B.G. Pehrson

Section J. Role of specific fatty acids in animal health, reproduction, and performance

Using conjugated linoleic acids for healthier animal products and as a management tool 244
M.A. McGuire, E.E. Mosley, S.A. Mosley, A. Nudda and A. Corato

Fatty acid composition of phospholipids and levels of alpha-tocopherol, total antioxidative capacity and malondialdehyde in liver and muscular tissue after dietary supplementation of various fats in cattle 255
J. Rehage, O. Portmann, R. Berning, M. Kaske, W. Kehler, M. Hoeltershinken, R. Duehlmeier, M. Coenen and H.-P. Sallmann

A field sample study investigating possible indicators of undernutrition in cattle 256
S. Agenäs, M.F. Heath, R.M. Nixon, J.M. Wilkinson and C.J.C. Phillips

Section K. Mastitis

Activation of immune cells in bovine mammary gland secretions by zymosan treated bovine serum 260
K. Kimura and Jesse P. Goff

The effect of milk yield at dry-off on the likelihood of intramammary infection at calving 262
P.J. Rajala-Schultz, K.L. Smith and J.S. Hogan

Mastitis therapy for persistent **Escherichia coli** on a large dairy 263
P.M. Sears, C.E. Ackerman, K.M. Crandall and W.M. Guterbock

Diagnostic key data for lactate dehydrogenase activity measurements in raw milk for the identification of subclinical mastitis in dairy cows 264
A. Neu-Zahren, U. Müller, S. Hiss and H. Sauerwein

Evaluation of leukocyte subset for occurrence of mastitis on dairy herd 265
H. Ohtsuka, M. Kohiruimaki, T. Hayashi, K. Katsuta, R. Abe and S. Kawamura

Antimicrobial treatment strategies for Streptococcal and Staphylococcal mastitis 266
K.D. Crandall, Philip M. Sears and Walter M. Guterbock

Subclinical mastitis in dairy cows in farms with organic and with integrated production: prevalence, risk factors and udder pathogens 267
M. Roesch, E. Homfeld, M.G. Doherr, W. Schaeren, M. Schällibaum, J.W. Blum

Evaluation of a novel on-farm test for antibiotic susceptibility determination in mastitis pathogens 268
B.C. Love and P. Rajala-Schultz

Section L. Rumen digestion and metabolism

Physically effective fiber and regulation of ruminal pH: more than just chewing 270
M.S. Allen, J.A. Voelker and M. Oba

High dietary cation difference induces a state of pseudohypoparathyroidism in dairy cows resulting in hypocalcemia and milk fever 279
J.P. Goff and R.L. Horst

Effects of two different dry-off strategies on metabolism in dairy cows 280
M. Odensten, K. Person Waller and K. Holtenius

Ruminal pH, concentrations and post-feeding pattern of VFA and organic acids in cows experiencing subacute ruminal acidosis 281
G.R. Oetzel and K.M. Krause

Characterization of the Na^+/Mg^{2+} exchanger as a major Mg^{2+} transporter in isolated ovine ruminal epithelial cells 282
M. Schweigel, B. Etschmann, F. Buschmann, H.S. Park and H. Martens

High potassium diet, sodium and magnesium in ruminants: the story is not over 284
F. Stumpff, I. Brinkmann, M. Schweigel and H. Martens

Functional characterization of the time course of rumen epithelium adaptation to a high energy diet 286
B. Etschmann, A. Suplie and H. Martens

Characterization of an ovine vacuolar H^+-ATPase as a new mechanism for the energization of ruminal transport processes 288
M. Schweigel, B. Etschmann, E. Froschauer, S. Heipertz and H. Martens

The absorptive capacity of sheep omasum is modulated by the diet 289
O. Ali, H. Martens and C. Wegeler

Section M. Application of genomics to production diseases

Microarray analysis of bovine neutrophils around parturition: implications for mammary gland and reproductive tract health 292
Jeanne L. Burton, Sally A. Madsen, Ling-Chu. Chang, Kelly R. Buckham, Laura E. Neuder and Patty S.D. Weber

Functional genomics analysis of bovine viral diarrhea virus-infected cells: unraveling an enigma 307
John D. Neill and Julia F. Ridpath

Early postpartum ketosis in dairy cows and hepatic gene expression profiles using a bovine cDNA microarray 308
J.J. Loor, H.M. Dann, D.E. Morin, R.E. Everts, S.L. Rodriguez-Zas, H.A. Lewin and J.K. Drackley

Genetic improvement of dairy cattle health   309
J.B. Cole, P.D. Miller, and H.D. Norman

Parturition-induced gene expression signatures in bovine peripheral blood
mononuclear cells   310
L-C. Chang, R. van Dorp, P.S.D. Weber, K.R. Buckham, and J.L. Burton

Mechanisms of glucocorticoid-induced L-selectin (CD62L) down-regulation in bovine
blood neutrophils   311
Patty S.D. Weber, Ling-Chu. Chang, Trine Toelboell and Jeanne L. Burton

T-cell receptor Vb gene repertoire analysis of the mammary gland T-cells on the
Staphylococcus aureus causing bovine mastitis   312
T. Hayashi, H. Ohtsuka, M. Kohiruimaki, K. Katsuda, Y. Yokomizo, S. Kawamura and R. Abe

A new sensitive microarray for studying metabolic diseases in cattle   313
B.E. Etchebarne, W. Nobis, M.S. Allen, and M.J. VandeHaar

Ontogenetic development of mRNA levels and binding sites of hepatic beta-adrenergic
receptors in cattle   314
J. Carron, H.M. Hammon, C. Morel and J.W. Blum

Bovine lactase messenger RNA levels determined by real time PCR   315
E.C. Ontsouka, B. Korczack, H.M. Hammon and J.W. Blum

Design and application of a bovine metabolism long oligonucleotide microarray   316
B.E. Etchebarne, W. Nobis, K.J. Harvatine, M.S. Allen, P.M. Coussens and M.J. VandeHaar

Meeting summaries

Meeting summary and synthesis: new scientific directions for production medicine
research   332
J.K. Drackley

Meeting summary and synthesis: practical applications and new directions for applied
production medicine research   337
Walter M. Guterbock

Author index   341

The History and Influence of the ICPD

Thomas H. Herdt
Department of Large Animal Clinical Sciences, College of Veterinary Medicine, Michigan State University, East Lansing, MI 48823, USA

The International Conference on Production Diseases in Farm Animals (ICPD) began at the University of Illinois in 1968 with a conference on parturient hypocalcemia of dairy cows. The initial conference brought together by invitation researchers who were active in the investigation of this important metabolic disease of dairy cows. The success of that initial meeting spawned the creation of the ICPD, with the first meeting by that name occurring in England and hosted by Reading University in 1972. At that meeting the theme was broadened to include a variety of topics related to production diseases.

The stated objective of the 2nd ICPD was to *"enable scientists of many disciplines such as nutrition, biochemistry, physiology, endocrinology, radiobiology and veterinary science to meet and make a unified approach to the solution of production disease"*. This has remained the objective of subsequent ICPD meetings, which have been held at regular intervals as listed in Table 1.

Topics of major interest throughout the series of ICPD meetings have been:

* the classical metabolic/deficiency diseases such as hypocalcemia, ketosis, fatty liver, and nutritional myopathy;
* the relationship between nutrition and fertility;
* the relationship between nutrition and immunity;
* the modification of animal production efficiency by feed additives, antibiotics and hormones;
* welfare and animal behavior in intensive animal production systems.

Table 1. Locations and dates of previous International Conferences on Production Disease in Farm Animals.

Year	ICPD #	Location
1968	I	Urbana-Champaign, Illinois
1972	II	Reading, England
1976	III	Wageningen, The Netherlands
1980	IV	Munich, Germany
1983	V	Uppsala, Sweden
1986	VI	Belfast, Northern Ireland
1989	VII	Cornell, New York
1992	VIII	Bern, Switzerland
1995	IX	Berlin, Germany
1998	X	Utrecht, The Netherlands
2001	XI	Copenhagen, Denmark
2004	XII	East Lansing, Michigan

The continued importance of the classical metabolic diseases in ICPD discussions is shown in Figure 1, which illustrates over the course of the thirteen ICPD meetings the number of papers presented on topics related to electrolyte, macro-mineral, energy and protein metabolism, and associated diseases. The expansion of ICPD discussions into new and overlapping areas is illustrated in Figures 2 and 3, which show the increasing emphasis on such topics as immunity and immunophysiology, genetics and genomics, ethology, and epidemiology in the study of production diseases. These dynamics in program content demonstrate the vibrant and contemporary nature of the ICPD as it continues to promote the incorporation of even broader disciplines and modes of investigation into the resolution of production diseases in farm animals.

The growth of the ICPD and its intensely international character is demonstrated in Figure 4 and Table 2, respectively. The ICPD is the premier international and interdisciplinary conference for the discussion of production disease across all livestock species. Other conferences have a more limited focus on either animal production or animal disease, and typically attract primarily either animal scientists or veterinary scientists. The ICPD, in contrast, creates a unique cross section of many disciplines and backgrounds.

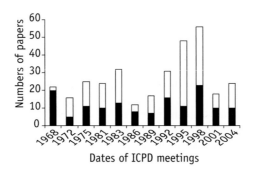

Figure 1. Numbers of ICPD papers addressing topics related to metabolism, or metabolic disease. White indicates electrolyte and macro-mineral metabolism, black indicates energy and protein metabolism.

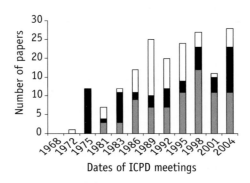

Figure 2. Number of ICPD papers addressing the relationship of metabolic modifiers (white), genetics and genomics (black) and immunity and immunomodulators (grey) to metabolism and production diseases.

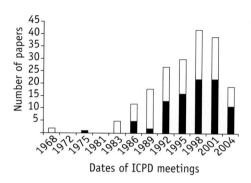

Figure 3. Numbers of ICPD papers addressing the topics of economics and epidemiology of production diseases (white) or the relationship of animal management, behavior, or welfare to production disease (black).

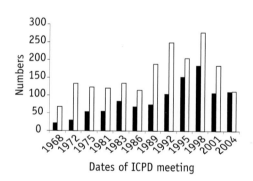

Figure 4. Numbers of papers (black) and attendees (white) at past ICPD meetings.

Table 2. Countries having one or more participants at either of the two most recent ICPD meetings.

Albania	Germany	Poland
Argentina	Hungary	Romania
Belgium	India	Russia
Bulgaria	Iran	Slovak Republic
Canada	Ireland	Slovenia
China	Israel	Sweden
Costa Rica	Italy	Switzerland
Croatia	Japan	Thailand
Czech Republic	Kazakhstan	The Netherlands
Denmark	Libya	Turkey
Egypt	Lithuania	Ukraine
Estonia	New Zealand	United Kingdom
Finland	Norway	Uruguay
France	Pakistan	USA

Section A.
Transition cow biology and management

Advances in transition cow biology: new frontiers in production diseases

J. K. Drackley
Department of Animal Sciences, University of Illinois, Urbana, IL 61801, USA

Abstract

Emphasis should be returned to the interaction of demands for productivity with the impacts of suboptimal management systems on animals as the root cause of production diseases. Much has been learned about the temporal regulation of feed intake and metabolism in liver and adipose tissue of dairy cows during the periparturient period. However, triggers for production disease likely lie at the interface of environmental stressors and productivity. Frontiers for further research progress may lie in the areas of "psychoneuroimmuno-metabolism", the "new biology" of adipose tissue and obesity, and applications of the various "–omics" technologies for high-throughout screening of expression of genes and proteins. Further progress demands a blurring of traditional disciplines such as nutrient metabolism, immunology, stress physiology, and neuroendocrine physiology, along with continued interaction with those clinicians and scientists working at the farm level.

Keywords: psychoneuroimmunology, cytokines, stress, intermediary metabolism, genomics

Introduction

The continued success of the International Conference on Production Diseases (ICPD) unfortunately indicates that production diseases remain a critical problem for the livestock industry. The first conference in Urbana, Illinois in 1968 was focused on parturient hypocalcemia in dairy cows. During the 10 additional conferences between its inception and the current 12th meeting, the scope of the conference has grown substantially to address other diseases and disorders in dairy cows, as well as production diseases in other livestock species. There is no doubt that considerable progress has been made in unraveling the etiology and pathology of production diseases. However, the fundamental question of what causes production diseases remains unanswered.

It is an honor and privilege to be asked to present the keynote address at the 12th ICPD. My objective is to highlight some significant recent advances in understanding the causes of production disease and to point out remaining challenges from a biological scientist's perspective. I also discuss three areas that may constitute "frontiers" of basic science investigation that may shed light on development of production disease. Because of my background and research focus my comments will be centered on production diseases of dairy cows; however, many of the concepts can easily be extrapolated to other species as well.

"Production disease" revisited

The concept of production diseases was defined by J. M. Payne (Payne, 1972) as *"... a group of conditions hitherto known as metabolic disorders, some of which are well known but others less so. All of them are due to imbalances between input and output, or, in other words, to inadequate intake of the various nutrients needed for production. In broader terms, production disease is a man-made problem; it consists of a breakdown of the various metabolic systems of the body under the combined strain of high production and modern intensive husbandry."*

The framing of production disease as a "man-made" problem was important in subsequent development of research. At the 1976 ICPD, Ekesbo (1976) discussed the impact of intensive agriculture and the resulting environment of the animal on development of production disease, which he expanded to include various economically important infectious diseases such as mastitis in dairy cows, respiratory disease in calves, and enteritis in pigs. The implicit links between demands for productivity and the constraints of intensive management are qualitatively attractive and deserve further attention, as discussed in the next sections. That these links are real is evidenced by the fact that, for example, common production diseases that affect dairy cows, such as ketosis, fatty liver, and displaced abomasum, rarely if ever affect beef cows on pasture.

Thus, a framework was created for defining production disease broadly to include infectious and non-infectious disease situations that have their roots in the interaction between the animal's physiology and the environment in which it lives. The diversity introduced by this broad definition is easily grasped by examining the topics of papers presented at this meeting. In addition to the traditional metabolic disorders in dairy cows (milk fever and ketosis), production disease now can be considered to encompass subclinical acidosis, retained placenta, endometritis and other reproductive disorders, and feet and leg abnormalities inducing lameness, such as digital dermatitis and laminitis.

As discussed by Reid in his keynote address at the 1989 ICPD (Reid, 1989), the concept of production disease has been useful in providing a framework for research into the interactions among nutritional, metabolic, and environmental stressors that ultimately lead to production disease. The complex interactions and associated relationships among health disorders in periparturient dairy cows that have been identified in epidemiological studies (*e.g.*, Curtis *et al.*, 1985; Markusfeld, 1987; Correa *et al.*, 1993; Emmanuelson *et al.*, 1993; Peeler *et al.*, 1994) emphasize that these factors cannot be studied in isolation. The concept of production disease has promoted interdisciplinary research among nutritionists, clinicians, physiologists, biochemists, immunologists, and pathologists. However, this interaction must continue to expand and be nurtured by administrators of research institutions if further progress is to be made. Moreover, the best new disciplines of integrative biology should be brought to bear on the problem of production diseases.

Does high milk yield cause production disease?

It is noteworthy that Payne's 1972 indictment of production disease as a man-made problem did not implicate "strain of high milk production" as a primary cause of production disease, but only one component of a two-way interaction. Nevertheless, the push to attain high milk yield has been a common evil invoked by many to explain production diseases over the subsequent three decades. In general, however, the scientific literature describing the epidemiological relationships between high milk yield and increased production disease is inconclusive at best. The question has recently been addressed in an elegant review by Ingvartsen *et al.* (2003). In their review, mastitis was the only disease whose incidence clearly increased with increasing milk yield. Despite the fact that available data were inconclusive, Ingvartsen *et al.* (2003) concluded that continued selection for milk yield *"will continue to increase lactational incidence rates"* for ketosis and lameness.

Studies comparing health data between lines of dairy cows selected for high milk yield and control lines maintained at average milk yield generally show higher health care costs for the high-yielding cows (*e.g.*, Dunklee *et al.*, 1994; Jones *et al.*, 1994). Hansen (2000) argued that selection for body traits, especially those related to udder conformation, body size, and angularity, may be placing cows at greater risk for production disease. For example, placing favorable emphasis on cows that appear sharper might result in cows that are more prone to metabolic problems such as ketosis and displaced abomasum.

On the other hand, selection for high milk yield must also concurrently increase the ability of support systems (*e.g.*, gastrointestinal tract, liver, adipose tissue, skeletal muscle) to provide and process nutrients for the mammary gland to manufacture high milk yield. This principle is embodied in the concepts of "homeorhesis" as described by Bauman and Currie (1980), in which needs for the developing late-term fetus or copious milk secretion in early lactation assume top priority for available nutrients. Adaptations among all these systems ensure that the priorities of the animal are met while still maintaining homeostasis. Given these principles, it is illogical that high milk yield *per se* would result in the various metabolic systems being more likely to fail and cause production disease. Rather, it is more likely that focused genetic selection has made the animal more susceptible to environmental influences that may impinge on the ability of the coordinated systems to maintain homeostasis in the face of management deficiencies or environmental stressors. In other words, failure of the adaptive mechanisms may lead to production diseases.

As pointed out by Ingvartsen *et al.* (2003), the greatest incidence of production diseases is very early after calving and occurs concurrently with the greatest rate of increase (acceleration) of daily milk yield, not with peak milk yield itself. The rate of acceleration of milk yield is greater for higher yielding cows than lower yielding cows. Peak disease incidence also coincides with the nadir in energy balance and the greatest rate of body fat mobilization, as reflected in peak NEFA concentrations in blood. Adaptations to stressors and immune responses to pathogens are responses by the animal to potential threats to homeostasis, and require diversion of scarce nutrients to other systems that must deal with the stressors or immune processes. If individual stressors are severe enough, or in the face of multiple stressors, such responses may become the highest priority for the animal in order to ensure survival. Given the exquisite matching of nutrient demands and supply in healthy high-producing dairy cows, it is reasonable to expect that stressors which place additional nutrient or adaptive demands on the cow may disrupt her ability to maintain homeostasis. Thus, the conceptual view that high-producing cows are more susceptible to environmental insults that in turn disturb homeostasis and cause production disease is more attractive than the notion that production disease is caused by high production *per se.*

Energy balance, dry matter intake, and production disease

In this context, therefore, factors affecting energy balance (and balance of other nutrients as well) likely are highly related to development of production disease. While body reserve mobilization and negative balances of energy and nutrients are part of the normal adaptation to lactation in most mammalian species (Friggens, 2003), extreme deficits likely contribute to increased susceptibility to disease either directly or indirectly. Energy balance during late gestation is largely a factor of dry matter intake (DMI), because the variation in energy requirements is relatively small. An exception to this statement may be in cows carrying twins, which are notoriously susceptible to production diseases.

After parturition, available data are consistent that the extent of postparturient energy balance is more highly correlated with DMI than with milk yield. For example, Figure 1 demonstrates that the yield of solids-corrected milk during week 3 postpartum was not related to energy balance. In contrast, the relationship between DMI and energy balance in the same cows is strong (Figure 2). Thus, nutritional, environmental, infectious, or management-imposed limitations on dry matter intake increase the extent of negative energy balance and may predispose the cow to disease. Similar scenarios also might be created for nutrient drains after parturition in swine.

Given the importance of postpartum DMI in determining the extent of negative energy balance, knowledge of the regulation of DMI is important. Considerable progress has been made in recent years in furthering the understanding of feed intake regulation by periparturient cows, and excellent current reviews are available (Allen, 2000; Ingvartsen and Andersen, 2000). Diet formulation during the dry period impacts periparturient DMI (Hayirli *et al.*, 2002, 2003). Based on classic research by Bertics *et al.* (1992), much emphasis was placed on maximizing prepartum DMI as a tool to avoid postpartum problems. However, research in our laboratory (Grum *et al.*, 1996; Douglas *et al.*, 1998; Dann *et al.*, 2003) as well as by other groups (Kunz *et*

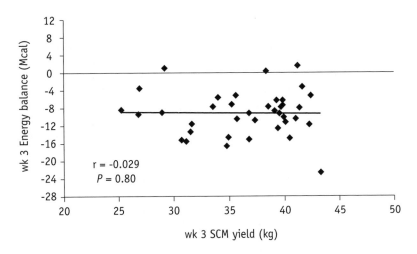

Figure 1. Relationship between enegy balance and yield of solids-corrected milk (SCM) for individual cows during week 3 postpartum. Unpublished analysis from data in Drackley et al. *(1998).*

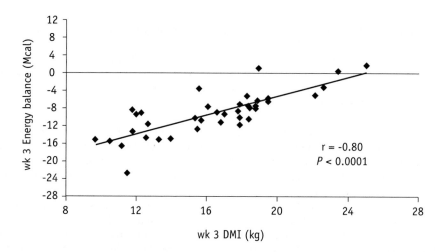

Figure 2. Relationship between energy balance and DMI during week 3 postpartum for individual cows. Unpublished analysis from data in Drackley et al. *(1998).*

al., 1985; Holcomb *et al.*, 2001; Agenäs *et al.*, 2003) has shown that postpartal DMI actually may be improved by moderate feed restriction or limit feeding before parturition. Recently, we (Drackley, 2003a) and Grummer's group (unpublished data, 2003) have demonstrated that it is more likely the constancy of prepartum DMI, or preventing large prepartal decreases in DMI, that is more important in preventing lipid accumulation in the liver and in improving postpartal DMI. High-starch pre-calving diets may accentuate these prepartal decreases in DMI (Rabelo *et al.*, 2003). A major recent advance has been the quantitative description of nutrient supply from the portal-drained viscera and its modification by the liver of cows during the peripartal period (Reynolds *et al.*, 2003).

In response to the inevitable negative energy and nutrient balance after parturition, dairy cows like other mammals undergo several classic adaptations. These include mobilization of body lipid stores as NEFA, decreased oxidative use of glucose, increased gluconeogenesis, increased mobilization of muscle protein, and shuttling of peripheral carbon as alanine to the liver for glucose synthesis. Much has been learned over the last two decades about the temporal regulation of these pathways, which has certainly been a major advance in our understanding of how production diseases develop. While not dealt with here, the topics have been extensively reviewed elsewhere (Grummer, 1993, 1995; Drackley, 1999; Bell *et al.*, 2000; Herdt, 2000; Drackley *et al.*, 2001).

Environmental and infectious stressors and production disease

Payne's definition of production disease as a "man-made problem" consisting of "*a breakdown of the various metabolic systems of the body under the combined strain of high production and modern intensive husbandry*" (Payne, 1972) likely still holds the key to understanding and preventing production diseases. Earlier conferences brought to light the likelihood that environmental stressors present in intensive management systems by definition play a role in the etiology of production diseases. However, research attention in this area seems to have lagged behind other avenues of pursuit. It is likely that too much emphasis has been given to narrow aspects of the topic, such as focusing on precalving diet formulation, individual nutrient deficits, or environmental pathogen infectivity rather than focusing on the interactions of the high-producing animal with stressors in its environment. Viewed in this way, then, a high producing animal may be more susceptible to a "trigger" for disease, whether infectious or environmental, when management and nutrition are not optimal.

In recent years, there has been an explosion of research in integrative disciplines that are tackling complex human diseases, in many cases using animal models. The emerging understanding of the complex multidirectional interactions among pathways of nutrient metabolism, the immune system, the central nervous system, and the neuroendocrine system has been instrumental in recent progress in elucidating the etiology of production diseases or, at the very least, creating new hypotheses that can be tested. An example of such interactions is the recent accumulation of evidence that retained placenta in dairy cattle may be the result of a failure of the immune system to recognize the placenta as "foreign" and trigger the expulsion process (Kimura *et al.*, 2002). Thus, nutritional or environmental factors that impinge on the ability of the immune system to carry out this function lead to occurrence of retained placenta and its undesirable sequelae.

Increasing evidence demonstrates that traditional disciplinary boundaries drawn among nutrient metabolism, stress physiology, immunology, and endocrine physiology are artificial and thus may impede progress in understanding complex phenomena such as production disease. Capitalizing on knowledge gained from medical research may open important new avenues for investigation into production diseases of animals, and finally enable true progress to be made in understanding the interaction of the "strain of high milk production" with the stressors possible in "modern intensive husbandry". This topic has been introduced previously by the author (Drackley, 1999; Drackley *et al.*, 2001).

Frontiers in basic sciences for investigation of the biology of production diseases

Based on the principles discussed above, in my opinion there are at least three major frontiers that should be addressed with greater effort by researchers interested in production diseases. These three frontiers lie in areas I will refer to as "psychoneuroimmuno-metabolism", the "new biology" of adipose tissue, and the "age of the -omics". Each is discussed in the following sections.

"Psychoneuroimmuno-metabolism"

The *"combined strain of high production and modern intensive husbandry"* as origin of production disease naturally lends itself to principles of stress physiology. Stress responses, inflammatory responses, and classic immune responses share in common many of the same mediators and response factors. In recent years, acceptance and understanding of the functioning of the neuroendocrine and immune systems as a single sensory unit has led to development of a new field called "psychoneuroimmunology" (Kelley, 2001; 2004). Effects of cytokines released in response to these divergent stimuli, including neurogenic stress (Black, 2002), are broadly similar to many responses observed in various production diseases. Such effects have recently been demonstrated in dairy cows administered endotoxin by Waldron *et al.* (2003a, b). Evidence for involvement of inflammation in development of fatty liver and possibly other production disease-related pathologies has been presented by several groups (*e.g.*, Bertoni *et al.*, 1996, 1998, 2001; Ametaj *et al.*, 2002; Calamari *et al.*, 2002; Janovick *et al.*, 2004). Because of the effects of these factors on nutrient metabolism and maintenance of the appropriate homeorhetic state, one could coin the term "psychoneuroimmuno-metabolism" to connote the interactions among the central nervous system, neuronendocrine system, immune system, and metabolic regulation. Such interactions are an important arena for further investigation relative to production disease.

Dairy cows during the periparal period potentially may be subjected to a host of stressors, including the inevitable metabolic adaptive stressors but also heat and cold stress, overcrowding, changes in grouping and social structure, changes in housing, uncomfortable stalls or footing, infectious challenges, and mycotoxins, among others. Physiological mediators of responses to these stressors include the classic hypothalamic-pituitary-adrenal axis leading to glucticoid release, norepinephrine from the sympathetic nervous system, epinephrine in acute stress responses, and proinflammatory cytokines. Recent discoveries that adipose tissue, muscle, and liver can elaborate cytokines has raised new questions about their release in response to energy or nutrient intake and environmental stressors. Responses to these stressors (discussed by Drackley, 2003b) may include decreased DMI, which may be a primary factor in creating abnormally negative energy balance. Other potential responses may include diversion of scarce nutrients away from milk synthesis and maintenance, as discussed earlier, interference with lactogenic hormones, increased body fat mobilization, further suppression of the immune system, and hypocalcemia.

One of the continued challenges of this line of research is to identify appropriate "currencies" to quantify effects of stressors (Ingvartsen *et al.*, 2003). An example of recent progress in this area is the report by Holtenius *et al.* (2004), in which it was found that cows in herds with high incidence of mastitis had higher postpartum concentrations of NEFA and lower glutamine concentrations than cows in low-incidence herds. Moreover, the authors stated that "...*there were differences between the two types of herds in management variables...*" that perhaps could explain these finding based on differences in stressor load.

The field of psychoneuroimmunology is an integrative discipline, and emphasizes interactions of disease agents with other stressors faced by animals (Kelley, 2004). Such principles seem ideally suited to tackling the complex nature of production diseases in livestock.

The "new biology" of adipose tissue

In the last few years it has become apparent that adipose tissue is not an inert storehouse of energy in the form of fat, but rather is a key metabolic tissue in communication with the rest of the body and brain via nervous innervation and elaboration of numerous proteins that act locally and/or in distant tissues. The discovery of leptin was the beginning of a long line of signaling molecules since discovered. The study of the human problems of obesity and type II diabetes has provided exciting discoveries on regulation of metabolism in adipose tissue and liver, and potential roles of insulin resistance and cytokines in such pathologies (*e.g.*, Lewis *et al.*, 2002).

It has long been known that overconditioned or fat cows are highly susceptible to a complex of metabolic disorders and infectious diseases after calving. Fat cows have poor appetites, mobilize large amounts of NEFA from adipose tissue triglycerides, and accumulate triglyceride in the liver.

However, we and others are accumulating evidence that cows that consume excesses of energy over their requirements during the dry period, even if not overconditioned, display many of the same metabolic characteristics as cows that are too fat. Our recent work (Dann, 2004) showed that cows that overconsumed energy throughout the early dry period had signs of insulin resistance and changes consistent with greater likelihood for deposition of triglyceride in liver. These cows also had lower DMI and milk yield during the first 10 days postpartum. Moreover, cows that were overfed during the dry period, although not different immediately after calving from cows fed restrictedly during the dry period, were more susceptible to a feed-restriction induced ketosis beginning at day 5 postpartum (Dann, 2004). Others also have shown that overfeeding in the dry period results in insulin resistance (Holtenius *et al.*, 2003). These data fit well with the Holtenius classification of type I and type II ketosis, as reviewed recently by Herdt (2000). Development of theories and concepts from the "new biology" of adipose tissue in relation to feeding diets that are excessively energy dense thus has the potential for furthering the understanding of production diseases.

The "age of the –omics"

Metabolic adaptations to physiological state and to external stressors involve changes in gene transcription and the translation of mRNA to proteins. The last few years has seen an explosion of new research tools for study of global gene expression patterns in different physiological states (functional genomics). DNA microarrays have been constructed using sequences derived from large-scale projects expressed sequence tags. Coupled with efforts to sequence and map the genomes of major agricultural species, the ability now exists to determine the relative expression of large portions of the animal's genome. The importance of this capability lies not so much in the idea that some animals may be genetically predisposed to production diseases (although that certainly is a possibility), but rather in the ability to determine environmental (*e.g.*, nutritional, thermal, behavioral, and microbial) impacts on the degree to which genes or groups of genes are expressed during key stages of the production cycle (*e.g.*, early lactation).

Examples of the potential power of functional genomics to transform our understanding of production diseases are found in presentations by Burton *et al.* and Loor *et al.* in this proceedings. In data from our group (Loor *et al.*, 2004 and this volume), we have shown that high-energy diets fed during the dry period result in marked changes in gene expression patterns in liver, and that cows with an induced ketosis also have greatly altered expression patterns in liver.

When coupled with large-scale characterization of protein translation (proteomics) and changes in metabolite pools (metabolomics), these techniques offer a powerful new set of capabilities to elucidate changes during development of production diseases. Thus, the field has never had more tools at its disposal.

Challenges for future research on production diseases

Many challenges remain for researchers in the field. Among these potentially the most serious is the deterioration of funding for livestock oriented research, both fundamental and applied, world-wide. Much of the "frontier" research described above is particularly expensive yet could be enormously fruitful. Other challenges relate to development of suitable models to understand the interactions between stressors and development of production disease.

To continue to make progress at the basic level requires that scientists studying production diseases continue to exploit the capabilities offered by epidemiology and applied research. Clues must be sought from veterinarians and animal scientists working at the farm level, whose observations of "real-world" management and disease incidence are vital for formation of testable hypotheses. Given the complex interactive nature of the multiple systems involved, multidisciplinary basic research that blurs the traditional fields of metabolism, immunology, physiology, and neuroendocrinology are necessary for further progress in understanding the etiology of production diseases.

References

Agenäs, S., E. Burstedt and K. Holtenius, 2003. Effects of feeding intensity during the dry period. 1. Feed intake, bodyweight, and milk production. Journal of Dairy Science 86, 870-882.

Allen, M.S., 2000. Effects of diet on short-term regulation of feed intake by lactating dairy cattle. Journal of Dairy Science 83, 1598-1624.

Ametaj, B.N., B.J. Bradford, G. Bobe, Y. Lu, R. Nafikov, R.N. Sonon, J.W. Young and D.C. Beitz, 2002. Acute phase response indicates inflammatory conditions may play a role in the pathogenesis of fatty liver in dairy cows. Journal of Dairy Science 85 (Suppl. 1), 189. (Abstr.)

Bauman, D.E. and W.B. Currie, 1980. Partitioning of nutrients during pregnancy and lactation: a review of mechanisms involving homeostasis and homeorhesis. Journal of Dairy Science 63, 1514-1529.

Bell, A.W., W.S. Burhans and T.R. Overton, 2000. Protein nutrition in late pregnancy, maternal protein reserves and lactation performance in dairy cows. Proceedings of the Nutritions Society 59, 119-136.

Bertics, S.J., R.R. Grummer, C. Cadorniga-Valino and E.E. Stoddard, 1992. Effect of prepartum dry matter intake on liver triglyceride concentration and early lactation. Journal of Dairy Science 75, 1914-1922.

Bertoni, G., M.G. Maianti, L. Calamari and V. Cappa, 1996. The acute phase proteins variation, liver activity and lipid metabolism of dairy cows around calving. Proceedings of the VIIth International Symposium of Veteriary Laborotary Diagnosticians, Jerusalem, Israel, pp 59.

Bertoni, G.E. Trevisi, L. Calamari and R. Lombardelli, 1998. Additional energy and protein supplementation of dairy cows during early lactation: milk yield, metabolic-endocrine status and reproductive performances. Zootecnica e Nutrizione Animale 24, 17-29.

Bertoni, G., E. Trevisi and X.T. Han, 2001. Relationship between the liver activity in the puerperium and fertility in dairy cows. 52nd Annual Meeting EAAP, Budapest.

Black, P.H., 2002. Stress and the inflammatory response: A review of neurogenic inflammation. Brain, Behavior, Immunity 16, 622-653.

Calamari, L., F. Librandi, E. Trevisi and G. Bertoni, 2002. Transition period in dairy cows: immune system, inflammatory conditions and liver activity. Journal of Dairy Science 85 (Suppl. 1), 246-247. (Abstr.)

Correa, M.T., H. Erb and J. Scarlett, 1993. Path analysis for seven postpartum disorders of Holstein cows. Journal of Dairy Science 76, 1305-1312.

Curtis, C.R., H.N. Erb, C.H. Sniffen, R.D. Smith and D.S. Kronfeld, 1985. Path analysis of dry period nutrition, postpartum metabolic and reproductive disorders, and mastitis in Holstein cows. Journal of Dairy Science 68, 2347-2360.

Dann, H.M., 2004. Dietary energy restriction during late gestation in multiparous Holstein cows. Ph.D. Diss., Univ. Illinois, Urbana.

Dann, H.M., N.B. Litherland, J.P. Underwood, M. Bionaz and J.K. Drackley, 2003. Prepartum nutrient intake has minimal effects on postpartum dry matter intake, serum nonesterified fatty acids, liver lipid and glycogen contents, and milk yield. Journal of Dairy Science 86(Suppl. 1), 106.

Douglas, G.N., J.K. Drackley, T.R. Overton and H.G. Bateman, 1998. Lipid metabolism and production by Holstein cows fed control or high fat diets at restricted or ad libitum intakes during the dry period. Journal of Dairy Science 81 (Suppl. 1), 295. (Abstr.)

Drackley, J.K., 1999. Biology of dairy cows during the transition period: the final frontier? Journal of Dairy Science 82, 2259-2273.

Drackley, J.K., 2003a. Interrelationships of prepartum dry matter intake with postpartum intake and hepatic lipid accumulation. Journal of Dairy Science 86 (Suppl. 1), 104-105. (Abstr.)

Drackley, J.K. 2003b. Physiological and pathological adaptations in dairy cows that may increase susceptibility to periparturient metabolic diseases. 54th EAAP Meeting, 31 August – 3 September 2003, Rome, Italy.

Drackley, J.K., D.W. LaCount, J.P. Elliott, T.H. Klusmeyer, T.R. Overton, J.H. Clark and S.A. Blum, 1998. Supplemental fat and nicotinic acid for Holstein cows during an entire lactation. Journal of Dairy Science 81, 201-214.

Drackley, J.K., T.R. Overton and G.N. Douglas, 2001. Adaptations of glucose and long-chain fatty acid metabolism in liver of dairy cows during the periparturient period. Journal of Dairy Science 84 (E. Suppl.), E100-E112.

Dunklee, J.S., A.E. Freeman and D.H. Kelley, 1994. Comparison of Holsteins selected for high and average milk production. 2. Health and reproductive response to selection for milk. Journal of Dairy Science 77, 3683-3690.

Ekesbo, I., 1976. Possible ways of fighting environmentally evoked production diseases. In: Proceedings Third International Conference on Production Disease in Farm Animals. Centre for Agricultural Publishing and Documentation, Wageningen, pp. 18-21.

Emanuelson, U., P.A. Oltenacu and Y.T. Grohn, 1993. Nonlinear mixed model analyses of five production disorders of dairy cattle. Journal of Dairy Science 76, 2765-2772.

Friggens, N.C., 2003. Body lipid reserves and the reproductive cycle: towards a better understanding. Livestock Production Science 83, 219-236.

Grum, D.E., J.K. Drackley, R.S. Younker, D.W. LaCount and J.J. Veenhuizen, 1996. Nutrition during the dry period and hepatic lipid metabolism of periparturient dairy cows. Journal of Dairy Science 79, 1850-1864.

Grummer, R.R., 1993. Etiology of lipid-related metabolic disorders in periparturient dairy cows. Journal of Dairy Science 76, 3882-3896.

Grummer, R.R., 1995. Impact of changes in organic nutrient metabolism on feeding the transition dairy cow. Journal of Animal Science 73, 2820-2833.

Hansen, L.B., 2000. Consequences of selection for milk yield from a geneticist's viewpoint. Journal of Dairy Science 83, 1145-1150.

Hayirli, A., R.R. Grummer, E.V. Nordheim and P.M. Crump, 2002. Animal and dietary factors affecting feed intake during the prefresh transition period in Holsteins. Journal of Dairy Science 85, 3430-3443.

Hayirli, A., R.R. Grummer, E.V. Nordheim and P.M. Crump, 2003. Models for predicting dry matter intake of Holsteins during the prefresh transition period. Journal of Dairy Science 86, 1771-1779.

Herdt, T.H., 2000. Ruminant adaptation to negative energy balance. Influences on the etiology of ketosis and fatty liver. Veterinary Clinics of North America - Food Animal Practice 16, 215-230.

Holcomb, C.S., H.H. Van Horn, H.H. Head, M.B. Hall and C.J. Wilcox, 2001. Effects of prepartum dry matter intake and forage percentage on postpartum performance of lactating dairy cows. Journal of Dairy Science 84, 2051-2058.

Holtenius, K., S. Agenäs, C. Delavaud and Y. Chilliard, 2003. Effects of feeding intensity during the dry period. 2. Metabolic and hormonal responses. Journal of Dairy Science 86, 883-891.

Holtenius, K., K. Persson Waller, B. Essen-Gustavsson, P. Holtenius and C. Hallen Sandgren, 2004. Metabolic parameters and blood leucocyte profiles in cows from herds with high or low mastitis incidence. Veterinary Journal 168, 65-73.

Ingvartsen, K.L. and J.B. Andersen, 2000. Integration of metabolism and intake regulation: a review focusing on periparturient animals. Journal of Dairy Science 83, 1573-1597.

Ingvartsen, K.L., R.J. Dewhurst and N.C. Friggens, 2003. On the relationship between lactational performance and health: is it yield or metabolic imbalance that cause production disease in dairy cattle? A position paper. Livestock Production Science 83, 277-308.

Janovick, N.A., J.J. Loor, H.M. Dann, H.A. Lewin and J.K. Drackley, 2004. Characterization of changes in hepatic expression of inflammation-associated genes during the periparturient period in multiparous Holstein cows using quantitative real time-PCR. Journal of Dairy Science 87 (Suppl. 1), 460. (Abstr.)

Jones, W.P., L.B. Hansen and H. Chester-Jones, 1994. Response of health care to selection for milk yield of dairy cattle. Journal of Dairy Science 77, 3137-3152.

Kelley, K.W., 2001. It's time for psychoneuroimmunology. Brain Behavior Immunity 15, 1-6.

Kelley, K. W., 2004. From hormones to immunity: the physiology of immunology. Brain Behavior Immunity 18, 95-113.

Kimura, K., J.P. Goff, M.E. Kehrli, Jr. and T.A. Reinhardt, 2002. Decreased neutrophil function as a cause of retained placenta in dairy cattle. Journal of Dairy Science 85, 544-550.

Kunz, P.L., J.W. Blum, I.C. Hart, H. Bickel and J. Landis, 1985. Effects of different energy intakes before and after calving on food intake, performance and blood hormones and metabolites in dairy cows. Animal Production 40, 219-231.

Lewis, G.F., A. Carpenter, K. Adeli and A. Giaca, 2002. Disordered fat storage and mobilization in the pathogenesis of insulin resistance and type 2 diabetes. Endocrine Reviews 23, 201-229.

Loor, J.J., N.A. Janovick, H.M. Dann, R.E. Everts, S.L. Rodriguez-Zas, H.A. Lewin and J.K. Drackley, 2004. Microarrary analysis of hepatic gene expression from dry-off through early lactation in dairy cows fed at two intakes during the dry period. Journal of Dairy Science 87 (Suppl. 1), 196. (Abstr.)

Markusfeld, O., 1987. Periparturient traits in seven high dairy herds. Incidence rates, association with parity, and interrelationships among traits. Journal of Dairy Science 70, 158-166.

Payne, J.M., 1972. Production disease. Journal of the Royal Agricultural Society of England 133, 69-86.

Peeler, E. J. M.J. Otte and R.J. Esslemont 1994. Inter-relationships of periparturient diseases in dairy cows. Veterinary Record 134, 129-132.

Rabelo, E., R.L. Rezende, S.J. Bertics and R.R. Grummer, 2003. Effects of transition diets varying in dietary energy density on lactation performance and ruminal parameters of dairy cows. Journal of Dairy Science 86, 916-925.

Reid, I. M., 1989. Production disease in farm animals – past, present and future. In: Proceedings Seventh International Conference on Production Disease in Farm Animals. Cornell University, Ithaca, pp. 162-171.

Reynolds, C.K., P.C. Aikman, B. Lupoli, D.J. Humphries and D.E. Beever, 2003. Splanchnic metabolism of dairy cows during the transition from late gestation through early lactation. Journal of Dairy Science 86, 1201-1217.

Waldron, M.R., T. Nishida, B.J. Nonnecke and T.R. Overton, 2003a. Effect of lipopolysaccharide on indices of peripheral and hepatic metabolism in lactating cows. Journal of Dairy Science 86, 3447-3459.

Waldron, M.R., B.J. Nonnecke, T. Nishida, R.L. Horst and T.R. Overton, 2003b. Effect of lipopolysaccharide infusion on serum macromineral and vitamin D concentrations in dairy cows. Journal of Dairy Science 86, 3440-3446.

Research priorities from a producer's point of view

Walter M. Guterbock, DVM, MS,
Sandy Ridge Dairy, Scotts, MI

The context in which producers work

Owners and managers of livestock farms are concerned with running a business. Animals create the profit but there are many aspects to running a livestock business besides managing the animals. Often the managers of large livestock enterprises delegate the direct care and management of the animals to employees. Whether a farm has one cow or ten thousand, feed must be obtained, products must be marketed, forages have to be harvested on time, labor needs to be hired and managed, equipment and facilities have to be repaired and maintained, financing has to be obtained, bankers and regulators need to be satisfied, and paperwork taken care of. Often diseases that seem very important to veterinarians or other advisers are not the top priority of the owner/manager.

It is a common misconception that as units get bigger, less attention is paid to the welfare and health of the animals. In fact, as units get bigger, problems of the same relative size look bigger because of sheer numbers. A 10% rate of displaced abomasum (where the denominator is the number of cows that calve in a period) on a 40-cow dairy would be 4 cows a year, which might appear acceptable. On a 4000 cow dairy, it is 400 cows a year, or more than one daily, which would appear unacceptable although it is the same rate.

There is an opportunity here for the advisers of producers to become advocates for animal health and well being, to call attention to the needs of animals in modern livestock management systems. The advisers can also help train employees and develop systems to help maximize the well being and performance of the animals.

Many advisers assume that maximal output per animal equates with maximal profit. While this is often true, it is also true that the owner/manager must look at cost of production. In adopting a costly innovation, there is always the risk that it may not yield the promised gains, or that it may be solving a problem that is not the main constraint to increased productivity. In that event the producer incurs a new cost without getting the promised benefit. If a cost is cut, one is certain that it is cut, although one is not sure of the negative consequences of making the cut. This explains in part why many producers are more interested in cutting costs than in trying promising new interventions.

Innovations often are difficult to adopt because of the peculiarities of cow grouping, management, or feeding strategy on a given farm. For example, a feed additive that is meant just for transition cows may be easy to adopt on a tie-stall dairy where cows are fed individually, but difficult on a medium-sized farm (say, 300 cows) where making a special mix for recent postpartum cows would mean mixing a very small batch that the mixer truck may not be able to handle well. If the additive is too expensive to feed to all the cows, it may not be feasible to use it.

Researchers and consultants love data. Unfortunately, the data collected in the production context is often confounded, incomplete, unaudited, and full of errors. For example, culling reasons in DHIA records are notoriously unreliable, because many cows that leave the herd for other reasons, such as failure to get pregnant or lameness, may get coded as "low production," which is the immediate reason they finally are sold. Displaced abomasum (DA) may not be recorded, or may be recorded only for cows that are presented for surgery, not for those that are sold immediately after diagnosis because they are considered a poor risk for surgery. Managers vary in their enthusiasm for keeping detailed records, and often on-farm records contain only the minimum needed to run the farm. Archives on events in the past lactation or cows that have left the herd may be inaccessible or lost.

Multiple factors are always at work in the production environment. In the dairy industry, weather, timing of somatotropin injections, variations in calving cycles, and day-to-day variations in feedstuffs can affect milk yield. The effect of an intervention or additive must be separated from the noise of short-term variations. Small changes in production caused by an intervention, on the order of one liter a day, are almost impossible to see. An intervention that only affects milk yield in one group of cows, say those that calved recently, may not cause any noticeable increase in overall daily milk yield of the herd. Yet it is ultimately an increase in the amount of milk sold by the farm (or a decrease in the cost of producing it) that must pay for any innovation. Seeing effects in reproduction or in health events of low prevalence (like DA) may require large numbers of cows over extended periods, and may not be possible in small herds.

Research workers are wedded to the .05 probability level for the reporting of "significant" results. Results that are statistically significant may not mean much in terms of interventions on real farms. This is especially true when logistic regression is used to analyze data and significant interactions are found. Statistically significant results may also be of little biological importance in the herd. For example, a management change that changes the prevalence of pinkeye may be of little concern to a producer who does not have much pinkeye in his herd. Producers operate in an uncertain environment and almost always have to make decisions based on incomplete information. As a producer, if I had a 70% chance of making the right decision, I would be satisfied.

Recommendations to producers should consider the time value of money. Benefits far in the future have to be discounted back to their present value. While the benefits of a Johne's disease eradication program may seem obvious to a Johne's researcher, such a program will take many years to achieve results, and there are many opportunities for mistakes. The benefit may not be seen for ten years. As a producer I am less interested in spending money on such a program than in something with a more immediate payoff.

Researchers like to be able to explain mechanisms at the cellular or molecular level. Many research reports focus on enzymes or molecules, not even on whole cells. Producers deal with whole animals and groups of animals. If research is to have practical application, it must in the end be translated to the animal or population level. It is hard to get funding for population-based research, but more of it would be desirable for producers. Also, we are short of research on the outcomes of interventions. For example, increased milking frequency (IMF) in early lactation has looked attractive in several small studies, but producers need to know what has happened when IMF has been tried in commercial herds. Unfortunately, it is hard to tease out the effect from the noise of milk yield data from the whole herd.

Veterinarians and veterinary researchers tend to focus too much on microbes as causes of production problems. Large dairy farms and feedlots are rarely if ever run as closed herds. One can assume that almost all of the important diseases of cattle are represented in the herd. The aim of management is to assure that most of the animals are healthy and strong enough to resist the inevitable exposure to pathogens, and that the spread of pathogens is minimized. Eradication is rarely a realistic goal for diseases that are not covered by government eradication programs. The swine and poultry industries have made greater strides in biosecurity than beef and dairy.

General needs of livestock industries

The price of the product and the ability to sell it freely are still the major determinants of the profitability of a livestock farm. So producers are naturally interested most in research that can affect these. Food safety scares, like that surrounding *E. coli* 0157H7 in hamburger, are of great concern to producers. This explains why producer groups, particularly in the dairy industry, are more willing to fund research that improves food safety and develops novel products than research that increases milk production per cow or per herd. All producers would like their herds to be more productive, but maintaining price and markets is a more pressing need.

All livestock species are increasingly impacted by environmental regulations. Current (2004) regulatory efforts are concentrated mainly on large units, but they will eventually affect farms of all sizes. Producers need research that will help them dispose of manure in ways that do not hurt the environment and that meet environmental regulations. Research may also help regulators devise regulations that producers can live with. Environmental regulation can be seen as a competitive advantage for producers who are already operating because it raises the barriers for entry to the industry by new competitors.

Livestock industries are also under pressure from their customers to assure the public that animals are not abused in the production process. It appears that many large buyers of animal products in the US, such as chain restaurants and food wholesalers, will require animal welfare certification from suppliers. Animal welfare regulations are already more stringent in Europe than in the US. Research that will help bring a scientific basis to welfare regulations will help producers. Perceptions of what is actually stressful or abusive to animals vary with the observer. Many people who might make rules may not have a deep understanding of production processes. On the other hand, producers may become desensitized to animal suffering because they think it is an inevitable part of the production process.

As livestock units get bigger and ownership becomes more concentrated, producers begin to see each other as competitors, rather than collaborators. Vertical integration and proprietary products that may command premium prices are supplanting the old co-operative model, where all producers belonged to a co-op and all sold a commodity product together. Similarly, a more proprietary model may replace the old co-operative extension model of providing advice, where knowledge was shared among producers. Producers may be unwilling to share with competitors innovations and knowledge that they believe to be their own. Suppliers or consultants with proprietary products or services are usually unwilling to educate their competitors. This means that more production-oriented research may be funded privately and conducted within large enterprises, and less in publicly funded research stations and universities. As public funding decreases, pharmaceutical and feed companies will fund proportionately more research with an interest in testing or selling a specific product. This will increase the need for objective, independent research, although funding for it may be hard to find.

It is critical that researchers in production diseases maintain close communication with producers to assess research needs in specific industries. Producers need to understand that quality research takes time and money, and in the future it appears that more of the money is going to have to come from producers or producer groups than it has in the past.

Specific needs of the dairy industry

In this article I will be devoting more space to the dairy industry than to others because it is the industry I know best. This should not imply that it is more important than other livestock industries.

The dairy cow is a metabolic athlete. Few other animals, perhaps only the lactating sow and the laying hen, are asked to take in three to four times their maintenance feed requirement and process the nutrients into usable products. The transition from the dry period to the peak of lactation and feed intake requires large adaptations of the liver and digestive tract. Most of the economically important diseases of cattle and much of the premature loss to death and disease occur during the late dry period and early lactation. In general, if a cow gets through the transition well, she will have a successful lactation (provided that she gets pregnant).

It is heartening to see more research being conducted on transition cows. It is inherently a difficult period to study. Exact calving dates cannot be known a priori. It is impossible to have a pretrial covariate period. Milk yield is not very repeatable from lactation to lactation, so it is difficult to have convincing controls. Events around calving, such as twinning or dystocia, can have a great effect on the outcome of the lactation. Feed intake decreases and then increases rapidly, and interventions to measure intake (like Calan gates) may affect intake. High-fiber, low-energy dry cow diets appear promising but need to be investigated further, as do manipulations of dry period length.

More research is needed on the adaptation of the rumen microflora to the feed changes around calving. In particular, more work needs to be done on subacute rumen acidosis. It appears empirically that the disease exists, but it is difficult to define precisely. It appears that healthy cows can produce well with rumen pH below 5.0, and that cows have self-protective mechanisms (mainly going off feed) that protect them from acidosis. It also appears to me that our models for predicting feed intake in cattle are far too simple. While fiber digestibility and rate of passage are clearly important, there also must be neuroendocrine mechanisms working that are similar to those that have been described in monogastric models like the rat. The role of adipose tissue in metabolic signaling needs to be defined further. The amount of adipose tissue appears to affect both appetite and metabolism, perhaps through leptin.

Culling and death rates are simply too high in the dairy industry. In fact, the rates are unsustainable when compared to the rate at which new heifers can be raised to replace the cows leaving the herd. Certainly improving the nutrition and management of transition cows can reduce loss rates, but on the dairies I manage only about half of the losses are attributable to transition problems. The most frequent culling reason for most dairy farms is still failure to become pregnant in time.

Reproductive performance in the US dairy industry as a whole has not improved over more than forty years of veterinary involvement in reproductive management programs. Mastitis, lameness, and injury are also major causes of loss. One has to ask if the dairy cow we currently have is adapted to the production system in which we have placed her.

Mastitis remains the most common disease of dairy cows. It is my opinion that efforts to improve milking hygiene and milking equipment function have, in better-managed herds, been carried as far as they practically can. Yet some of these well-managed herds continue to have unacceptably high clinical mastitis rates. The existing paradigm of mastitis consultants has been that the normal state of the udder is sterile, and that efforts must focus on keeping pathogens out. What if bacteria normally enter and leave the udder with some frequency, and only some cows get clinical mastitis as a result? Why do a third or a half of cows with clinical mastitis in many modern herds have no bacteria isolated from their milk? More mastitis research should focus on the immune response of the cow and the events that lead to the secretion of abnormal milk and less on the pathogens.

Reproductive failure, lameness, and digital dermatitis (hairy heel warts) are also common causes of economic loss to dairy producers and need more research. The development of estrus synchronization programs has helped the industry tremendously, but the widening use of ultrasound for early pregnancy diagnosis has made it clear that the rates of early embryonic loss are much higher than we previously thought. Some of this may be due to genetic abnormalities in the fetus, but some may be preventable through management or nutrition changes. Unfortunately, seeing differences or changes in reproductive performance requires many animals, studied over one or more years. Current research on the reproductive effects of fatty acid nutrition and metabolism appears to be very promising. Research on improving animal welfare and housing systems should help prevent injury and lameness.

The dairy industry, like other livestock industries, faces significant challenges in environmental regulations. We view coming regulation of air quality, especially of the discharge of nitrogenous compounds from manure and urine, with great concern. Research that will help producers conform to environmental regulations and help regulators develop realistic regulations would help the industry greatly. Better definition of nutritional requirements may help avoid overfeeding of nutrients of regulatory interest and reduce the amounts in manure.

Specific research needs of the beef industry

Respiratory disease continues to be an enormous challenge to the beef industry. Currently no effective preventive or therapeutic strategies are available for atypical interstitial pneumonia in feed yards. Respiratory disease caused by *Haemophilus* and *Mycoplasma spp.* continue to be challenges to the industry.

Minimizing transportation stress will continue to be a big welfare issue for the beef industry, as will industry practices in marketing and commingling young cattle. Improvement of handling practices in feed yards and slaughter facilities will also continue to be priorities.

Section A.

Specific research needs of the swine industry

The loss rate of sows is similar to that in dairy cows. This is an economic issue for the industry but is also likely to become an animal welfare issue. The fact that animals can only tolerate our management systems for two to three years is not easily justified to the public.

Weaning is a period of unavoidable stress to baby pigs that continues to be a challenge to management systems and to nutritionists.

Porcine Respiratory and Reproductive Syndrome (PRRS) is a very costly disease in the swine industry and is currently a high research priority for producers.

Some pork industry practices, such as keeping sows confined constantly either in farrowing or gestation crates, have come under public pressure and have even been the subject of legislation. The industry will continue to require help form research to formulate realistic animal welfare standards.

Specific research needs of the poultry industry

Recent adverse publicity about the treatment of chickens in a slaughter plant points out that there is still a long way to go in animal welfare in the poultry industry. The emphasis growing young animals to heavy weights has led to problems with pathological fractures and locomotion in young chickens and turkeys. A production system that does not allow animals to live more or less normal lives, however short, will probably not be acceptable to the consuming public.

Public pressure to stop forced moulting of laying hens may force changes in the management of those animals or development of less drastic measures to extend hens' useful life.

The dry and concentrated nature of poultry manure means that methods used for environmental management in cattle and swine are probably not applicable. Since poultry manure can no longer be fed to cattle, other disposal methods must be found.

Acknowledgements

I am indebted to Drs Bob Smith and Mark Spire for providing information on the beef industry, Dr. Barbara Straw for providing information on the swine industry, and Jim Brown for information on poultry.

Metabolic profiling to assess health status of transition dairy cows

Robert J. Van Saun
Penn State University, University Park, Pennsylvania, USA

Use of blood chemistries in the form of metabolic profiles to determine nutritional status has been advocated, but acceptance has been limited by economics and interpretation difficulties. Different blood metabolite criteria are needed to evaluate disease potential compared to disease diagnosis. Laboratory blood metabolite reference ranges are often based on mid to late lactation cow populations and may not be appropriate for evaluating transition cows. Objectives of this study were to determine effects of time relative to calving and health status on blood metabolite concentrations and determine if any diagnostic relationships are present between prepartum metabolite concentrations and postpartum health status.

Metabolic profiles were performed on plasma samples collected from 113 cows housed at 15 commercial dairies over three time periods. These periods were defined as: Early dry (ED), >30 days precalving; Close-up Dry (CU), 3 to 21 days precalving and Fresh (FR), 3 to 30 days postcalving. Metabolites measured included urea nitrogen (PUN), creatinine (Cr), glucose (Glu), total protein (TP), albumin (Alb), total bilirubin (TB), alkaline phosphatase (ALP), creatine kinase (Ck), γ-glutamyltransferase (GGT), aspartate aminotransferase (AST), sorbitol dehyrdogenase (SDH), sodium (Na), potassium (K), chloride (Cl), calcium (Ca), phosphorus (P), magnesium (Mg), total cholesterol (Chol), triglycerides (TG), nonesterified fatty acids (NEFA) and β-hydroxybutyrate (BHB). Disease diagnosis and treatment events were recorded. Blood metabolites were evaluated by ANOVA for repeated measures with period, health and their interaction as main effects and herd as a covariate. Relative risk of postpartum disease was determined using contingency tables of metabolite concentration categories and health status.

Of all cows, 53% had one or more disease events postcalving. Percentage healthy calvings varied greatly between herds. Herd was significant in all metabolite models, except NEFA and Ck. Time period influenced ($P<0.05$) all metabolite concentrations, except Ca, P and K. Health status influenced ($P<0.02$-0.002) NEFA, BHB, TG, GGT and AST independent of time period. An interaction ($P<0.02$-0.0002) between time period and health status was found for Alb, PUN, Glu, Chol, TG, AST, BHB and NEFA. Sick cows had lower Alb, PUN, Glu and Chol and higher AST, BHB and NEFA compared to healthy cows in the FR period. Fresh cow Alb concentration was stratified into three groups: <30 g/l, 30 to 35 g/l and >35 g/l and associated with health status. Percentage of FR cows experiencing a health event within each group was 67, 61 and 32, respectively. Cows with CU Alb concentrations <32.5 g/l were 1.5 ($P<0.04$; 1.04-2.04, 95% CI) times more likely to experience a postpartum disease event. Cows with FR Alb concentration <33.0 g/l were 1.8 ($P<0.003$; 1.2-2.7, 95% CI) times more likely to have a disease event. If NEFA values were >0.4 mM in either CU or FR samples, cows were 1.6 ($P<0.03$) and 1.5 ($P<0.04$) times more likely to have a disease event, respectively. Disease risk was greater if NEFA concentration was >0.6 mM at CU (1.7, $P<0.02$) and FR (1.9, $P<0.001$) periods. No metabolites measured in the ED period were associated with disease risk.

Based on these findings, reference ranges for diagnostic interpretation of blood metabolite concentrations should be adjusted to time periods relative to calving. Interactions between time period and health status suggest prepartum blood metabolite concentrations may provide some indication to postpartum disease risk and can be useful as a herd monitoring tool. Preliminary data suggest Alb and NEFA concentrations in CU and FR periods can be used to predict potential disease risk.

Acknowledgements

This research is supported in part by funding from Pennsylvania Department of Agriculture.

A consideration about the energy supply in peripartum of dairy cows on the basis of change in plasma free amino acid concentration

S. Kawamura., K. Shibaro., R. Hakamada., M. Otsuka, S. Sat and H. Hoshi
Internal Medicine, Department of Veterinary Clinical Sciences, Kitasato University School of Veterinary Medicine and Animal Sciences, Japan

Recently, management of dry phase of high yielding dairy cows has become extremely important to evaluate nutrition in close up transition. It is obvious that metabolic response in dam site for drastic and large nutritional demand to cope with delivery and lactation depends greatly on nutritional condition in close up transition. Normal or morbid metabolism of amino acids in peripartum of dairy cows hasn't been cleared yet.

The authors studied the changes of plasma free amino acids (PFAA) concentration in dry and lactating dairy cows and the effect on plasma components by supplying several kinds of amino acids before parturition.

The changes in concentrations of PFAA from 20 days before to 90 days after parturition were studied. Leu, Ile, Tyr, Ala, Pro, Arg, Asp, Asn, Orn, Val, Trp and Cys declined from the middle of close up transition and then recovered until 30 days after delivery. Gly and Ser maintained the tendency of a rise gradually until after one month of delivery. Methionine, Thr, Lys, Phe Glu and His declined from before the delivery and did not show recovery or maintained the decrease after parturition. Supplementation of rumen protected methionine brought about the recovery of concentration in plasma of several kinds of amino acids, however it was to slight extent.Six pregnancy cows were injected intravenously with several kinds of amino acids from before five days of expected delivery date until delivery day. Plasma parameters related to diagnosis of a metabolism diseases in these cows have improved remarkably.

These results suggest that the changes of PFAA of dairy cows in peripartum could be a reflection of the elucidation in amino acid levels of protein metabolisms and mobilization of amino acids from muscle tissue. Furthermore, supply of several kinds of amino acid in prepartum satisfies the deficiency of protein and energy in the cows and may have an preventive effect to the metabolism disease in peripartum.

Metabolic predictors of displaced abomasum in transition dairy cows

S. LeBlanc, T. Duffield and K. Leslie
Department of Population Medicine, University of Guelph, Ontario, Canada

Risk factors for displaced abomasum (DA) have been described, but significant gaps remain in understanding its pathogenesis. The objective of this field study was to identify metabolic tests available in clinical practice that identified cows at increased risk of DA. A technician visited 1044 cows in 20 herds weekly from one week before expected calving until one week postpartum. Cows were assigned a BCS and samples were collected at each visit for complete blood count and for measurement of serum non-esterified fatty acids (NEFA), cholesterol, β-hydroxybutyrate (BHB), glucose, urea, calcium and phosphorus. Prepartum, collection of a free-flow urine sample was attempted, and postpartum a milk sample was collected, for measurement of ketones with Acetest (Bayer, Etobicoke, Canada) and Keto-Test (Elanco Animal Health) tests, respectively. All disease events before 30 DIM were recorded. Test results were screened for association with disease occurrence using contingency tables. Multivariable logistic regression models were built to account for covariates and the correlation of cows with herds.

There were 53 cases of DA (incidence rate = 5.1%) and the median time to diagnosis was 11 DIM. In cows with DA, mean NEFA concentrations began to diverge from the mean in cows without DA 14 days before calving, whereas mean serum BHB concentrations did not diverge until the day of calving. Among the variables studied, prepartum, only NEFA was associated with risk of subsequent DA. Examined retrospectively, 0 to 6 days before the actual day of calving, the optimum put-point was 0.5 mEq/l. Cows with NEFA concentration $\geq$ 0.5 mEq/l were 3.5 times (95% CI, 1.9 – 7.1) more likely to develop DA after calving (P<0.0001). For prospective application, among cows sampled between 4 and 10 days before their expected calving date, the optimum NEFA cut-point remained 0.5 mEq/l. The sensitivity, specificity and likelihood ratio (LR) were 64%, 66%, and 2.6, respectively. Exclusion of samples taken within 2 days of actual calving did not change the optimum cut-point and only increased the LR to 3.3. Urine samples were obtained from 46% of cows. Among these, cows with a positive urine ketone test were 11.8 times more likely to develop DA (P<0.0001), but the sensitivity and specificity of the test were 39% and 96%, respectively. Serum calcium concentration up to and including the day of calving was not associated with DA.

In a separate model for samples between 1 and 7 days postpartum (accounting for the effects of retained placenta and metritis) increasing serum concentrations of both BHB and NEFA were associated with increased risk of subsequent DA. However, considered separately, postpartum, serum BHB was a more specific test than NEFA concentration. The odds of DA were 7 times greater in cows with serum BHB $\geq$ 1200 µmol/l (P<0.0001; LR = 3.5). Cows with milk BHB concentration $\geq$ 200 µmol/l measured by the cowside test were 3.5 times (P<0.001) more likely to develop DA; the sensitivity, specificity, and LR of this test alone were 47%, 80%, and 2.4, respectively. The timing and severity of negative energy balance, and the success of cows' adaptation to it, are key elements in the pathogenesis of DA. Strategic use of metabolic tests to monitor transition cow programs in dairy herds should focus on NEFA in the last week prepartum, and BHB in the first week postpartum.

Evaluation of a rapid test for NEFA in bovine serum

L. Gooijer, K. Leslie, S. LeBlanc and T. Duffield
Department of Population Medicine, Ontario Veterinary College, University of Guelph, Guelph, Ontario, Canada, N1G 2W1

Excessive or prolonged periparturient negative energy balance (NEB) is an important issue for dairy producers, and may be associated with increased risk of clinical disease and impaired production and reproductive performance. Affected cows commonly have elevated circulating levels of non-esterified fatty acids (NEFA) prior to calving and increased beta-hydroxybutyrate (BHB) postpartum. Monitoring the incidence of subclinical ketosis postpartum has been the recommended method of surveillance for this problem. Prepartum, blood NEFA concentration may be used to detect cows at risk for problems with severe NEB. Serum NEFA greater than 0.4 mEq/l NEFA has been proposed to identify excessive prepartum NEB. Measuring NEFA has traditionally involved submission of serum to a diagnostic laboratory. The DVM NEFA test (Veterinary Diagnostics. Newburg, Wisconsin, USA) is a new, rapid, spectophotometry method to determine NEFA concentration in serum through light absorbance. The objective of this study was to determine the test characteristics of the DVM NEFA test and its usefulness as a method of identifying problems with NEB in prepartum dairy cows.

Primaparous and multiparous animals were enrolled between 7 and 4 days prior to their expected calving date. Blood was collected by coccygeal venapuncture and serum harvested. Cows were re-sampled twice weekly until calving. In addition, each cow was sampled in week 1 and week 2 postpartum for BHB and other indicators of metabolism. NEFA concentration was measured using the DVM NEFA test. An aliquot of each serum sample was submitted to the Animal Health Laboratory (AHL) at the University of Guelph for analysis by a Hitachi 911 automated analyzer (Roche, Laval, Quebec). The AHL NEFA concentration was considered the gold standard for this evaluation. A total of 491 samples from 256 cows from eight farms in the Guelph, Ontario area were utilized in this study.

The Pearson correlation coefficient between the DVM NEFA and the AHL NEFA determination was 0.75. Using 350 samples drawn within 14 days prepartum, and NEFA > 0.4 mEq/l from the AHL test as the gold standard, the sensitivity and specificity of the DVM NEFA test were 84% and 96%, respectively. It is noteworthy that changing the NEFA cut-off level to > 0.5 mEq/l resulted in a similar sensitivity and specificity of 85% and 97%, respectively. There was a significant association between prepartum DVM NEFA and postpartum BHB in the first week after calving (n=280) (P<0.01 for Corrected Yates). The odds ratio was 2.33, meaning that cows detected with high prepartum DVM NEFA values are 2.3 times more likely to have subclinical ketosis (BHB > 1400) in the first week after calving. Furthermore, the association between prepartum DVM NEFA and postpartum BHB for both week 1 and week 2 postpartum were also statistically significant (P<0.02 for Pearson's Chi Square and the Corrected Yates). In this case, the odds ratio indicates that cows with detected high DVM NEFA prepartum were 2.0 times more likely to have subclinical ketosis (BHB > 1400) in both of the first two weeks after calving.

It was concluded that the DVM NEFA test characteristics were satisfactory for detection of cows with elevated prepartum NEFA, and useful for prediction of cows with increased risk of postpartum subclinical ketosis.

Association of rump fat thickness and plasma NEFA concentration with postpartum metabolic diseases in Holstein cows

N.P. Joshi[1], T.H. Herdt[1] and L. Neuder[2]
[1]Dept. Large Animal Clinical Sciences, Michigan State University; [2]Green Meadows Farm, Inc., Elsie, MI, USA

Real-time ultrasound measurement of subcutaneous fat on live animals is extensively used in the beef and swine industries to monitor body composition, carcass yield, and quality. Changes in subcutaneous fat thickness represent changes in energy stores. Subcutaneous fat thickness is typically measured 2 to 3 cm off the midline, at the level of the 12 – 13th rib, however measurement of fat thickness over the rump area may be more advantageous in dairy cattle due to easier access and greater fat thickness variability in that area. The objective of this study was to determine the association of prepartum rump fat thickness (RFT) and plasma NEFA concentrations with peripartum diseases including; ketosis, metritis and mastitis.

Ten animals per week (group) for four weeks were selected randomly as they entered the close-up pen at three weeks prior to their expected date of calving. During this period, a totally-mixed ration was continuously available and fresh feed was added daily. Ultrasound images and EDTA-blood samples were taken weekly until all cows had calved. Images were captured 5 cm cranial to the right *tuber ischium* using a portable ultrasound instrument (Sonosite – 180) with a 5.2 MHz "curved array" transducer. Images were digitized and subcutaneous fat thickness measured using Dicomworks software (www.dicomwork.com). Plasma nonesterified fatty acid (NEFA) concentrations were determined from coccygeal blood samples drawn on the same day as the ultrasound images were taken. Logistic regression models were formulated using a "step-down" technique. Parity, group, days prepartum, dystocia, RFT, and NEFA were the initial independent variables in the models. Rump fat thickness and plasma NEFA were found to be correlated ($r=0.32$) therefore, separate models were formulated. Metabolic morbidity events (metritis, mastitis and ketosis) were the dependent variables.

Of the 40 cows, 15 developed ketosis, 5 metritis, and 5 mastitis within 30 days postpartum. The mean days from blood collection and RFT measurements until calving were 17, 10 and 3 for each respective sampling. The mean weekly prepartum RFT were (mean ±SE) 1.04 ± 0.07, 1.02 ± 0.07 and $0.95+0.06$ cm for cows that were normal after calving and 1.28 ± 0.18, 1.46 ± 0.16 and 1.37 ± 0.16 cm for cows that became ketotic within the first three weeks after calving. Similarly, mean weekly prepartum NEFA concentrations were 0.11 ± 0.01, 0.10 ± 0.01 and 0.17 ± 0.04 mEq/l and 0.14 ± 0.01, 0.16 ± 0.02 and 0.31 ± 0.07 mEq/l in cows that were normal or ketotic after calving, respectively. Logistic regression analysis demonstrated a 5.32 ($P<0.03$) and 2.31 (<0.08) increase in ketosis risk per log unit increase in plasma NEFA concentration and a 5.65 ($P<0.05$) and 8.04 ($P<0.04$) increase in ketosis risk per log unit increase in RFT during week -2 and week -1 prepartum, respectively.

Monitoring of rump fat thickness and NEFA concentrations during last 2 weeks before calving may be quick and reliable method of determining risk of postpartum ketosis. These observations do not demonstrate superiority of one test over the other, although RFT values are available in real time. Rump fat thickness measures were strongly associated with ketosis. Metritis and mastitis were not associated with prepartum rump fat thickness or prepartum plasma NEFA concentration.

Section A.

Effect of pre-partum feeding intensity on postpartum energy status of Estonian Holstein cows

H. Jaakson, K. Ling, H. Kaldmae, J. Samarütel, T. Kaart and O. Kärt
Institute of Animal Science, Estonian Agricultural University1, Kreutzwaldi 1, 51 014 Tartu, Estonia

In early lactation, due to the discrepancy between the amount of energy required and that can be ingested, the negative energy balance (NEB) is developing in dairy cows. Several investigations report the dry period feeding to affect cows' *post-partum* ability to use their depot energy and to manage NEB. In order to find out the effect of *pre-partum* feeding intensity on the *post- partum* energy status an experiment with eight 2nd to 4th lactation cows divided into two groups (A and B; 4 cows per group) with different *pre-partum* and the same *post-partum* feeding strategy was carried out on the experimental farm of Estonian Agricultural University in 2002/2003. Experimental period included two *pre-partum* and four *post-partum* weeks. Cows daily rations consisted of grass silage (*ad-libitum* throughout the experimental period), barley meal (group A - 1 kg and group B - 5 kg *pre-partum*) and concentrates (for both groups 5 kg during the first five *post-partum* days, thereafter increasing the daily amount by 1 kg to the 10th day, and 10 kg from the 10th day onwards).

The following parameters were registered, calculated and/or determined on the14^{th}, 7th and 1st *pre-partum* day and on the 1st, 7th, 14th, 21st and 28th *post-partum* day: energy intake (EI); energy balance (EB); milk yield and composition (protein, lactose and fat content); body condition score (BCS) and blood glucose (GLC), ketone bodies (KB), non-esterified fatty acids (NEFA), insulin (INS) and glucagon (GLUC) concentration. The repeated measures general linear models analyses with SAS system MIXED procedure were performed to estimate the influence of *pre-partum* feeding and lactation day on registered parameters, the least square means were used to compare the two groups of cows on different lactation days. For positively skewed traits the logarithm transformation was applied.

Ration as a factor influenced the following investigated parameters: milk fat content, blood GLC, KB and NEFA concentrations. Lactation day had influence on BCS, EI, EB, milk fat, protein and lactose content; blood KB, TG and NEFA concentration. Factors interaction had effect on EI and blood KB concentration.

Post-partum EI did not differ between the groups. EB differed only on the 1st *post-partum* day being more negative in group A, apparently due to higher ECM production. *Post-partum* dynamics of GLC (tended to be lower in group A on the 7th day), NEFA (tended to be higher in group A on the $1^{st[}$, 14th and 21st day) and KB (tended to be higher in group A on the 1st day and higher on 7th day) concentrations may point to the possibly more negative EB in group A during longer period, however, similar dynamics of INS and GLUC concentrations in both groups did not support that. Our results do not definitely support the statement that raised *pre-partum* feeding intensity may result in deeper *post-partum* NEB due to depressed intake and lack of glyconeogenic capacity.

Acknowledgements

The study was supported by Estonian Science Foundation.

Using a pooled sample technique for herd metabolic profile screening

Robert J. Van Saun
Penn State University, University Park PA, USA

Use of metabolic profiles in evaluating herd nutritional or health status has not be widely accepted in the United States. Costs associated with metabolic profiling have been a primary deterrent. To profile a herd or animal group, 8 to 12 individuals are sampled for evaluation. Profile results are interpreted as a mean value or percent of individuals deviating from expected values. Use of pooled samples was evaluated as a method to collect usable information on herd metabolic status without the high cost of individual sampling. The objective of this study was to determine if blood metabolite concentrations from pooled serum samples were different from arithmetic mean results of individual samples. Metabolic profiles were performed on serum samples collected from 111 cows on 15 different farms for three defined time periods relative to calving (Early Dry, Close-up Dry, Fresh). Pooled samples (n=48, 16 in each time period) containing between 5 and 12 individuals were randomly composited by blending equal volumes of individual serum.

Metabolic profile analyses included urea nitrogen (BUN), creatinine (Cr), glucose (Glu), total protein (TP), albumin (Alb), total bilirubin (TB), alkaline phosphatase (ALP), creatine kinase (Ck), gamma-glutamyltransferase (GGT), aspartate aminotransferase (AST), sorbitol dehyrdogenase (SDH), sodium (Na), potassium (K), chloride (Cl), calcium (Ca), phosphorus (P), magnesium (Mg), total cholesterol (Chol), triglycerides (TG), beta-hydroxybutyrate (BHB), nonesterified fatty acids (NEFA), total CO_2 (TCO) and anion gap (AG, calculated). Individual pooled sample results (n=1056) were compared to arithmetic means of individuals by a one sample T-test. Difference between mean and pooled value, percent mean difference and difference as a proportion of the sample population standard deviation were tested by T-test. Effect of period and herd were tested by ANOVA. Significance was defined as $P<0.05$.

Overall mean-pooled difference and difference as a ratio to sample standard deviation was significantly different from zero for Ck, TCO, SDH and AG. Mean difference as a percent of mean was significantly different from zero for TG, TB, Ck, TCO, SDH and AG. Early dry pooled samples had more significant differences than close-up dry or fresh cow pooled samples. Mean difference as a proportion of sample standard deviation was 0.2 or less for most (72.7%, 16 of 22) metabolites. For individual samples, 192 (18.2%) pooled metabolite values were different from mean metabolite concentrations. Pooled samples for TCO and AG accounted for the greatest number, 14.1% (27 of 192) and 16.1% (31 of 192), respectively, of significant differences with individual sample arithmetic means. Another 8.9, 7.3 and 7.3% of the differences were accounted for by Cl (17 of 192), Na (14 of 192) and Ca (14 of 192), respectively. Of the individual mean-pool comparisons, herd of origin accounted for 73% ($P<0.003$) of the variation attributed with significantly different values. Pooled samples from 3 herds accounted for 39.6% (76 of 192) of the significantly different comparisons. These data suggest pooled samples may be used to assess metabolic status of a group of cows. Most important measures of metabolic status showed minimal differences between pooled and individual samples. Effect of herd on sample differences may suggest poor sample handling practices. Further work to determine how best to interpret pooled samples should be explored.

Acknowledgements
Research supported in part by funding from Pennsylvania Department of Agriculture.

Section B.
Metabolic effects of immune mediators

Metabolic effects of immune mediators

Kirk C. Klasing
Department of Animal Science Dept., University of California, Davis, California 96616, USA

Infected animals do not grow or reproduce as quickly or efficiently as healthy animals and have greater incidence of metabolic diseases. These changes are orchestrated by pro-inflammatory cytokines and other mediators released by a stimulated immune system[1,2,3,4,5]. Injection of these mediators causes characteristic behavioral, endocrine, metabolic, and cellular responses that duplicate many of the classical signs of infection. Pro-inflammatory cytokines, including interleukin 1, interleukin 6, and tumor necrosis factor are especially important. These cytokines coordinate local immunity to pathogens and also act systemically to alter metabolic homeostasis, decreased food intake, and slower growth rate. Many metabolic diseases are also exacerbated by a pro-inflammatory response. In chickens, for example, the incidence of tibial dyschondroplasia is increased and the density of bones is markedly decreased within days of an inflammatory response to bacterial lipopolysaccharides. Similarly, morbidity from heat stress is magnified by the acute phase sequelae of an inflammatory response. Both genetic background of the bird and dietary characteristic modify the extent of the systemic components of the pro-inflammatory response.

References

[1] Humphrey, B.D. and K.C. Klasing, 2004. Modulation of nutrient metabolism and homeostasis by the immune system. World Poult. Sci. 60, 90-101.

[2] Barnes, D.M., Z. Song, K.C. Klasing and W. Bottje, 2002. Protein metabolism during an acute phase response in chickens. Amino Acids 22, 15-26.

[3] Leshchinsky, T.V. and K.C. Klasing, 2001. Divergence of the inflammatory response in two types of chickens. Dev. Comp. Immunol. 25, 629-38.

[4] Klasing, K.C., 1998. Nutritional modulation of resistance to infectious diseases. Poult. Sci. 77, 1119-1125.

[5] Klasing, K.C., 1998. Avian macrophages: regulators of local and systemic immune responses. Poult. Sci. 77, 983-989.

The effect of bovine respiratory disease on carcass traits

R.L. Larson, DVM, PhD
Commercial Agriculture Program, VeterinaryExtension and Continuing Education, University of Missouri, Columbia, MO, USA

Evidence suggests that bovine respiratory disease influences carcass traits such as carcass weight, marbling, and subcutaneous fat cover [1,2,3,4]. Steers with lung lesions had lighter hot carcass weight, lower dressing percent, less internal fat, and lower marbling scores than steers without lesions [2]. They also tended to have less external fat and smaller longissimus muscle area than steers without lung lesions [2].

The regulation of muscle growth and fat deposition in beef cattle has not been fully elucidated. Hormones such as insulin, insulin-like growth factor I (IGF-I), growth hormone (GH), thyroid hormone, and leptin contribute to the partitioning of nutrients for tissue growth. There is also evidence that disease mediators such as interleukin-1α, interleukin-1β, and tumor necrosis factor α (TNFα) (cytokines) can also play a role in nutrient partitioning. A clear mechanistic pathway linking disease to changes in carcass traits has not been made, three possibilities have been identified. First, metabolic signals such as cytokines and cortisol could have an effect on carcass composition through modification of hypothalamic secretions of thyrotropin-releasing hormone, by inhibition of IGF-I and insulin actions on muscle and fat tissues, and by direct protein catabolism and lipolysis. Second, disease-induced anorexia could decrease serum IGF-I and increase serum GH which causes irreversible change in the partitioning of nutrients for tissue deposition. Cortisol may be involved in anorexia-associated decreases in carcass weight and fatness through decreased thyroid hormone activity and increased protein catabolism. Third, cytokines and endotoxin induce various behavioral symptoms of sickness including lethargy, adipsia, and reduced social interactions. The result may be an indirect effect of anorexia on growth and carcass traits in that sick cattle are effectively on feed for fewer days than healthy penmates.

References
[1] McNeill J.W., J.C. Paschal, M.S. McNeill, *et al.*, 1996. J. Anim. Sci. 74 (Suppl. 1), 135.
[2] Gardner B.A., H.G. Dolezal, L.K. Bryant, *et al.*, 1999. J. Anim. Sci. 77, 3168-3175.
[3] Stovall T.C., D.R. Gill, R.A. Smith, *et al.*, 2000. OSU Anim. Sci. Res. Report 82.
[4] Roeber D.L., N.C. Speer, J.G. Gentry, *et al.*, 2001. Prof. Anim. Sci. 17, 39.

Does calf health during the feedlot period affect gain and carcass traits?

L.R. Corah[1], W.D. Busby[2], P. Beedle[2], D. Strohbehn[2], J.F. Stika[1]
[1]*Certified Angus Beef LLC, Wooster, Ohio;* [2]*Iowa State University, Ames, IA, USA*

Research studies[1,2], designed for other objectives, have inferred health problems lower quality grade at time of carcass harvest, but to date the authors are unaware of previous data sets that evaluate the effect of health on subsequent carcass traits. A data set (n=6697) involving calves from 12 states, consigned to the eight Iowa feedlots feeding cattle for the Tri-County Carcass Futurity, were used to determine impact of health (treated 0, 1, 2+ times) on feedlot gain and carcass traits. Calves were individually weighed upon arrival, at re-implant, and pre-harvest.

The predominant health problems were respiratory and occurred within the first 30 days of arrival at the yard. Calves were marketed when visually evaluated to have one centimeter of fat cover with complete carcass data gathered on each animal. Calf sex, origin of calf (Southeast *vs.* Midwest), and season of delivery (fall *vs.* spring *vs.* summer) all affected feedlot gain (kg/d). Feedlot gain, individual treatment costs, and mortality rate were 1.47, $0, and 0.1%; 1.42, $19.14, and 3.7%; 1.39, $44.47, and 8.7% for calves treated 0, 1, 2+ times, respectively. Calf mortality rates were not impacted by calf sex, origin of calf, or season of delivery. The percent USDA Prime, Ch0 and Ch+, Ch-, Select, and Standard for 0, 1, 2+ calf carcasses were 1.9, 21.5, 48.8, 25.3, and 2.6; 1, 19.5, 43.4, 30.1, and 5.9; and 0.9, 15.2, 42.8, 30.6, and 10,6, respectively. Yield grades 1 and 2, 3, and 4 and 5 were 52.3, 44.9, and 2.8; 65.9, 32.7, and 1.4; and 71.7, 28.1, and 0.2 for calves treated 0, 1, 2+ times, respectively.

About 68 percent of the calves were Angus-type calves eligible for *Certified Angus Beef*® (CAB®) acceptance with CAB® acceptance for 0, 1, 2+ treated calves being 28.1%, 26.3%, and 22.4%, respectively. As the percentage of "Angus" in the calves increased, quality grade increased, yield grade increased, and percent health treatments decreased. CAB® acceptance rates were impacted by calf sex (steers=25.1% and heifers=34.6%) and season of feedlot delivery (spring=18.2%, summer=24.6%, and fall/winter=29.4%), but not origin of the calf (Southeast *vs.* Midwest). Individual animal treatment costs were influenced by origin of calf (Southeast=$4.18 and Midwest=$9.11), time of arrival at feedlot (spring=$3.94, summer=$3.14, and fall/winter=$6.77), and calf sex (steers=$6.50 and heifers=$3.32). Calves treated two or more times upon feedlot arrival had reduced feedlot gain, reduced quality grade, reduced CAB® acceptance rates, increased treatment costs, and increased mortality rates.

Acknowledgements

Appreciation is expressed to Mike King, Colorado State University biometrician, for his statistical analysis of the data. Appreciation is expressed to the Tri-County Steer Carcass Futurity board members, feedlots, and consignors.

References

[1] Gardner B.A., H.G. Dolezal, L.K. Bryant, F.N. Owens and R.A. Smith, 1999. Health of finishing steers: Effects of performance, carcass traits, and meat tenderness. J. Anim. Sci. 77, 3168-3175.
[2] McNeill J.W., J.C. Paschal, M.S. McNeill, and W.W. Morgan, 1996. Effect of morbidity on performance and profitability of feedlot steers. J. Anim. Sci. 74 (Suppl. 1), 135. (Abstr.).

Effects of dexamethasone on mRNA levels and binding sites of hepatic β-adrenergic receptors in neonatal calves and dependence on colostrum feeding

H.M. Hammon, J. Carron, C. Morel, and J.W. Blum
Division of Animal Nutrition and Physiology, Vetsuisse Faculty, University of Berne, Switzerland

Plasma glucose concentrations in neonatal calves are influenced by colostrum (C) feeding and by glucocorticoids. We have recently shown that dexamethasone (DEXA) increased plasma glucose concentrations in neonatal calves, but did not stimulate key enzymes of hepatic gluconeogenic pathways[1]. As catecholamines are involved in regulation of hepatic glucose metabolism, we have investigated mRNA levels and receptor binding of β-adrenergic receptors in liver of neonatal calves and have tested the hypothesis that DEXA treatment and C feeding affect gene expression and/or receptor binding of β-adrenergic receptors.

Calves (n = 7, each group) of FD- and FD+ were fed a milk-based formula (F) that had the same nutrient composition as C, but almost no hormones and growth factors, whereas calves of CD- and CD+ were fed C. DEXA (30 μg/[kg body weight d]) was injected in calves of FD+ and CD+. Calves were fed C or F for the first 3 d, milk replacer on d 4, and were euthanized on d 5 of life. Liver samples were taken and mRNA concentrations of β1-, β2- and β3-adrenergic receptors were measured by real-time PCR[3]. Relative quantification was done using glyceraldehyde-3-phosphate dehydrogenase as reference gene transcript[1]. Binding of β-adrenergic receptors was evaluated by radioreceptor binding assays[3] using 3H-CGP-12177 (β-antagonist) as ligand. Competitive binding studies were performed to evaluate β-adrenergic subtypes with alprenolol, propranolol (both β1- and β2-antagonists), ICI-188,551 (β2-antagonsit), atenolol (β1-antagonist) and adrenaline as competitors. For quantification of receptor sites saturation binding assays were performed[3].

Abundance of mRNA for β2-adrenergic receptors in liver was highest (P<0.01) and for β1-adrenergic receptors was higher (P<0.01) than for β3-adrenergic receptors. DEXA treatment decreased (P<0.05) β1- and β2-adrenergic receptors, and β3-adrenergic receptors were higher (P<0.05) in CD- than in FD-. Competitive binding studies revealed highest affinities for alprenolol, propranolol and ICI-188,551, which did not significantly differ from each other. Atenolol in concentrations up to 10-5 M did not displace 3H-CGP-12177 from specific binding. Binding curve for adrenaline was best fitted by a two-receptor model. Dexamethasone treatment decreased (P<0.05) β-adrenergic binding sites in liver of neonatal calves, whereas the binding affinity of 3H-CGP-12177 was not affected by DEXA treatment or C feeding. Binding sites correlated positively with mRNA levels of β2-adrenergic receptors (r = 0.56; P<0.01). Although mRNA of all three β-adrenergic receptor subtypes was found in liver of neonatal calves, β2-adrenergic receptors were the dominant subtype. Binding studies indicated binding of 3H-CGP-12177 to β2-adrenergic receptors. DEXA treatment decreased β2-adrenergic binding sites and this was regulated at the transcription level. Reduced β2-receptors corresponded with reduced activities of gluconeogenic enzymes after DEXA treatment[1]. C feeding barely affected β-adrenergic binding sites in the liver of neonatal calves.

References

[1] Hammon H.M., S.N. Sauter, M. Reist, *et al.*, 2003. Dexamethasone and colostrum feeding affect hepatic gluconeogenic enzymes differently in neonatal calves. J. Anim. Sci. 81, 3095-3106.

Section B.

[2] Inderwies T,. M.W. Pfaffl, H.H.D. Meyer, *et al.*, 2003. Detection and quantification of mRNA expression of a- and b-adrenergic receptor subtypes in the mammary gland of dairy cows. Domest. Anim. Endocrinol. 24, 123-135.

[3] Hammon H.M., R.M. Bruckmaier, U.E. Honegger, *et al.*, 1994. Distribution and density of a- and b-adrenergic receptor binding sites in the bovine mammary gland. J. Dairy. Res. 61, 47-57.

Pathogenesis of avian growth plate dyschondroplasia

N.C. Rath, J.M. Balog, G.R. Huff, and W.E. Huff
Poultry Production and Product Safety Research Unit, Agricultural Research Service, USDA, Fayetteville, AR 72701, USA

Tibial dyschondroplasia (TD) is a metabolic cartilage disease characterized by the presence of an avascular plug of cartilage in the proximal tibia and tibiotarsal bones. It is a leading cause of lameness in meat-type poultry. The pathology of TD is well characterized but the etiology of the naturally occurring disease is unclear. Although genetic predisposition is considered to be a factor for the susceptibility to TD, it is more likely that exogenous factors trigger its pathogenesis. Growth plate development is a complex process rendered through a multitude of growth and metabolic interactions which culminate in endochondral bone development. Consequently, the metabolic deficiency of some key nutritional, growth, and developmental factors have been suggested to cause the pathogenesis of TD. Accordingly, these deficiencies have been thought to prevent post-proliferative chondrocytes to undergo hypertrophic maturation and subsequent bone formation. Studies in our lab have demonstrated that premature chondrocyte death is the cause for the retention of the avascular plug of cartilage which is metabolically inactive. It explains why the TD-affected tissues lack many of the growth and metabolic factors. The majority of the studies on TD are done comparing normal growth plate versus tissues in TD lesions; therefore, it is difficult to determine what initiates its pathogenesis.

To address this issue we have developed an experimental model of TD by feeding one week-old broiler chicks 100 ppm thiram, a dithiocarbamate pesticide, for 24-48 h which causes severe TD lesion in more than 85% of birds. The model allows comparison of changes in the growth plate induced by thiram which may be consequential to the subsequent development of TD lesions. To determine what changes are brought about by thiram, we compared several cellular and biochemical profiles of growth plate tissues of control and treated birds using parameters of cell multiplication, gene expression associated with chondrocyte development and maturation, and cell death.

Our results show that thiram does not diminish the expression of genes that would indicate impairment of chondrocyte maturation and arrest of the cells in a prehypertrophic state. Neither was there any cell multiplication that would lead to cartilage accumulation, and thickening in post-proliferative zones. However, thiram treatment caused capillary endothelial cell and chondrocyte death. Histological observation showed a gradual shrinkage in the cytoplasmic and nuclear volume of the chondrocytes which lead to their appearance resembling, prehypertrophic, not hypertrophic chondrocytes.

In light of our past studies which showed the chondrocytes of TD lesions failed to grow in culture and the fact that the avian growth plate is a well vascularized tissue, we surmise that thiram exerts its TD-producing effect through its ability to cause endothelial cell death by preventing the vascularity of the growth plate. The loss of blood vessels secondarily causes metabolic deficiency preventing the maturational process in prehypertrophic chondrocytes. Because several dithiocarbamate and their congeners are widely used to control insects, fungi, and rodents, it is possible that an inadvertent exposure to such chemicals during the period of intense growth may be a cause of tibial dyschondroplasia in poultry.

Concentrations of haptoglobin and fibrinogen during the first ten days after calving in dairy cows with acute endometritis

M. Drillich, D. Voigt, D. Forderung and W. Heuwieser
Faculty of Veterinary Medicine, Clinic for Reproduction, Free University of Berlin, Germany

Haptoglobin and fibrinogen are acute phase proteins indicating acute inflammatory processes[1,2,3]. The aim of this study was to analyze the blood levels of these proteins during the first ten days post partum (pp) in cows with and without acute endometritis.

The study was conducted on a commercial dairy farm in Germany, housing 1000 dairy cows in free stall facilities with rubber mats and slotted floor. Blood samples were taken from cows at parturition, at day 4 to 5, day 6 to 7 and day 10 to 11 pp. Haptoglobin was measured by Phase[TM] Range Haptoglobin Assay (BioRepair) based on by photometric extinction. Fibrinogen was quantified with a coagulometer (Biomatic 2000, Dade Behring). All cows were examined 4 to 5 days pp for signs of acute endometritis, defined as fetid discharge and a rectal temperature >39.5°C [4,5]. In group A, cows with endometritis were treated with an antibiotic drug (1.0 mg/kg Ceftiofur, s.c.) on three to five consecutive days. In group B, cows with endometritis received an initial treatment with 2.2 mg/kg of Flunixin-Meglumin (i.v.) and the same antibiotic treatment regime as group A. Cows without acute endometritis were regarded as healthy control group (group C).

At parturition, haptoglobin was slightly higher in cows diagnosed with acute endometritis at day 4 to 5 pp than in healthy cows. For cows in groups A and B, concentrations of haptoglobin and fibrinogen were higher at day 4-5, day 6-7, and day 10-11 pp compared to healthy cows in group C at the same time. For all groups, concentrations of haptoglobin and fibrinogen decreased significantly from day 4 to 5 to day 10 to 11 pp. No differences were found between groups A and B.

In conclusion, haptoglobin and fibrinogen are useful parameters to monitor acute endometritis early post partum. The additional treatment with the non-steroidal anti-inflammatory drug Flunixin-Meglumin at day 4 to 5 pp for cows with acute endometritis showed no effect on concentrations of haptoglobin or fibrinogen.

References
[1] Ek N., 1972. Acta Vet. Scand. 13, 175-184.
[2] Alsemgeest S.P., *et al.*, 1994. Vet. Quart. 1, 21-23.
[3] Hirvonen J, *et al.*, 1998. Theriogenology 51, 1071-1083
[4] Smith B.I., *et al.*, 1998. J. Dairy Sci. 81, 1555-1562.
[5] Drillich M, *et al.*, 2001. J. Dairy Sci. 84, 2010-2017.

Extracellular pH alters innate immunity by decreasing the production of reactive oxygen and nitrogen species, but enhancing phagocytosis in bovine neutrophils and monocytes

D. Donovan
Department of Large Animal Medicine, College of Veterinary Medicine, Athens, GA, USA

Acidic pH has an effect on both humoral and cellular immunity. The objective of this experiment was to determine the effects of extracellular pH on phagocytosis, and on the production of nitric oxide (NO) and reactive oxygen species (ROS) by bovine leukocytes.

Sixty milliliters of blood was obtained by jugular venipuncture from cows at least 250 days in milk with an average body condition score of 3.2 for use in ROS (n = 10), phagocytosis (n= 12), and NO (n = 4) assays. One medium used in these studies was composed of Phosphate Buffer Saline, with the addition of 0.5% Bovine Serum Albumin and 5 mM Glucose (PBG). PBG was aliquoted, and individual samples adjusted to pH 6.0, 6.4, 6.8, 7.2, 7.6, 8.0 with HCl or NaOH. PBG medium containing 3×10^6 total leukocytes per ml was used for measurement of ROS and phagocytosis. ROS was assessed in quadruplicate (100µl of cells in 96 well plates) after stimulation with 10^{-6}, 10^{-7}, 10^{-8}, and 10^{-9} M phorbol myristate acetate (PMA) for 1 hr by measuring the conversion of dihydrorhodamine 123 to its fluorescent form in comparison with PBG controls. Phagocytosis was assessed by incubation of 200µl of cells with commercial bodipy labeled *S. aureus* or *E. coli* particles for 1 hr. The number of bacteria associated with the leukocytes was evaluated by flow cytometry. Minimal Essential Medium with addition of 10% Fetal Bovine Serum, 2 mM L-Glutamine, 2 mM sodium pyruvate and 50 µg/ml of Gentamycin sulfate (MEMG) was aliquoted and individual samples adjusted to pH 6.0, 6.4, 6.8, 7.2, 7.6, 8.0 with HCl or NaOH. One hundred microliters of MEMG, after adjustment to the desired pH, was added to quadruplicate wells with 6×10^5 mononuclear cells. To induce nitric oxide, 10 µg/ml, 1 µg/ml, 0.1 µg/ml, or 0.01 µg/ml *E. coli* 055 LPS was added to wells at each pH. Supernatants were removed after sixty hours of incubation and nitric oxide production was evaluated using the Greiss reaction. Data were analyzed by mixed procedures of SAS 8.2 (2002).

It was determined that pH (P<0.01) greatly effected the production of ROS, and acidic pH decreased (P=0.031) the production of ROS relative to alkaline conditions. Phagocytosis of multiple *E. coli or S. aureus* particles showed a tendency to be increased under acidic medium pH relative to basic pH conditions. Alkaline conditions appeared to favor nitric oxide production over acidic conditions (P<0.05). Acidosis appears to hinder the functions of innate immunity, which could result in a delayed response to bacterial infections. In other studies, lymphocyte recall response to antigen and circulating virus neutralizing antibody titers were diminished in acidotic animals.

High nitrite/nitrate status in neonatal calves is associated with increased plasma levels of S-nitrosoalbumin and other S-nitrosothiols

I. Cattin[1], S. Christen[2], S.G. Shaw[3] and J.W. Blum[1]
[1]*Division of Nutrition and Physiology, Vetsuisse Faculty;* [2]*Institute for Infectious Diseases;*
[3]*Department of Clinical Research, Medical Faculty, University of Berne, Berne, Switzerland*

Neonatal calves show high, endogenously produced plasma nitrite/nitrate (NO_2/NO_3) levels that rapidly decrease to low levels in mature cattle[1]. NO_2/NO_3 are the principal breakdown products of nitric oxide (NO), which has important physiological functions such as control of blood flow. Binding of NO to proteins and non-protein thiols (S-nitrosylation) is thought to prolong its action. The aim of the present study was to determine whether high plasma NO_2/NO_3 levels in newborn calves are associated with high S-nitrosothiol (RSNO) levels, and whether albumin was one of the targets for nitrosylation.

Blood plasma obtained from calves (n=13) on day 0 (before first feeding), 1, 7 and 112 after birth and from adult cows (n=8) were analysed for RSNO by measuring $CuSO_4$-catalyzed NO release with DAF-2. The same plasma samples were analyzed for NO_2/NO_3, total protein, albumin, and the water-soluble antioxidants ascorbate and urate. S-nitrosylated plasma proteins were studied using the biotin switch method in which protein-bound RSNO are specifically labelled with biotin after reduction with ascorbate.

As reported previously[1], newborn calves had high plasma NO_2/NO_3 levels that declined to low levels in adulthood. This change in NO_2/NO_3 status was associated with high levels of plasma RSNO in newborn calves that progressively declined with increasing age. There was a large decrease in ascorbate and urate concentrations between day 0 and 1, which could indicate a possible involvement of these two molecules in the mobilization of NO during early development. Albumin was amongst the most prevalent S-nitrosylated proteins in plasma of newborn calves. S-nitrosoalbumin levels rapidly declined during early development and were barely detectable in mature animals. S-nitrosylation of immunoglobulin G was also observed.

To study the potential physiological role of changing RSNO levels during development, we assessed the effect of blood serum collected from cattle at different ages (d 0 and d 7 after birth, as well as 3-5 y) on rat blood vessel reactivity *in vitro*. Authentic S-nitrosoalbumin dose-dependently reduced norepinephrine-induced vasoconstriction in denuded rat aorta segments. Addition of serum increased vasoconstriction initially, followed by a prolonged period of progressive relaxation. Inhibition of vasorelaxation by the NO scavenger oxyhemoglobin in the presence, but not in the absence of serum, strongly suggests that serum of calves contains an NO-releasing factor. However, serum samples from adults did not differ in their ability to inhibit norepinephrine-induced vasoconstriction compared to samples from young animals.

High NO_2/NO_3 levels in neonatal calves are associated with high levels of RSNO, among which S-nitrosoalbumin is a major molecular species, that decline with increasing age. RSNO such as S-nitrosoalbumin may be important for the development and function of the vascular system in young calves.

References
[1] Blum J.W., C. Morel, H.M. Hammon, *et al.*, 2001. High constitutional nitrate status in young cattle. Comp. Biochem. Physiol. A 130, 271-282.

Effects of 0.03% dietary β-glucans on nonspecific/specific immunity, oxidative/antioxidative status and growth performance in weanling pigs

S. Schmitz, S. Hiss and Helga Sauerwein
Institute of Physiology, Biochemistry and Animal Hygiene, University of Bonn, Germany

Immunomodulatory feed additives offer alternatives to anti-microbial growth promoters in swine production. β-1,3/1,6-glucans are branched carbohydrates derived from yeast cell walls. Based on the reports about stimulatory effects of these substances on the immune system, the present study aimed to determine the influence of β-glucans on specific and nonspecific immune defense mechanisms, oxidative stress, antioxidant capacity and growth performance in weaned pigs in order to evaluate the potential benefit of the application in practise.

30 piglets were randomly allocated to either the control group (C; n = 10) or the treatment group (T; n = 20). After weaning at 25 – 28 days of age, the two groups were fed the same conventional feed with 0 or 0.03% β-glucans (Antaferm-M, Dr. Eckel GmbH, Niederzissen, Germany). After 4 weeks of treatment all animals were fed a diet without β-glucans. Oxidative stress was monitored in whole blood directly after sampling using the D-ROM® system (detection of reactive oxygen metabolites)[1]. Nonspecific immune parameters recorded were phagocytic activity and oxidative burst of neutrophile granulocytes in whole heparinised blood via flowcytometry and also haptoglobin serum concentrations (Hp) by enzyme immuno assay. The specific immune defense was investigated by means of Immunoglobulin G and A serum concentrations (IgG and IgA) by ELISA. The total antioxidant capacity was quantified using the TEAC assay[2]. Blood samples were taken 1 week before weaning and in weekly intervals thereafter. Measurements in whole blood or from blood cells were limited to the samples obtained 1 week before weaning, 3 weeks after weaning and 2 weeks after the end of β-glucan feeding. Individual body weights were recorded weekly and feed intake was quantified daily per pen. All data were analyzed with the PROC MIXED procedure of SAS. The model included terms for the fixed effects of dietary treatment, time and the appropriate interactions.

For the parameters of the nonspecific defense, increased (P<0.05) oxidative burst was observed 2 weeks after the end of glucan feeding in the treatment group when compared to the controls. Phagocytic activity and Hp concentrations were not affected by dietary treatment. A β-glucan feeding x time interaction was observed for IgA (P<0.05) and TEAC levels (P<0.05). The two parameters increased over time in both groups. IgG and oxidative stress showed no glucan-related differences. Moreover, alterations related to weaning were observed within the groups mainly for Hp and oxidative stress, whereas both increased (P<0.0001) after weaning. These observations support that weaning implies stress for the animals. There was a treatment x time interaction (P<0.05) for the average daily gains (ADG). Weaning resulted in a reduction in ADG in C and T group (P<0.005 and P<0.0001, respectively). There were no differences in food intake between the groups.

The results of this study indicate that dietary β-glucans may affect the oxidative burst response of neutrophile granulocytes in young pigs; however the impact of the β-glucan preparation used herein on immune function seems marginal and offers no benefit for growth performance.

References

[1] Cesarone M.R., G. Belcaro, M. Carratelli, *et al.*, 1999. A simple test to monitor oxidative stress. Int. Angiol. 2, 127-130.

[2] Miller N.J., C. Rice-Evans, M.J. Davies, *et al.*, 1993. A novel method for measuring antioxidant capacity and its application to monitoring the antioxidant status in premature neonates. Clin. Sci. 84, 407-412.

Influence of organic nutrition and housing on selected immunological and metabolic properties in fattening pigs

S. Millet[1], E. Cox[2], M. Van Paemel[1], B.M. Goddeeris[2] and G.P.J. Janssens[1]
[1]Laboratory of Animal Nutrition, Ghent University, Heidestraat 19, B-9820 Merelbeke, Belgium;
[2]Laboratory of Immunology, Ghent University, Salisburylaan 133, B-9820 Merelbeke, Belgium

Major differences in organic pig farming are management measures and the use of organic nutrition. The lower stocking density, the presence of an outdoor area and a rooting and lying area with straw, lead to changed climatic conditions in comparison with pigs housed in a conventional barn. The present trial evaluated the effect of housing and nutrition on haptoglobin concentration during the growing-finishing period and lactate concentration at slaughter linked to initial pH of the meat.

A group of 64 pigs of a commercial breed (Seghers Hybrid) was divided over either organic or conventional housing. All pigs were born in a conventional barn. The pigs assigned to the organic barn were moved to this housing system 1 week after weaning. Both housing systems consisted of 8 groups of 4 animals (2 barrows and 2 sows), with each time 4 pens receiving a conventional diet and 4 pens receiving an organic diet. Feeds were isocaloric. Organic feed and housing were according to European guidelines on organic farming. Serum haptoglobin was measured at the start of the experiment, 3, 6 and 9 weeks later and at slaughter. Lactate concentration was determined in fresh stabbing blood at the slaughterhouse. At 40 min post mortem, pH was measured in the loin around the 13th costa (*musculus longissimus thoracis et lumborum*) of both carcass sides.

Organic housing induced lower haptoglobin levels when compared to conventional housing, but only at slaughter ($P=0.019$; Table 1). Lactate concentration in the stabbing blood was significantly higher in the pigs of the conventional housing. The higher haptoglobin concentration at slaughter coincided with higher lactate concentrations in the stabbing blood and a slightly higher initial pH in the *m. longissimus thoracis*.

Due to the lack of difference in haptoglobin concentration during the trial, we cannot conclude that housing type affected health status. The higher haptoglobin concentration at slaughter in the pigs of the conventional barn might have resulted from preslaughter handling. The larger space allowance probably increased the spontaneous activity of the organic pigs. With increasing fitness, muscles generate less ATP through anaerobic pyruvate catabolism[1]. Pigs from the organic house had lower blood lactate levels at slaughter and might have spent less energy during the preslaughter treatment. Indeed, the lower plasma lactate concentration in the OH pigs might be due to improved coping with stress around slaughtering, which can affect meat quality. However, the changes in initial pH were subtle and were fairly good in all animals.

Organic nutrition does not affect haptoglobin concentration over a growing-finishing period, but organically housed pigs can have the ability for improved coping with stress at slaughtering, which was reflected in slightly higher pH values 40' post mortem.

Acknowledgements

The Ministry of the Flemish Community supported this study. Data have partly been published in The Veterinary Journal. Doi information 10.1016/j.tvjl.2004.03.012

Section B.

Table 1. Haptoglobin and lactate concentration at slaughter and pH in the m. longissmius thoracis (LT) 40' post mortem.

	OH		CH			P		
	OF	CF	OF	CF	SEM	Housing	Nutrition	Hous. x Nutr.
Haptoglobin (log hapt concentration)	-0.66	-0.57	-0.31	-0.01	0.07	<0.001	0.098	0.389
Lactate (mmol/l)	5.3	5.5	8.3	7.3	0.5	0.015	0.683	0.502
pH LT 40' pm	6.164	6.163	6.000	6.075	0.024	0.008	0.421	0.415

OH= organic housing, CH= conventional housing, OF= organic feeding, CF= conventional feeding

References

[1] Jorgensen P.F. and J.F. Hyldgaard-Jensen, 1975. The effect of physical training on skeletal muscle enzyme composition in pigs. Acta Vet. Scand. 16, 368-78.

Section C.
Animal behaviour and welfare in intensive production systems

Associations of cow comfort indices with total lying time, stall standing time, and lameness

K.V. Nordlund, N.B. Cook and T.B. Bennett
School of Veterinary Medicine, University of Wisconsin, Madison, WI, USA

Dairy consultants use several indices of cows standing in freestall barns to evaluate freestall comfort on dairy farms[1]. The objective of this study was to measure several standing indices for one 24-hour period in the high group pens of a selection of Wisconsin dairy herds, and determine the relationship of the herd comfort index to actual lying time in stalls and lameness prevalence in the herd.

Twelve Wisconsin dairy herds were selected to include six sand freestall herds and six mattress freestall herds. For one 24-hour period, the mature cow high group pen on each farm was video filmed. For 10 cows in each herd, daily times spent lying down in the stall and times spent standing in the stall were obtained. All cows in the high group pen were locomotion scored using the four-point system described by Cook[2]. For each herd, standing indices were determined each hour of the 24-hour period. Three standing indices were calculated; Cow Comfort Index (CCI: proportion of cows in stalls that are lying down), Stall Standing Index (SSI: proportion of cows in stalls that are standing, *i.e.* 1-CCI) and the Proportion Eligible Lying (PEL: proportion of cows not eating that are lying down).

The start time of the morning milking was used as a reference point to align the hourly data for each farm. Differences in cow comfort indices between sand and mattress herds were examined using repeated measures in the mixed procedure of SAS. $P<0.05$ was used to determine significance. The association between hourly cow comfort indices for each herd and the mean 10-cow daily lying and stall standing times and the mean pen lameness prevalence was examined using PROC REG and PROC GLM in SAS. The optimal hourly relationship between the indices and the outcome variables was selected based on an optimal combination of adjusted R^2 and P value. A P value of <0.002 was used to determine significance in order to reduce the chances of making an erroneous conclusion due to the multiple comparisons being made.

There was a significant effect of sand or mattress base on CCI and SSI (P=0.002) and on PEL (P=0.003). The hourly effect was also significant for each variable (P<0.001). More CCI variability was observed in mattress herds compared to sand herds, and the average over the 24h period was 14% lower. All of the indices were poor predictors of mean daily lying time at all hours of the day. There was a significant relationship between CCI/SSI and mean daily standing time at 5h and 2h before the morning milking. At 2h before the morning milking, there was a significant relationship between CCI/SSI and lameness prevalence in the pen (R^2 0.89, P=0.0005), but there was also a significant effect of base (P=0.0008). SSI greater than 24% was uniformly associated with pen lameness prevalence rates greater than 20%. Discussion: Traditional indices of cow comfort do not predict mean daily lying times of individuals within the pen. However, the CCI or SSI do predict time spent standing in the stall. Cook *et al.*[3] have documented that lame cows spent increased time standing in the stall in mattress facilities, but not in sand barns. We therefore propose the use of the SSI, taken 2h before the morning milking as a predictor of standing behavior and related lameness prevalence rates.

References

[1] Overton, M.W., D.A. Moore and W.M. Sischo, 2003. Comparison of commonly used indices to evaluate dairy cattle lying behavior. In: Proc. Dairy Housing Conf., Fort Worth, TX. Amer. Soc. Agric. Engineers, St Joseph, MI, pp. 125-130.
[2] Cook, N.B., 2003. Prevalence of lameness among dairy cattle in Wisconsin as a function of housing type and stall surface. JAVMA 223, 1324-1328.
[3] Cook N.B., K.V. Nordlund and T.B. Bennett, 2004. Effect of free stall surface on daily activity patterns in dairy cows, with relevance to lameness prevalence. J. Dairy Sci. (Accepted).

Interaction of lameness with sand and mattress surfaces in dairy cow freestalls

K.V. Nordlund, N.B. Cook and T.B. Bennett
School of Veterinary Medicine, University of Wisconsin, Madison, WI, USA

Cook[1] found a significantly lower prevalence of lameness in a group of Wisconsin dairy herds using sand stalls compared to other stall surfaces. The objective of this study was to identify behavioral differences between cows housed in free stalls bedded with deep sand (SAND) and cows housed in free stalls with rubber crumb mattresses (MAT).

Twelve Wisconsin dairy herds were selected to include six sand stall herds (SAND) and six mattress stall herds (MAT). All lactating cows in the entire herd were locomotion scored and the prevalence of clinical lameness was calculated. For one 24-hour period, the mature cow high group pen on each farm was video filmed. Ten cows in the pen were randomly selected, locomotion scored, and color marked. Using the video recordings, location in the pen (alley or stall), activity (standing, lying, feeding, drinking) and time spent performing each activity (to the nearest minute) was recorded for each marked cow. The data were analyzed using the mixed procedure of SAS. One way ANOVA was used to compare cow and herd level data and a mixed effect model was created to investigate differences in cow behavior between SAND cows and MAT cows.

Mean (SE) lameness prevalence was significantly higher in MAT herds (24.0%, 2.1) than in SAND herds (11.1%, 1.3), (P<0.001). There were no significant differences in parity, days in milk at the day of visit, last DHIA recorded daily milk yield and last DHIA recorded ME305 milk yield between filmed cows in SAND herds and those in MAT herds. Cows with normal ambulation scores behaved similarly in both types of barn with respect to time spent milking, eating, socializing and lying down. However, there were significant differences in the amount of time spent standing in the stalls at all ambulation scores. Normal cows in MAT herds stood in stalls for 2.4h/d compared to cows in SAND herds that stood for 1.7h/d. Time up in stall for slightly lame cows in MAT herds was 4.4h/d compared to 2.1h/d in SAND herds (P<0.0001) and for moderately lame cows in MAT herds it was 6.1h/d compared to 1.8h/d in SAND herds (P=0.0183).

Altered daily activity time budgets were identified in lame cows on mattress free stalls. Increased time spent standing in the stall compressed the time available to lame cows for other activities such as socializing, feeding and lying down. Changes in time budgets were not observed in lame cows on sand free stalls. We speculate that the behavior of lame cows changes on mattresses due to either increased pain or decreased security during either the lying or rising motions, in contrast to the unchanged behavior of lame cows on sand.

References
[1] Cook, N.B., 2003. Prevalence of lameness among dairy cattle in Wisconsin as a function of housing type and stall surface. JAVMA 223, 1324-1328.

Table 1. Effect of locomotion score and stall base type (SAND v MAT) on daily time budgets.

Mean (SD) activity, hr/day	Locomotion score					
	1 (normal)		2 (slight lame)		3 (moderate lame)	
Stall base type	Mat	Sand	Mat	Sand	Mat	Sand
Lying time	12.0 (2.1)	12.0 (1.6)	11.7 (2.3)	12.0 (2.8)	10.0 (4.9)	12.8 (0.2)
Standing in stall	2.4 (1.4)	1.7 (0.8)	4.4 (2.2)	2.1 (1.9)	6.1 (3.9)	1.8 (1.6)
Time standing in alley	2.8 (1.7)	2.3 (1.2)	1.6 (0.5)	2.2 (1.4)	1.4 (0.7)	1.8 (0.1)
Time Up feeding	4.3 (1.0)	4.7 (1.0)	3.8 (1.1)	4.6 (1.3)	3.5 (1.1)	5.1 (0.4)
Time Up milking	2.5 (0.7)	3.3 (1.2)	2.6 (0.6)	3.2 (0.8)	3.0 (0.9)	2.7 (1.3)

Section C.

Electric and magnetic fields (EMF) affect milk production and bevavior of cows

D. Hillman[1], C.L. Goeke[1], D. Stetzer[2], K. Mathson[2], M.H. Graham[3], H.H. VanHorn[4], C.J. Wilcox[4]
[1]Michigan State U, E. Lansing; [2]Stetzer Electric, Blair, WI; [3]U. Cal, Berkeley; [4]U. Florida, Gainesville, FL, USA

Dairy farmers complained that animal behavior and milk production were affected by electricity below 0.5 Volt or 1.0 mA at cow contact. Power quality (PQ) was investigated on 12 farms to determine the relationship to animal performance.

Milk yield was estimated from daily tank weights and number of cows milked[1]. Event recorders plugged into wall outlets measured number of transient events, transients, voltage (peak-peak), waveform degree angle, *etc.* on farms. Data from 1705 cows and 939 data points were analyzed by multi-herd least-squares multiple regression and SAS-ANOVA statistical programs.

Milk/cow decreased as transient events increased: $Y = -0.028$ kg/transient event (range 0 to 122/d), and as N-G transients increased/d. Step-potential voltage and frequency of currents were measured by oscilloscope from electrodes grouted into the floor of milking stalls. Milk decreased as the number of 3rd, 5th, and 7th harmonics and triplen harmonics increased ($P<0.003$). EMF exceeded 10 kV/m. EMF decreased melatonin, increased Insulin-like growth factor (IGF-1), and electrolytes in blood and CSF of cows[2] and decreased melatonin in humans[3] exposed to EMF, and impaired health of humans exposed to 180-Hz current and 3rd,5th, and 7th harmonics[4]. Harmonics are generated by nonlinear loads, *i.e.*, capacitor switching to balance loads on lines, variable speed motors, and switch-mode devices in electronic equipment on single-phase lines, *e.g.* computers, printers, MRI, *etc.* When harmonics are combined in a three-phase neutral, they produce odd numbered, integer multiples of the fundamental 60-Hz, which return exclusively through the neutral, causing heating of circuits, motors, panels, and may cause fires. Installing a shielded neutral isolating transformer between the utility and farm secondary removed EMF, and mitigated health and production problems on dairy farms. Transients and harmonics are pollution on power lines, as microbes are in milk.

References
[1] Hillman, D., D. Stetzer, M.H. Graham, C.L. Goeke, K. Mathson, H.H. VanHorn and C.J. Wilcox, 2003. Relation of Electric Power Quality to Milk Production of Dairy Herds. Presentation Paper 033116, ASAE, St. Joseph, MI. www.pq.goeke.net.
[2] Burchard, J.F. and Electric and Magnetic Field Research at McGill University, 2003. Stray Voltage on Dairy Farms, Proceedings 2003, Coop. Extension, Ithaca, NY.
3. Burch J.M., J.S. Reif, C.W. Noonan and M.G. Yost, 2000. Melatonin Metabolite Levels in Workers Exposed to 60-Hz Magnetic Fields: Work in Substations and with 3-Phase Conductors. J. Occup. Environ. Med. 42, 136-142.
4. Kaune, W.T., T. Dovan, R.I. Kavet, D.A. Savitz and R.R. Neutra, 2002. Study of High- and Low-Current-Configuration Homes From the 1988 Denver Childhood Cancer Study, Bioelectromagnetics 23, 177-188.

Effect of roughages added to the milk replacer diet of veal calves on behavior and gastric development

A.M. van Vuuren, N. Stockhofe, J.J. Heeres-van der Tol, L.F.M. Heutink and C.G. van Reenen
Animal Sciences Group, Wageningen UR, PO Box 65, 8200 AB Lelystad, The Netherlands

A factorial study was performed, to evaluate the effect of type, amount and particle size of roughages on oral behavior and pathology of rumen and abomasums in veal calves. Three roughages were tested: wheat straw, dried corn silage and dried corn cob silage at two particle lengths (chopped at 4 cm or ground at 1 cm) and at two intake levels (250 or 500 g per day). Three control treatments were included: milk replacer only, milk replacer with iron and milk replacer with hay ad libitum. Each treatment was allocated to 4 pens of 5 calves. Abnormal oral behavior was recorded at the end of the fattening period (wk 22). Calves were slaughtered at 290 kg LW. The reticulo-rumen was emptied and subsequently weighed. The development of the rumen mucosa was estimated by a qualitative macroscopic evaluation. Presence and number of alterations in the abomasal mucosa were recorded as erosions or abomasal ulcers.
At wk 22, calves fed milk only or with iron showed the highest frequency of abnormal oral behavior like tongue playing and tongue rolling (Table 1). Adding hay or straw to the diet reduced the frequency of abnormal oral behavior and increased the frequency of rumination compared to the milk only treatment. Particle size and amount of roughage had no additional effect on abnormal behavior. Amount of roughage but not particle length increased eating time from 90 to 192 min per day (se = 30 min). Highest weight of the reticulo-rumen and best-developed ruminal mucosa were observed in calves that received milk with hay. Compared to milk only, rumen mucosa was better developed in calves additionally fed with corn silage and corn cob silage, but less when compared with hay-fed calves. Feeding of straw was not very effective in initiating ruminal mucosa development: in 80% of the straw-fed calves mucosal development was comparable to that achieved with milk only. Amount of roughage but not particle length influenced weight and development of the reticulo-rumen. On average, the increase in roughage intake from 250 to 500 g increased the weight of the reticulo-rumen by 260 g. Abomasal erosions and ulcers were nearly exclusively observed in the pyloric region of the abomasums and occurred in all treatment groups. However, the proportion of calves with ulcers or erosions and also the number of lesions were lowest in the milk only groups and in the group fed with hay *ad libitum*. Adding roughage to the diet increased the number of lesions.
From these results we concluded that administration of high fiber roughages may have beneficial effects on oral behavior of veal calves, but may have detrimental effects on the integrity of the mucosa of the abomasum. Thus, administration of roughage to veal calves as required by EU legislation, will not always improve the animal's well-being.

Section C.

Table 1.

Parameter	None	Fe	Hay	Straw	Corn silage	Corn cob silage
Abnormal oral behavior[1]	24.8[cd]	31.4[d]	5.4[a]	10.5[b]	20.4[c]	19.3[c]
Ruminating[1]	6.9[a]	5.1[a]	16.6[b]	12.0[b]	6.6[a]	5.6[a]
Low rumen mucosa development[2]	100	95	0	80	24	25
Rumen weight, kg	1.45[c]	1.50[c]	2.50[d]	1.80[a]	2.00[b]	2.10[b]
Abomasal erosions[2]	15[a]	30[a]	30[ab]	39[b]	25[a]	20[a]
Abomasal ulcers[2]	25[a]	35[a]	40[a]	73[b]	65[b]	78[b]

[1]Proportion of observations, %
[2]Proportion of calves with observation, %

Transportation stress of cattle by railway

Z. Cui[1], B. Han[2], G. Gao[1] and D. Cui[1]
[1]College of Animal Science & Veterinary Medicine, Inner Mongolia Agricultural University, Huhhot 010018, P.R. of China; [2]College of Veterinary Medicine, China Agricultural University, Beijing 100094, P.R. of China

The objectives of this study were to evaluate the epidemiological investigations of transportation stress in cattle by railway transport from Inner Mongolia to Hong Kong, China and the control methods of the stress condition.

509 native yellow commercial cattle from Inner Mongolia were selected as a transportation stress model, which were transported to Hong Kong during 48h in the train between 1989 and 1993. The proportionate clinical and biochemical parameters were carried out.

Cattle were transported by railway for purposes of breeding, fattening or slaughter. In terms of volume transported and economic importance, the most significant trade was the railway transportation of cattle to slaughter. Recent research has attempted to identify situations in the transport chain that were stressful or hazardous to the cattle, or that lower carcass and meat quality. Epidemiological investigations indicated that the total number of cattle from Inner Mongolia to Hong Kong between 1989 and 1993 was 82088, and the loss on the way was 5090, It is, currently, obviously that the number increased year by year. The direct loss was over 304 thousands U.S.$., and the indirect loss was hard to estimate, which included breaking a promise and exporting trade credit which was influenced by the loss of fat and weight, meat quality and instable number of cattle produced by the loss on the way. Other major influences on cattle welfare during railway transport are the quality of stockmanship and stockcar. The most common hazard on the moving train was overloading, which greatly increases the risk of animal injury and occurrence of carcass bruising. Therefore, it seriously influenced the health development of foreign trade. From the point view of clinical signs, the cattle showed severe horrible dyspnea, the body temperature increased more than 40 °C, heart rate and cortisol were elevated as a part of adaptation to the transport environment, but indicating high physical and emotional loads on the animals with no resting possibilities. When the train has run for 2-4 hours, they will tremble, lay down, breathe heavier, and have no response to the driving, and less than 7% were feed and water deprivation during intervals of transportation, and they often die resulted from crowding and trampling by each other in the same compartment. Although, some ill cattle have survived with escort men's care, they will be disabled when arrived at Hong Kong, so have been discarded. When the fur skin striped, the rear muscle showed pale and soft (PSE), therefore, the edible value was lower. Experimental evidence suggests that the most stressful aspects of the transportation chain for cattle affected the plasma total protein, albumin and lactic acid concentrations, which were higher ($P<0.01$, $P<0.01$, $P<0.01$), and transported animals possesed a lower plasma concentration of potassium ($P<0.01$).

However, there were no effective control methods on transportation stress in cattle so far. Good animal handling, Chinese traditional medicine taking, considerate driving technique, and by using correctly designed pens, loading ramps and stock crates were necessary. There are several control approaches during transportation as described below:

1. To promote education of the relative persons, improve management of commercial cattle.
2. To improve installations.
3. To strictly carry out railway transport regulation of fresh and live commodity by the railway transport department.

4. To take Chinese traditional medicine, such as radix acanthopanacis senticosi solution, Huoxiang Zhengqi Shui, and with some other antipyretic analgestic such as antondin, and antipyrine.

The transportation stress is still a problem, which seriously restricts the commercial development. Based on the results, pathogenesis on the condition should be continued.

Section D.
Infection and infectious diseases associated with production systems

Postweaning Multisystemic Wasting Syndrome (PMWS) in pigs

Matti Kiupel
Diagnostic Center for Population and Animal Health, Department of Pathobiology and Diagnostic Investigation, College of Veterinary Medicine, Michigan State University, East Lansing, MI, USA

Abstract

Postweaning Multisystemic Wasting Syndrome (PMWS) is a newly recognized disease in swine that is clinically characterized by progressive weight loss and enlarged lymph nodes and less frequently dyspnea, tachypnea, icterus and diarrhea in weaned pigs of approximately 6-12 weeks of age. Morbidity and mortality are variable and may be as high as 50%, but are typically less than 10%. PMWS was first described in Western Canada, but has now been recognized worldwide, indicating its significant economic importance.

Gross lesions in pigs with PMWS consist of generalized lymphadenopathy that can occur in combination with cutaneous pallor and/or icterus, interstitial pneumonia, hepatic necrosis and atrophy, renomegaly, splenomegaly, fluid-filled, thin-walled sections of lower intestine, with occasional edema of the wall of the cecum, and gastric ulcers and/or edema of the gastric wall. The most distinctive microscopic lesions in affected pigs are lymphoid cell depletion and granulomatous inflammation in lymphoid organs with inconsistently occurring clusters of magenta inclusion bodies in the cytoplasm of macrophages.

Porcine circovirus type 2 (PCV2) is a small, single stranded, circular DNA virus of the *Circoviridae* family. Based on current knowledge, PCV2 is an indisputable pathogen and the principal infectious cause of PMWS. However, serologic surveys suggest that endemic infection with PCV2 is common in swine worldwide and PCV2 antigen and specific IgG titers can be detected in healthy and diseased pigs. Furthermore, most experimental studies required co-infection with PCV2 and another virus or immune stimulation to reproduce PMWS. This paper will review the pathology, epidemiology, experimental reproduction and diagnosis of PMWS.

Postweaning Multisystemic Wasting Syndrome (PMWS)

Postweaning Multisystemic Wasting Syndrome (PMWS), a newly recognized disease of nursery and grower pigs, was first described in 1995/96 by Clark and Harding in Canada (Clark, 1996; Clark, 1997; Harding and Clark, 1997). Since than, PMWS has been diagnosed with increasing frequency in the US (Allan *et al.*, 1998a; Kiupel *et al.*, 1998; Morozov *et al.*, 1998) and Canada (Ellis *et al.*, 1998; Larochelle *et al.*, 1999b; Nayar *et al.*, 1997), and has now been recognized worldwide (Allan *et al.*, 1998a, 1999b; Choi *et al.*, 2000; Harding, 2004; Kennedy *et al.*, 1998a,b; Mankertz *et al.*, 2000; Onuki *et al.*, 1999; Rosell *et al.*, 2000b; Segales *et al.*, 1997). Retrospective studies have confirmed that PMWS existed prior to 1990 (Rosell *et al.*, 2000b; Sato *et al.*, 2000), but there has been a significant increase of cases since 1997 (Sato *et al.*, 2000).

PMWS is clinically characterized by progressive weight loss, enlarged lymph nodes, and less frequently dyspnea, tachypnea diarrhea, pallor and icterus in postweaned pigs. (Clark, 1997).

Morbidity and mortality are variable and may be as high as 50%, but are typically less than 10%. Higher death rates appear to be more common in Europe, particularly in the UK. Gross lesions in pigs with PMWS consist of generalized lymphadenopathy (Figure 3) in combination with one or a multiple of the following lesions: (1) cutaneous pallor and icterus (Figure 1), (2) diffusely noncollapsed, firm, rubbery lungs that have surface mottling with pronounced grayish lobules and scattered reddish-brown areas (Figure 2), (3) yellow atrophic and/or fibrotic livers (Figures 4 and 6), (4) grossly enlarged, edematous kidneys with a semitranslucent appearance or diffusely scattered white foci on the subcapsular surface (Figure 5), (5) an enlarged, meaty, noncongested spleen, (6) fluid-filled, thin-walled sections of lower intestine with occasional edema of the wall of the cecum, and (7) gastric ulcers and/or edema of the gastric wall (Clark, 1997; Harding and Clark, 1997). Hepatic disease has been characterized as the major cause of icterus, wasting and death in natuarally occurring and experimentally reproduced cases of PMWS (Krakowka *et al.*, 2000, 2001; Rosell *et al.*, 2000a).

The most distinctive microscopic lesions in affected pigs are lymphoid cell depletion and granulomatous inflammation in lymphoid organs with inconsistently occurring clusters of intensely basophilic intracytoplasmic inclusion bodies (Figures 7 and 8) in macrophages (Clark, 1997; Ellis *et al.*, 1998; Kiupel *et al.*, 1998; Morozov *et al.*, 1998; Rosell *et al.*, 1999). Other lesions include interstitial lympho-histiocytic patchy or diffuse interstitial pneumonia (Figure 9), lympho-histiocytic periportal hepatitis with occasional hepatocellular necrosis (Figures 11 and 12), lympho-histiocytic and eosinophilic peripelvic nephritis with nonsuppurative perivasculitis (Figure 10), lympho-histiocytic infiltration of gastric, cecal, and colonic mucosa with marked submucosal edema, and lympho-histiocytic interstitial pancreatitis (Clark, 1997; Ellis *et al.*, 1998; Kiupel *et al.*, 1998; Morozov *et al.*, 1998; Rosell *et al.*, 1999).

Hepatic lesions have been characterized in more detail and classified into four stages of hepaticdamage (Rosell *et al.*, 2000a). In severe cases such lesions are characterized by massive, eneralized hepatocellular necrosis followed by fibrosis and nodular regeneration.

Porcine circovirus type 2 (PCV2) nucleic acid or antigen is most commonly demonstrated within the cytoplasm of epithelioid macrophages and multinucleated giant cells (Figures 13, 14 and 15), but can be found in histiocytic cells in a wide range of tissues, such as alveolar macrophages (Figure 16), Kupffer cells (Figure 17) and follicular dendritic cells (Kiupel *et al.*, 1998; Rosell *et al.*, 1999; Segales *et al.*, 2004). Viral antigen and/or nucleic acid has also been demonstrated in the nuclei and cytoplasm of various epithelial cells, including hepatocytes (Figure 17), renal tubular epithelium, bronchiolar epithelial cells, enterocytes, pancreatic acinar and ductular cells as well as endothelial cells (Figure 18), smooth muscle cells, T- and B-lymphocytes and small and large neurons, Purkinje cells and few oligodenrocytes in the central nervous system and spinal cord (Kiupel *et al.*, 1999; McNeilly *et al.*, 1999; Rosell *et al.*, 2000b; Segales *et al.*, 2004; Stevenson *et al.*, 2001).

There has been only limited characterization of the ultrastructural changes in different organs of pigs with PMWS. In one study, a lymph node with granulomatous inflammation and characteristic intracytoplasmic inclusions in epithelioid macrophages was examined. Ultrastructurally, the intracytoplasmic inclusions in macrophages were electron dense and round to ovoid with sharp margins. The matrix was heterogeneous, with different areas being granular, crystalline in a herringbone pattern or crystalline in cross-sectional arrays of non-enveloped, small, icosahedral, viral particles, approximately 17 nm in diameter (Kiupel *et al.*, 1998). No intranuclear virus particles were observed.

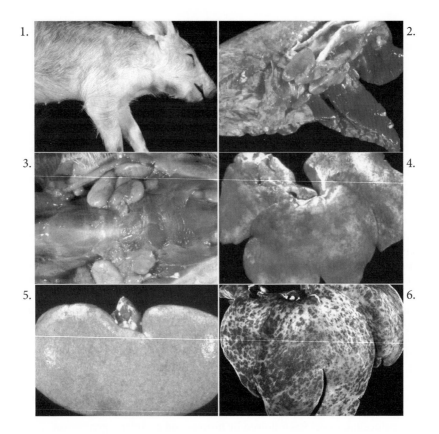

Figure 1. Severely icteric and wasting pig with PMWS.

Figure 2. Diffusely edematous, mottled, non-collapsed lungs with multifocal dark red areas primarily in the cranioventral lobes and enlarged bronchial lymph nodes from pig with PMWS.

Figure 3. Ventral view of the intermandibular area with the mandibular and retropharyngeal lymph nodes enlarged up to 4 times normal size from a pig with PMWS.

Figure 4. Swollen liver with a mottled yellow pattern due to severe necrosis from a pig with PMWS.

Figure 5. Swollen kidney with a prominent cortical tubular pattern and large numbers of necrohemorrhagic foci diffusely distributed throughout the cortex from a pig with PMWS. Note the severly enlarged renal lymph node.

Figure 6. Liver with diffuse cirrhosis characterized by loss of hepatic parenchyma, interlobular fibrosis and regeneration from a pig that had exhibited clinical signs of PMWS including icterus 6 weeks e.a.

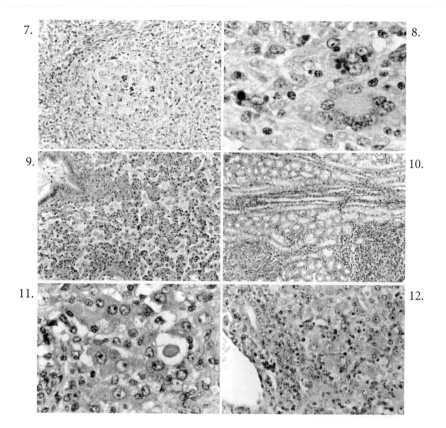

Figure 7. Diffuse lymphoid depletion and granulomatous lymphadenitis with large numbers of epithelioid macrophages containing characteristic intracytoplasmic inclusion bodies in a lymph node from a pig with PMWS.

Figure 8. Multinucleated giant cells and epitheloid macrophages with round, homogeneous, magenta-to-basophilic appearing inclusion bodies that form large single globules or botryoid clusters in a lymph node with granulomatous inflammation from a pig with PMWS.

Figure 9. Interstitial pneumonia with large numbers of interstitial and alvelor macrophages in a lung from a pig with PMWS.

Figure 10. Interstitial granulomatous nephritis in a kidney from a pig with PMWS.

Figure 11. Hepatocellular degeneration and single cell necrosis (apoptotic bodies) in a liver from a pig with PMWS.

Figure 12. Periportal granulomatous hepatitis in a liver from a pig with PMWS.

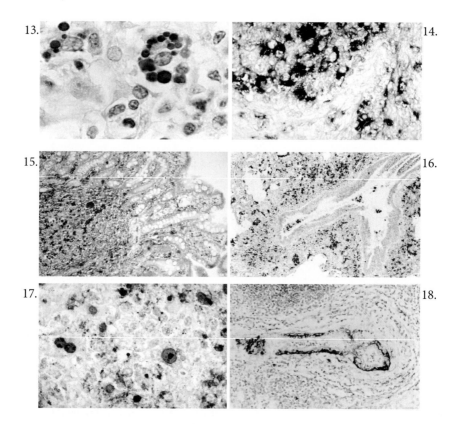

Figure 13. Immunohistochemistry demonstrating strong positive staining of intracytoplasmic circoviral inclusions in a lymph node from a pig with PMWS.

Figure 14. In-situ hybridization demonstrating large amounts of PCV2 nucleic acid in the cytoplasm of large numbers of macrophages in the follicular center of a lymph node from a pig with PMWS.

Figure 15. In-situ hybridization demonstrating large amounts of PCV2 nucleic acid in the cytoplasm of macrophages in the Peyer's patches and propria mucosa in the ileum from a pig with PMWS.

Figure 16. In-situ hybridization demonstrating large amounts of PCV2 nucleic acid in the cytoplasm of interstitial and alveolar macrophages in the lung from a pig with PMWS.

Figure 17. In-situ hybridization demonstrating PCV2 nucleic acid in the cytoplasm of large numbers of Kupffer cells and in nuclei of hepatocytes in the liver from a pig with PMWS.

Figure 18. In-situ hybridization demonstrating PCV2 nucleic acid in the necrotic wall of a muscular artery in the kidney from a pig with PMWS.

Porcine circovirus (PCV)

Circoviruses are the smallest animal viruses known so far. The viral genome consists of a single copy of circular single-stranded ambisense DNA genome. The size of the genome varies between 1.7 and 2.3 kb. Animal circoviruses have been isolated from chickens (chicken anemia virus, ChAV, Yuasa *et al.*, 1979), pigs (porcine circovirus, PCV, Tischer *et al.*, 1982), psittacines (psittacine beak and feather disease virus, PBFD, Pass and Perry, 1984) and pigeons (columbid circovirus, CoCV, Woods *et al.*, 1993). Phylogenetic analysis of PCV, chicken and psittacine circoviruses, plant geminiviruses and nanoviruses (previously known as plant circoviruses), classified PCV as most closely related to PBFDV. Both the PCV and PBFDV share sequence homology with and are intermediate between the 2 plant viral groups (Bassami *et al.*, 1998; Niagro *et al.*, 1998).

In 1982, the name porcine circovirus was first proposed for a small, 17 nm icosohedral virus with a circular DNA genome (Tischer *et al.*, 1982), which was previously described in 1974 by the same research group as a picornavirus-like noncytopathic contaminant of the PK-15 porcine kidney cell line (Tischer *et al.*, 1974). Previously an unidentified small cuboidal DNA virus had been identified in filtrates from primary kidney cell cultures derived from neonatal pigs with CT type A2 (Kanitz, 1972; Gustafson *et al.*, 1974). This virus was recently identified as a porcine circovirus (Stevenson *et al.*, 2001). Field strains of PCV have been identified in pigs with PMWS since 1996 (Allan *et al.*, 1998b; Clark, 1997; Daft *et al.*, 1996; Ellis *et al.*, 1998, Kiupel *et al.*, 1998; Morozov *et al.*, 1998; Nayar *et al.*, 1997; Segales *et al.*, 1997). Isolates of PMWS–associated PCV are genetically homogeneous, sharing at least 95% nucleotide homology (Allan *et al.*, 1998a; Hamel *et al.*, 1998; Meehan *et al.*, 1998; Morozov *et al.*, 1998). In contrast, PMWS-associated isolates of PCV are genetically and antigenically different from the PK-15 cell PCV, sharing less than 80% nucleotide homology with PK-15 PCV. Isolates of PCV that are genetically like PK-15 cell PCV are referred to as PCV1 and those like the first characterized PMWS isolates as PCV2 (Meehan *et al.*, 1998). Interestingly porcine circoviruses sharing at least 98% sequence homology with PCV1 from PK-15 cells have also been isolated from pigs with wasting disease (Le Cann *et al.*, 1997) and a stillborn piglet (Allan *et al.*, 1998a). More recently a field isolate of porcine circovirus from myoclonic piglets that had been used to experimentally reproduce congenital tremors by experimental inoculation of pregnant sows (Kanitz, 1972) has been identified as PCV1 (Stevenson *et al.*, 2001). These data suggest that field strains of PCV1 exist and may also be associated with various disease syndromes.

Porcine circoviruses have a high tenacity and are highly resistant to inactivation after exposure to pH3, chloroform and temperatures of 56°C and 70°C as well as a number of commonly used disinfectants (Allan *et al.*, 1994b). Infact, only Roccal D plus (Winthrop Labs) and One-stroke Environ (ConvaTec Labs) have been shown to be highly effective at inactivating the virus (Royer *et al.*, 2000). PCVs do not haemagglutinate erythrocytes from a variety of animals (Allan *et al.*, 1994b, 1998b) and no PCV-induced cytopathic effect has been observed in PCV-free PK-15 cells (Allan and Ellis, 2000) and primary fetal porcine kidney cells. PCV1 has also been shown to replicate in other cells including primary bovine kidney cells, semi-continuous bovine lung and testes cells, primary lamb kidney and semi-continuous lamb testes cells and Vero cells (Allan *et al.*, 1994b). Viral replication can be visualized by an indirect fluorescent antibody test and typically dense nuclear staining and perinuclear cytoplasmic stipling with occasional large cytoplasmic inclusions can be observed in PCV infected cell cultures (Allan *et al.*, 1994a). Ultrastructurally various types of inclusions can be found in the cytoplasm and nuclei of PK-15 cells that are persistently infected with PCV1 (Stevenson *et al.*, 1999). Small intracytoplasmic inclusions (0.1 – 0.5 µm) that are not surrounded by trilaminar membranes may be sites of viral assembly or maturation. Large intracytoplasmic inclusions (0.5 – 5.0 µm) that are surrounded by trilaminar membranes and contain paracrystalline arrays of aggregated virions, electron dense crystalline lamellae of 5 nm periodicity and/or whorls of myelinoid membranes are typical of autophagolysosomes. In contrast, intranuclear inclusions are not membrane bound and are 0.1 - 1.0 µm in diameter, round or ring shaped, dense and finely granular with sharply demarcated margins or consist of irregularly shaped aggregates of indistinct circular 10 - 12 nm in diameter virus-like particles (Stevenson *et al.*, 1999).

The role of PCV in PMWS

PCV2 is an indisputable pathogen and now recognized as the principal infectious cause of PMWS (Allan *et al.*, 2004; Ellis *et al.*, 2004; Harding, 2004; Segales *et al.*, 2004). So far, isolates of PCV from pigs with PMWS have been identified nearly exclusivly as PCV2 (Allan *et al.*, 1998a; Hamel *et al.*, 1998, 2000; Meehan *et al.*, 1998; Morozov *et al.*, 1998). Most reports have not associated PCV1 with PMWS and inoculation studies in pigs using PCV1 from PK-15 cells did not result in clinical disease (Allan *et al.*, 1995; Tischer *et al.*, 1986). One single case of an isolate of only PCV1 from pigs with PMWS has been reported (Le Cann *et al.*, 1997). In one study, at least 2 restriction enzyme types of PCV2 were detected in 5 of 15 pigs (Hamel *et al.*, 2000). It is uncertain whether these identified restriction enzyme types correspond to specific variants of PCV2 that concurrently infected pigs. In another study, 6 complete genomes of PCV2 isolates from North America were compared with genomic sequences of previously published PCV2 isolates (Fenaux *et al.*, 2000). The results indicate that although the genome of PCV2 is generally ≥90% homologous among different isolates, PCV2 isolates from different geographic regions generally differ in genomic sequences. This variation could have important implications for PCV2 virulence and associated disease syndromes.

Whereas the first reports of PMWS originated from farms free of major enteric and respiratory pathogens (PRV; Clark, 1996; Harding and Clark 1997), more commonly PMWS has been identified in pigs that also were affected with porcine respiratory and reproductive syndrome virus (PRRSV, Le Cann *et al.*, 1997; Sorden *et al.*, 1998; Rosell *et al.*, 1999; Segales *et al.*, 2002), pseudorabies virus (PRV, Rodriguez-Arrioja *et al.*, 1999), porcine parvovirus (PPV, Ellis *et al.*, 2000), *Hemophilus parasuis* (Madec *et al.*, 2000) and *Pasteurella multocida* (Madec *et al.*, 2000; Quintana *et al.*, 2001) and *Bordetella bronchiseptica* (Quintana *et al.*, 2001). Until the pathogenesis of PMWS has been clarified, it is important to identify swine pathogens that infect pigs with PMWS concurrent with PCV2.

Numerous experimental studies have partially clarified the role of PCV2 in PMWS. In the majority of these studies, experimental inoculation of gnotobiotic (GB), colostrum-deprived (CD) or conventional (CV) pigs with PCV2 alone has resulted in few or moderate gross and mild microscopic lesions typical of PMWS, but only mild or absent clinical disease (Allan *et al.*, 1999a; Balasch *et al.*, 1999; Kennedy *et al.*, 2000; Krakowka *et al.*, 2000; Magar *et al.*, 2000a, b). Experimental inoculation with a tissue homogenate from pigs with PMWS has failed to produce clinical disease in CV pigs (Balasch *et al.*, 1999). In one study, 1 of 4 CD pigs that had been inoculated with PCV2 alone became dull and lost weight (Allan *et al.*, 1999a; Kennedy *et al.*, 2000). Gross lesions were limited to pulmonary and mesenteric edema and mild microscopic lesions were observed in lymph nodes. Only one study has reproduced PMWS in 6 of 23 inoculated pigs, including clinical wasting disease, gross lesions and microscopic lesions, in CD pigs inoculated with PCV2 alone (Bolin *et al.,* 2001).

In contrast to the unsuccessful attempts to reproduce PMWS with PCV2 alone, the syndrome has been reproduced by dual inoculation of pigs with PCV2 and porcine parvovirus (PPV). Inoculation of GB pigs with PCV2 and PPV resulted in gross and microscopic lesions typical of PMWS, but no clinical disease (Ellis *et al.,* 1999). In a later study, clinical disease and lesions of PMWS were reproduced in GB pigs inoculated concurrently with PCV2 and PPV (Krakowka *et al.,* 2000). Based on IFA tests on frozen tissue sections, there were more PCV-infected cells in a wider variety of tissues in dually inoculated pigs than in pigs inoculated with PCV2 alone. In another study, CD pigs were inoculated with PCV2 alone, PPV alone or both PCV2 and PPV (Allan *et al.,* 1999a; Kennedy *et al.,* 2000). Only pigs inoculated with PCV2 and PPV concurrently, developed PMWS. Clinical signs included lethargy, anorexia, and reluctancy to move, icterus and hepatomegaly. Microscopic lesions were similar to those described in field cases of PMWS and PCV2 antigen was detected intralesionally in numerous tissues (Allan *et al.,* 1999a; Kennedy *et al.,* 2000). In all these studies dual inoculation with PPV and PCV2 clearly enhanced PCV2 replication. Large amounts of PCV2 were detected in lesions that were typical of PMWS, but only small amounts of PPV were detected and PPV was not specifically associated with lesions of PMWS (Choi and Chae, 2000). The mechanism of PPV associated enhancement of PCV2 replication is not known. Various authors speculated that PPV might induce immunosuppression or activate macrophages, thus enhancing PCV2 viral replication (Allan *et al.,* 2000b; Kennedy *et al.,* 2000; Krakowka *et al.,* 2000).

Other work demonstrated that concurrent infection with PRRSV and PCV2 also enhances PCV2 replication (Allan *et al.,* 2000a). Interestingly dual inoculation with PRRSV and PCV2 did not result in clinical disease and gross lesions typical of PMWS in one study, but produced microscopic lesions typical of PMWS (Allan *et al.,* 2000a). However, in another study severe clinical disease was produced in CD pigs inoculated with PRRSV and PCV2 (Harms *et al.,* 2001). Significantly larger amounts and a wider tissue distribution of PCV2, but not PRRSV, were observed in dually inoculated pigs compared to pigs inoculated with PCV2 or PRRSV alone.

More recently, PMWS has been reproduced in GB pigs inoculated with PCV2 that were also injected with keyhole limpet hemocyanin in incomplete Freund's adjuvant (Krakowka *et al.,* 2001). This study confirmed PCV2 as an indisputable pathogen and the principal infectious cause of PMWS. The mechanism by which injection with keyhole limpet hemocyanin in incomplete Freund's adjuvant enhances PCV2 replication and as a consequence causes PMWS is unknown. General immunstimulation including vaccination, as has been suggested by some researchers, seems an unlikely mechanism. The number of pigs developing PMWS is small in contrast to the number of pigs that are infected with PCV2 worldwide. One would assume that routine vaccinations and enteric and respiratory diseases, commonly found in nursery age pigs, should provide sufficient immunstimulation to cause PMWS in pigs infected with PCV2. A more likely hypothesis is that concurrent viral or bacterial infections provide a large pool of actively dividing cells that become a target for PCV2 infection. Further work is needed to elucidate the pathogenesis of PMWS.

Diagnosis of PMWS

Although PCV2 infection is necessary for the development of PMWS, infection with PCV2 alone will not cause PMWS in a large numbers of pigs. For this reason, demonstration of PCV2 antigen or nucleic acid and/or antibodies against PCV2 in a pig is insufficient to make a diagnosis of PMWS.

So far hematological parameters have not been demonstrated to be of diagnostic value for the diagnosis of PMWS (Segales *et al.,* 2000). In one study, hemograms of pigs with naturally occurring PMWS, pigs that had been inoculated with PCV2 alone and control pigs had statistic significant differences in the red blood cell count (RBC), the hemoglobin concentration (HGB), and the hematocrit value (HCT). Pigs with PMWS had lower values for these parameters than control and PCV2 inoculated pigs (Segales *et al.,* 2000). However, the detected microcytic and hypochromic anemia is characteristic of blood loss and its occurrence correlated with the presence of gastric ulcers (Segales *et al.,* 2000), an inconsistent finding in pigs with PMWS (Harding and Clark, 1997).

Available tests, recommended tissues for submission and interpretive guidelines for the diagnosis of PMWS have been published (Sorden, 2000). Based on these data a diagnosis of PMWS requires that a pig or a group of pigs exhibit all of the following:

- Clinical signs: wasting/weight loss/ill thrift/failure to thrive, with or without dyspnea or icterus.
- Histologic lesions: depletion of lymphoid organs/tissues and/or lymphohistiocytic to granulomatous inflammation in any organ (lungs, lymphoid organs, liver, kidney, intestine, pancreas).
- Detection of PCV2 within characteristic lesions. PCV2 may be demonstrated within microscopic lesions by observation of typical circoviral inclusions (approx. 10% of cases) in the cytoplasm of macrophages (Clark, 1997; Ellis *et al.,* 1998; Morozov *et al.,* 1998; Kiupel *et al.,* 1998; Rosell *et al.,* 1999) or by specific detection of PCV2 antigen or nucleic acid using immunhistochemistry (Ellis *et al.,* 1998; Kiupel *et al.,* 1998) or in-situ hybridization (Morozov *et al.,* 1998; Kiupel *et al.,* 1998; Rosell *et al.,* 1999), respectively.

None of the above listed criteria on its own is sufficient to make a diagnosis of PMWS.

Because the demonstration of PCV2 in microscopic lesions is required, tissues fixed in 10% neutral-buffered formalin are necessary for a diagnosis and the submission should include: lymph nodes (mesenteric, iliac, bronchial and others), spleen, tonsil, thymus, ileum, lungs, liver, kidney, heart and pancreas. Fresh samples are unnecessary for a diagnosis of PMWS, but should be submitted to rule out other infectious agents, such as PPV or PRRSV, both of which have been used in dual inoculation studies with PCV2 to cause PMWS. The isolation of PCV2 is not recommended for diagnostic purposes. It is a time consuming, labor-intensive process that appears to be less sensitive than IHC or ISH. A variety of methods exist to detect PCV2 within microscopic lesions. Because typical intracytoplasmic inclusions in macrophages can be found in only a small numbers of pigs with PMWS, immunohistochemistry (IHC; Clark, 1997; Allan *et al.,* 1998a; Ellis *et al.,* 1998; Kiupel *et al.,* 1998; McNeilly *et al.,* 1999) and in-situ hybridization (ISH; Daft *et al.,* 1996; Choi and Chae, 1999; Morozov *et al.,* 1998; McNeilly *et al.,* 1999; Rosell *et al.,* 1999; Stevenson *et al.,* 2001) are widely used to detect PCV2. Both polyclonal (Allan *et* al., 1998a; Ellis et al., 1998; Sorden et al., 1999) and monoclonal antibodies (Allan et al., 1999b) have been used for PCV2 IHC. It has been reported that ISH detects more positive cells than IHC (McNeilly et al., 1999), which is in contrast to comparative studies of IHC and ISH for the detection of PBFDV in formalin fixed tissues from birds (Latimer et al., 1992; Ramis et al., 1994). In general, ISH is more economical than IHC and using rate enhancement hybridization buffer systems (Sirinarumitr et al., 2000) and/or capillary suction systems, is even more rapid than IHC (personal data). Specific riboprobes (Nawagitgul et al., 2000) and oligoprobes (personal data) have been developed to identify PCV1 and PCV2. PCV2 infection has also been demonstrated with PCR (Allan et al., 1999b; Hamel et al., 2000; Hinrichs et al., 1999; Larochelle et al., 1999a,b; Morozov et al., 1998; Nayar et al., 1997; Stevenson et al., 2001) and multiplex PCR (Ouardani et al., 1999) and primer sets that are able to differentiate between PCV1 and PCV2 have been developed (Allan et al., 1999b; Larochelle et al., 1999a; Stevenson et al., 2001). Due to the lack of correlation between PCV2 infection and PMWS, PCR is not a useful technique for the diagnosis of PMWS, but might be used for large volumes of samples as a fast screening tool to identify an infection with PCV1 and/or PCV2. Competitive coamplification of a 502- or 506-bp region of the PCV type 1 (PCV1) or PCV2 ORF2, respectively, with a known concentration of competitor DNA that produces a 761- or 765-bp fragment, has been used to determine PCV genome copy numbers in infected cells (Liu et al., 2000). Interestingly, there was a significant difference between loads of PCV2 for clinically unremarkable pigs and pigs with PMWS, indicating that the development of PMWS may require a certain amount of PCV2 (Liu et al., 2000). In this study more than 50% of clinically healthy piglets could harbor both types of PCV, but PCV1 was detected in only 3 of 16 pigs with PMWS and PCV2 was detected in all pigs with PMWS.

Epidemiology of PCV2 and PMWS

On the basis of sero-surveillance, it appears that PCV infections are common in swine populations world wide (Dulac and Afshar, 1989; Edwards and Sands, 1994; Gerdes, 1993; Harding, 2004; Hines and Lukert, 1995; Horner, 1991; Tischer et al., 1986, 1995). In one study in Germany antibodies against PCV were observed in 35% of postweaned pigs, but 95% of older breeder pigs (Tischer et al., 1995), which indicates the ubiquitous nature of PCV infections. None of these studies distinguished between antibodies against PCV1 or PCV2. More recently, serum antibodies to PCV2 were reported in 55% to 100% of pigs tested in Northern Ireland (Walker et al., 2000) and in 100% of serum samples collected from varying age pigs from Belgium since 1985 (Mesu et al., 2000), suggesting that an endemic infection with PCV2 is common. A retrospective serological survey that determined the presence of antibodies to PCV1 and PCV2 in serum samples from sows collected at an abattoir in Canada in 1985, 1989 and 1997 identified serum samples positive to PCV1 and PCV2 in every year (Magar et al., 2000b). More serum samples were positive for PCV2 than for PCV1, suggesting PCV2 to be the main PCV type circulating in the Canadian pig population at least 10 years before PMWS was first reported. In all of these reports, most of the pigs with serum antibodies against PCV/PCV2 did not show signs of PMWS. In a recent epidemiological study from Canada, PCV2 was detected in 20 swine herds by PCR, but PMWS was diagnosed in only 2 of these herds (Dewey et al., 2000). Similar results were obtained in other surveys worldwide (Cotrell et al., 1999; Hamel et al., 2000; Mesu et al., 2000). On the other hand antibodies against PCV2 can be consistently detected in pigs with PMWS. A study in Spain demonstrated antibodies against PCV2 in 97.8% of 90 serum samples collected from necropsied pigs that had been diagnosed with PMWS (Rodriguez-Arrioja et al., 2000).

Seroconversion of conventional pigs to PCV2 has been demonstrated at 3-4 weeks postweaning in Canadian swine herds (Cotrell *et al.*, 1999; Harding *et al.*, 1998) and between 4 to 14 weeks of age in almost all pigs sampled in Belgium (Mesu *et al.*, 2000).

Farrow-to-finish, farrow-to-feeder pig, off-site nursery and grower pig herds have been affected by PMWS as well as operations ranging in size from 50 sows to more than 1000 sows (Harding and Clark, 1997). The syndrome most commonly affects pigs 2-3 weeks postweaning (Harding *et al.*, 1998), but has been reported in pigs up to 12 weeks-of-age (Kiupel *et al.*, 1998). Morbidity is typically 5%-15% and mortality may peak in an acute outbreak close to 10% (Harding *et al.*, 1998). A strong litter effect on disease susceptibility has been suspected, however no relationship to the parity of the sow was noticed (Madec *et al.*, 2000).

Excretion of PCV1 in feces and nasal mucus at 13 to 14 days post inoculation has been demonstrated in experimentally infected pigs (Tischer *et al.*, 1986). Excretion of PCV2 has been demonstrated in feces, saliva, and eye swabs at 35 days post inoculation in experimentally with PCV2 inoculated germ-free pigs (Krakowka *et al.*, 2000).

PMWS is now worlddwide recognized as an important economic disease of swine. Although PCV2 is accepted as the causal agent of PMWS, questions regarding transmission, pathogenesis and epidemiology remain. Especially factors triggering expression of clinical disease are poorly understood. There has also been an increased interest in other porcine circoviruses associated diseases, including dermatitis and nephropathy syndrome (PDNS), congenital tremors type AII, abortions and neonatal myocarditis and proliferative and necrotizing pneumonia (PNP). Continued research is necessary to better understand PMWS an d other PCV2 associated diseases.

Acknowledgements

The author wishes to acknowledge his mentor Gregory Stevenson and Suresh Mittal, Charles Kanitz and Harm HogenEsch for their guidance and support during his studies of PMWS and PCV2.

References

Allan, G.M. and J.A. Ellis, 2000. Porcine circoviruses: a review. J. Vet. Diagn. Invest. 12, 3-14.
Allan, G.M., Kennedy S., McNeilly F., Foster J.C., Ellis J.A., Krakowka S.J., Meehan B.M. and B. Adair, 1999a. Experimental reproduction of severe wasting disease by co-infection of pigs with porcine circovirus and porcine parvovirus. J. Comp. Path. 121, 1-11.
Allan GM, Mackie DP, McNair J, Adair B and MS McNulty, 1994a. Production, preliminary characterization and applications of monoclonal-antibodies to porcine circovirus. Vet Immunol Immunopathol 43, 357-371.
Allan GM, McNeilly F, Cassidy JP, Reilly GAC, Adair B, Ellis WA and MS McNulty, 1995, Pathogenesis of porcine circovirus; experimental infections of colostrum deprived piglets and examination of pig foetal material. Vet Micro 44, 49-64.
Allan GM, McNeilly F, Ellis J, Krakowka S, Botner A, McCullough K, Nauwynck H, Kennedy S, Meehan B and C Charreyre, 2004. PMWS: experimental model and co-infections. Vet Microbiol 98, 165-168.

Allan GM, McNeilly F, Ellis J, Krakowka S, Meehan B, McNair I, Walker I and S Kennedy, 2000a. Experimental infection of colostrum deprived piglets with porcine circovirus 2 (PCV2) and porcine reproductive and respiratory syndrome virus (PRRSV) potentiates PCV2 replication. Arch Virol 145, 2421-2429.

Allan GM, McNeilly F, Kennedy S, Daft B, Clarke EG, Ellis JA, Haines DM, Meehan BM and B Adair, 1998a. Isolation of porcine circovirus-like viruses from pigs with a wasting disease in the USA and Europe. J Vet Diagn Invest 10, 3-10.

Allan GM, McNeilly F, Meehan BM, Ellis JA, Connor TJ, McNair I, Krakowka S and S Kennedy, 2000b. A sequential study of experimental infection of pigs with porcine circovirus and porcine parvovirus: Immunostaining of cryostat sections and virus isolation. J Vet Med B 47, 81-94.

Allan GM, McNeilly F, Meehan BM, Kennedy S, Mackie DP, Ellis JA, Clarke EG, Espuna E, Saubi N, Riera P, Botner A and CE Charreyre, 1999b. Isolation and characterisation of circoviruses from pigs with wasting syndromes in Spain, Denmark and Northern Ireland. Vet Microbiol 66, 115-123.

Allan GM, Meehan BM, Todd D, Kennedy S, McNeilly F, Ellis JA, Clarke EG, Harding J, Espuna E, Botner A and C Charreyre, 1998b. Novel porcine circovirus from pigs with wasting disease syndromes. Vet Rec 143, 467-468.

Allan GM, Phenix KV, Todd D and MS McNulty, 1994b. Some biological and physico-chemical properties of porcine circovirus. J Vet Med B41, 17-26.

Balasch M, Segales J, Rosell C, Domingo M, Mankertz A, Urniza A and J Plana-Duran, 1999. Experimental inoculation of conventional pigs with tissue homogenates from pigs with post-weaning multisystemic wasting syndrome. J Comp Path 121, 139-148.

Bassami MR, Berryman D, Wilcox GE and SR Raidal SR, 1998. Psittacine beak and feather disease virus nucleotide sequence analysis and its relationship to porcine circovirus, plant circoviruses and chicken anemia virus. Virol 249, 453-459.

Bolin SR, Stoffregen WC, Nayar GPS and AL Hamel, 2001. Postweaning multisystemic wasting syndrome induced after experimental inoculation of ceasarian-derived, colostrums-deprived piglets with type 2 porcine circovirus. J Vet Diagn Invest 13, 185-194.

Choi C and C Chae, 1999. In-situ hybridization for the detection of porcine circovirus in pigs with postweaning multisystemic wasting syndrome. J Comp Path 121, 265-270.

Choi C and C Chae, 2000. Distribution of porcine parvovirus in porcine circovirus 2-infected pigs with postweaning multisystemic wasting syndrome as shown by in-situ hybridization. J Comp Pathol 123, 302-305.

Choi C, Chae C and EG Clark, 2000. Porcine postweaning multisystemic wasting syndrome in Korean pig: detection of porcine circovirus 2 infection by immunohistochemistry and polymerase chain reaction. J Vet Diagn Invest 12, 151-153.

Clark EG, 1996. Pathology of postweaning multisystemic wasting syndrome of pigs. Proc West Can Ass Swine Pract p21.

Clark EG, 1997. Post-weaning multisystemic wasting syndrome. Proceedings of the 28th Ann Meet Am Ass Swine Pract p499-501.

Cotrell TS, Friendship RM, Dewey CE, Josephson G, Allan G, McNeilly F and I Walker, 1999. Epidemiology of post-weaning multisystemic wasting syndrome in Ontario. Proc Ann Meet Am Ass Swine Pract, p.389-390.

Daft B, Nordhausen RW, Latimer KS and FD Niagro, 1996. Interstitial pneumonia and lymphadenopathy associated with circoviral infection in a six-week-old pig. Proc Ann Meet Am Ass Vet Lab Diagn, p32.

Dewey C, Johnston T, Oldford L and T Whiting T, 2000. Post weaning mortality in Manitoba swine. ADRI Project No: 98-074.

Dulac GC and A Afshar, 1989. Porcine circovirus antigens in PK-15 cell line (ATCC CCL-33) and evidence of antibodies to circovirus in Canadian pigs. Can J Vet Res 53, 431-433.

Edwards S and JJ Sands, 1994. Evidence of circovirus infection in British pigs. Vet Rec 134, 680-681.

Section D.

Ellis JA, Bratanich A, Clark EG, Allan G, Meehan B, Haines DM, Harding J, West KH, Krakowka S, Konoby C, Hassard L, Martin K and F McNeilly, 2000. Coinfection by porcin e circoviruses and porcine parvovirus in pigs with naturally acquired postweaning multisystemic wasting syndrome. J Vet Diagn Invest 12, 21-27.

Ellis J, Clark E, Haines D, West K, Krakowka S, Kennedy S and G. Allan, 2004. Porcine circovirus-2 and concurrent infections in the field. Vet Microbiol 98, 159-163.

Ellis J, Hassard L, Clark E, Harding J, Allan G, Willson P,Strokappe J, Martin K, McNeilly F, Mehan B, Todd D and D. Haines, 1998. Isolation of circovirus from lesions of pigs with postweaning multisystemic wasting syndrome. Can Vet J 39, 44-51.

Ellis J, Krakowka S, Lairmore M, Haines D, Bratanich A, Clark E, Allan G, Konoby C, Hassard L, Meehan B, Martin K, Harding J, Kennedy S and F McNeilly, 1999. Reproduction of lessions of postweaning multisystemic wasting syndrome in gnotobiotic pigs. J Vet Diagn Invest 11, 3-14.

Fenaux M, Halbur PG, Gill M, Toth TE and XJ Meng, 2000. Genetic characterization of type 2 porcine circovirus (PCV-2) from pigs with postweaning multisystemic wasting syndrome in different geographic regions of North America and development of a differential PCR-restriction fragment length polymorphism assay to detect and differentiate between infections with PCV-1 and PCV-2. J Clin Microbiol 38, 2494-2503.

Gerdes GH, 1993. Two very small viruses – a presumptive identification. Tydskr S Afr Vet Ver 64, 2.

Gustafson DP and CL Kanitz, 1974, Experimental transmission of congenital tremors in swine..Proc Ann Meet USAHA 78, 338-345.

Hamel AL, Lin LL and GPS Nayar, 1998. Nucleotide sequence of porcine circovirus associated with multisystemic wasting syndrome in pigs. J Virol 72, 5262-5267.

Hamel AL, Lin LL, Sachvie C, Grudeski E and GPS Nayar, 2000. PCR detection and characterization of type 2 porcine circovirus. Can J Vet Res 64, 44-52.

Harding JCS, 2004. The clinical expression and emergence of porcine circovirus 2. Vet Microbiol 98, 131-135.

Harding JCS and EG Clark, 1997. Recognizing and diagnosing postweaning multisystemic wasting syndrome (PMWS). Swine Health Prod 5, 201-203.

Harding JCS, Clark EG, Strokappe JH, Willson PI and JA Ellis, 1998. Postweaning multisystemic wasting syndrome: Epidemiology and clinical presentation. Swine Health Prod 6, 249-254.

Harms PA, Sorden SD, Halbur PG, Bolin SR, Lager KM, Morozow I and PS Paul, 2001. Experimental reproduction of severe disease in CD/CD pigs concurrently infected with type 2 porcine circovirus and porcine reproductive and respiratory syndrome virus. Vet Pathol 38, 528-539.

Hines RK and PD Lukert, 1995. Porcine circovirus: a serological survey of swine in the United States. Swine Health Prod 3, 71-73.

Hinrichs U, Ohlinger VF, Pesch S, Wang L, Tegeler R, Delbeck FEJ and M Wendt, 1999. Erster Nachweis einer Infektion mit dem porzinen Circovirus Typ 2 in Deutschland. Tieraerztl Umschau 54, 255-258.

Horner GW, 1991. Pig circovirus antibodies present in New Zealand pigs. Surveillance - Wellington 18, 23.

Kanitz CL, 1972. Myoclonia congenita: Studies of the resistance to viral infection of tissue culture cell lines derived from myclonic pigs. PhD dissertation, Purdue University, West Lafayette, IN.

Kennedy S, Allan G, Mc Neilly, Adair BM, Hughes A and P Spillane, 1998a. Porcine Circovirus infection in Northern Ireland. Veterinary record 145, 495-496.

Kennedy S, Meehan B and G Allan, 1998b. Porcine Circovirus infection in the republic of Ireland. Veterinary record 146, 511-512.

Kennedy S, Moffettt D, McNeilly F, Meehan B, Ellis J, Krakowka S and G Allan G, 2000. Reproduction of lesions of postweaning multisystemic wasting syndrome by infection of conventional pigs with porcine circovirus type 2 alone or in combination with porcine parvovirus. J Comp Pathol 122, 9-24.

Kiupel M, Stevenson GW, Kanitz CL *et al.*, 1999. Cellular localization of porcine circovirus in postweaning pigs with chronic wasting disease. Eur J Vet Pathol 5, 45-50.

Kiupel M, Stevenson GW, Mittal SK *et al.*, 1998. Circovirus-like viral associated disease in weaned pigs in Indiana. Vet Pathol 35, 303-307.

Krakowka S, Ellis JA, Meehan B, Kennedy S, McNeilly F and G Allan, 2000. Viral wasting syndrome of swine: Experimental reproduction of Postweaning multisystemic wasting syndrome in gnotobiotic swine by coinfection with porcine circovirus 2 and porcine parvovirus. Vet Path 37, 254-263.

Krakowka S, Ellis JA, McNeilly F, Ringler S, Rings DM and G Allan, 2001. Activation of the immune system is the pivotal event in the reproduction of wasting disease in pigs infected with porcine circovirus-2 (PCV-2). Vet Path 38, 31-42.

Larochelle R, Antaya M, Morin M and R Magar, 1999a. Typing of porcine circovirus in clinical specimen by multiplex PCR. J Virol Methods 80, 69-75.

Larochelle R, Morin M, Antaya M and R Magar, 1999b. Identification and incidence of porcine circovirus in routine field cases in Quebec as determined by PCR. Vet Rec 145, 140-142.

Latimer KS, Niagro FD, Rakich PM, Campagnoli RP, Ritchie BW, Steffens WL, Pesti DA and PD Luckert, 1992. Comparison of DNA-dot blot hybridization, immunoperoxidase staining and routine histopathology in the diagnosis of psittacine beak and feather disease in paraffin-embedded cutaneous tissues. J Assoc Avian Vet 6, 165-168.

LeCann P, Albina E, Madec F, Cariolet R and A Jestin, 1997. Piglet wasting disease. Vet Rec 141, 660.

Liu QiAng, Wang Li, Willson P, Babiuk LA, Liu QA and L Wang, 2000. Quantitative, competitive PCR analysis of porcine circovirus DNA in serum from pigs with postweaning multisystemic wasting syndrome. J Clin Microbiol 38, 3474-3477

Madec F, Eveno E, Morvan P, Hamon L, Blanchard P, Cariolet R, Amenna N, Morvan H, Truong C and D Mahe, 2000. Post-weaning multisystemic wasting syndrome (PMWS) in pigs in France: clinical observations from follow-up studies on affected farms. Livest Prod Scien 63, 223-233.

Magar R, Larochelle R, Thibault S and L Lamontagne, 2000a. Experimental transmission of porcine circovirus type 2 (PCV2) in weaned pigs: a sequential study. J Comp Pathol 123, 258-269.

Magar R, Muller P and R Larochelle, 2000b. Retrospective serological survey of antibodies to porcine circovirus type 1 and type 2. Can J Vet Res 64, 184-186.

Mankertz A, Domingo M, Folch JM, LeCann P, Jestin A, Segales J, Chmielewicz B, Plana-Duran J and D Soike, 2000. Characterization of PCV2 isolates from Spain Germany and France. Virus Res 66, 65-77.

McNeilly F, Kennedy S and D Moffett, 1999. A comparison of in-situ hybridization and immunohistochemistry for the detection of a new porcine circovirus in formalin-fixed tissues from pigs with post-weaning multisystemic wasting syndrome. J Virol Methods 80, 123-128.

Meehan BM, McNeillly F, Todd D, Kennedy S, Jewhurst VA, Ellis JA, Hassard LE, Clark EG, Haines DM, and GM Allan, 1998. Characterization of novel circovirus DNAs associated with wasting syndromes in pigs. J Gen Virol 79, 2171-2179.

Mesu AP, Labarque GG, Nauwynck HJ and MB Pensaert, 2000. Seroprevalence of porcine circovirus types 1 and 2 in the Belgian pig population. Vet Quart, Quart J Vet Scien 22, 234-236.

Morozov I, Sirinarumitr T, Sorden SD, Halbur PG, Morgan MK, Yoon KJ and PS Paul, 1998. Detection of a novel strain of porcine circovirus in pigs with postweaning multisystemic wasting syndrome. J Clin Microbiol 36, 2535-2541.

Nawagitgul P, Morozow I, Sirinarumitr T, Sorden SD and PS Paul, 2000. Development of probes to differentiate porcine circovirus types 1 and 2 *in vitro* by *in situ* hybridization. Vet Microbiol 75, 83-89.

Nayar GP, Hamel SA and L Lin, 1997. Detection and characterization of porcine circovirus associated with post-weaning multisystemic wasting syndrome in pigs. Can Vet J 38, 385-386.

Niagro FD, Forsthoefel AN, Lawther RP, Kamalanathan L, Ritchie BW, Latimer KS and PD Lukert, 1998. Beak and feather disease virus and porcine circovirus genomes: intermediates between the geminiviruses and plant circoviruses. Arch Virol 143, 1723-1744.

Section D.

Onuki A, Abe K, Togashi K, Kawashima K, Taneichi A and H Tsunemitsu, 1999. Detection of porcine circovirus from lesions of pigs with wasting disease in Japan. J Vet Med Sci 61, 1119-1123.

Ourdani M, Wilson L, Jette R, Montpetit C and S Dea, 1999. Multiplex PCR for detection and typing of porcine circoviruses. J Clin Microbiol 37, 3917-3924.

Pass DA and RA Perry, 1984. The pathology of psittacine beak and feather disease. Austral Vet J 61, 69-74.

Quitana J, Segales J, Rossell C, Calsamiglia M, Rodriguez-Arrioja GM, Chianini F, Folch JM, Maldonado J, Canal M, Plana-Duran J and M Domingo, 2001. Clinical and pathological observations on pigs with postweaning multisystemic wasting syndrome. Vet Rec 149, 357-361.

Ramis A, Latimer KS, Niagro FD, Campagnoli RP, Ritchie BW and D Pesti, 1994. Diagnosis of psittacine beak and feather disease (PBFD) viral infection, avian polyomavirus infection, adenovirus infection and herpesvirus infection in psittacine tissues using DNA in-situ hybridization. Avain Pathol 23, 643-657.

Rodriguez-Arrioja GM, Segales J, Balasch M, Rosell C, Quintana J, Folch JM, Plana-Duran J, Mankertz A and M Domingo, 2000. Serum antibodies to porcine circovirus type 1 and type 2 in pigs with and without PMWS. Vet Rec 146, 762-764.

Rodriguez-Arrioja GM, Segales J, Rosell C, Quintana J, Ayllon S, Camprodon A and M Domingo, 1999. Audjesky's disease virus infection concurrent with postweaning multisystemic wasting syndrome in pigs. Vet Rec 144, 152-153.

Rosell C, Segales J and M Domingo, 2000a. Hepatitis and staging of hepatic damage in pigs naturally infected with porcine circovirus type 2. Vet Pathol 37, 687-692.

Rosell C, Segales J, Plana-Duran J, Balasch M, Rodriguez-Arrioja GM, Kennedy S, Allan GM, McNeilly F, Latimer KS and M Domingo, 1999. Pathological, immunohistochemical, and in-situ hybridization studies of natural cases of Postweaning Multisystemic Wasting Syndrome (PMWS) in pigs. J Comp Path 120, 59-78.

Rosell C, Segales J, Rovira A and M Domingo, 2000b. Porcine circovirosis in Spain. Vet Rec 146, 591-592.

Royer R, Nawagitgul P, Paul P and P Halbur, 2000. Susceptibility of porcine circovirus to several commercial and laboratory disinfectants. Proc Ann Meet Am Assoc Swine Pract, p. 45.

Sato K, Shibahara T, Ishikawa Y, Kondo H, Kubo M and K. Kadota, 2000, Evidence of Porcine Circovirus Infection in Pigs with Wasting Disease Syndrome from 1985 to 1999 in Hokkaido, Japan. J Vet Med Sci 62, 627-633.

Segales J, Calsamiglia M, Rosell C, Soler M, Maldonado J, Martin M and M Domingo, 2002. Porcine reproductive and respiratory syndrome virus (PRRSV) infection status in pigs naturally affected with postweaning multisystemic wasting syndrome (PMWS) in Spain. Vet Microbiol 85, 23-30.

Segales J, Pastor J, Cuenca R and M Domingo, 2000 Haematological parameters in postweaning multisystemic wasting syndrome-affected pigs. Vet Rec 146, 675-676.

Segales J, Rosell C and M Domingo, 2004. Pathological findings associated with naturally acquired porcine circovirus type 2 associated disease. Vet Microbiol 98, 137-149.

Segales J, Sitjar M, Domingo M, Dee S, DelPozo M, Noval R, Sacristan C, DeLasHeras A, Ferro A and KS Latimer, 1997. First report of post-weaning multisystemic wasting syndrome in pigs in Spain. Vet Rec 141, 600-601.

Sirinarumitr T, Morozov I, Nawagitgul P, Sorden SD, Harms PA and PS Paul, 2000. Utilization of a rate enhancement hybridization buffer system for rapid *in situ* hybridization for the detection of porcine circovirus in cell culture and in tissues of pigs with postweaning multisystemic wasting syndrome. J Vet Diagn Invest 12, 562-565.

Sorden SD, 2000. Update on porcine circovirus and posrweaning multisystemic wasting syndrome (PMWS). Swine Health Prod 8, 133-136.

Sorden SD, Harms PA, Nawagitgul P, Cavanaugh D and PS Paul, 1999. Development of a polyclonal-antibody-based immunohistochemical method for the detection of type 2 porcine circovirus in formalin-fixed, paraffin-embedded tissue. J Vet Diagn Invest 11, 528-530.

Stevenson GW, Kiupel M, Mittal SK, Jiwon C, Latimer KS and CL Kanitz, 2001. Tissue distribution and genetic typing of porcine circoviruses in pigs with naturally occurring congenital tremors. J Vet Diagn Invest 13, 57-62.

Stevenson GW, Kiupel M, Mittal SK and CL Kanitz, 1999. Ultrastructure of porcine circovirus in persistently infected PK-15 cells. Vet Pathol 36, 368-378.

Tischer I, Bode L, Peters D, Pociuli S and B Germann, 1995. Distribution of antibodies to porcine circovirus in swine populations of different breeding farms. Arch Virol 140, 737-743.

Tischer I, Gelderblom H, Vettermann W and MA Koch, 1982. A very small porcine virus with circular single-stranded DNA. Nature 295, 64-66.

Tischer I, Mields W, Wolff D, Vagt M and W Griem, 1986. Studies on epidemiology and pathogenity of porcine circovirus. Arch Virol 91, 271-276.

Tischer I, Rasch R and G Tochtermann, 1974. Characterization of papovavirus and picornavirus-like particles in permanent pig kidney cell lines. Zentralbl Bakt Hyg A 226, 153-167.

Walker IW, Konoby CA, Jewhurst VA, McNair I, McNeilly F, Meehan BM, Cottrell TS, Ellis JA and GM Allan, 2000. Development and application of a competitive enzyme-linked immunosorbent assay for the detection of serum antibodies to porcine circovirus 2. J vet Diagn Invest 12, 400-405.

Woods LW, Latimer KS, Barr BC, Niagro FD, Campagnoli RP, Nordhausen RW and AE Castro, 1993. Circovirus-like infection in a pigeon. J Vet Diagn Invest 5, 609-612.

Yuasa N, Taniguchi T and I Yoshida, 1979. Isolation and some characteristics of an agent inducing anemia in chicks. Avian Dis 23, 366.

Influence of growth conditions on caprine nasal bacterial flora in the presence of *Mannhemia hameolytica*

D.R. McWhinney[1] and Kingsley Dunkley[2]
[1]Prairie View A&M University; [2]Texas A&M University, College Station TX, USA

Mannheimia (Pasteurella) haemolytica is the primary cause for acute pneumonic pasteurelosis (shipping fever) in goats and other ruminants. The organisms colonize the upper respiratory tract of healthy and unstressed goats and are found in low undetectable numbers among normal flora. *M. haemolytica* over-rides the normal flora in the upper respiratory tract in stressed goats and colonize the lower respiratory tract and lungs in the disease state. Despite significant advancement in molecular analysis of *M. haemolytica* virulence factors, shipping fever continues to plague the cattle, sheep and goat industries.

Some species of normal nasal microflora when grown in the presence of *M. haemolytica,* in enriched medium are suppressed or killed. Current studies were designed to grow nasal microflora *in vitro* in a minimal medium that is of similar biochemical constituent to the respiratory tract of healthy goats. A continuous-Flow Cultures (CFC) of microbes was developed from healthy non-challenged goats and goats challenged with live *M. haemolytica.* Samples were extracted from the CFC at 0hr, 1hr, 4hrs, 8hrs, 12hrs, and 24hrs intervals for a total of ten days. Samples were analyzed for pH, optical density, colony forming units, isolation and characterization of microbes, volatile fatty acid levels and inhibition studies.

Our results indicated a difference in the number of bacteria isolated from non-challenged healthy goats compared to challenged goats. A 50% loss in species was observed in the challenged sick goats. The pH, OD and volatile fatty acids showed significant differences (P=0.01) in the log phase of both CFC's. The 12-hrs sample from the non-challenged goats exibited a zone of inhibition of *M. haemolytica* on blood agar plates. This suggested that the combination of microbial flora and metabolites at that stage of the CFC inhibited *M. haemolytica* and is of importance in our effort to minimize the level of pneumonic pasteurelosis in goats.

References

Frank, G.H. and R.E. Briggs, 1994. Colonization of the tonsils and nasopharynx of calves by a rifampicin-resistant *Pasteurella haemolytica* and it inhibition by vaccination. Am. J. Vet. Res. 56, 866-868.

Hinton, A. Jr. and M.E. Hume, 1995. Inhibition of *Salmonella typhimurium* by the products of tartrate metabolism by a *Veillonella* species. Appl. Bacteriology 81, 188-190.

Acknowledgements

Work supported by USDA.

The relationship between exposure to bovine viral diarrhea virus and fertility in a commercial dairy herd

W. Raphael[1], S. Bolin[2], W. Guterbock[3], J. Kaneene[1,2], and D. Grooms[1]
[1]Dept. Large Animal Clinical Sciences; [2]Dept. Pathobiology and Diagnostic Investigation, Michigan State University, East Lansing, MI 48824-1314; [3]Den Dulk Dairy, Ravenna, MI 4945, USA.

The objective of this retrospective case-control study was to identify the effects of exposure to Bovine Viral Diarrhea Virus (BVDV) on fertility in a known Type II infected, vaccinated, 600-cow Holstein herd.

Serum was collected approximately every four weeks from all adult lactating cows calving over a nine month time period. Exposure to BVD was measured as the change in serum neutralization titer (<4 fold rise, ≥ 4 fold rise) to the Singer and 125C BVDV strains between the sample taken immediately prior to artificial insemination and the sample taken 62 ± 16 days later. Pregnancy was diagnosed by rectal palpation 32 – 38 days after insemination. All inseminations resulting in pregnancies were considered as cases (n = 85). A similar number of controls were randomly selected from all inseminations not resulting in pregnancy, with stratification by order of insemination post-calving to match the cases. One technician performed all inseminations.

In a logistic regression model, the risk of pregnancy was not affected by exposure to BVDV for either strain ($P > 0.10$). Covariates considered were parity (odds ratio lactation one *vs.* three or greater [Singer] = 2.56 (95% CI = 1.16 – 5.68); [125C] 2.22 (95% CI = 1.05 – 4.66)) and estrus detection method ($P > 0.10$). This study has the power (B = 0.19) at a level of significance of 5%, to detect a difference in probability of pregnancy of 0.24 in exposed (≥4 fold rise), from 0.55 in non-exposed (<4 fold rise). This study does not have adequate power to detect smaller effects of BVDV on fertility.

Tritrichomonas foetus cysteine protease CP30 induces cell death in host cells

B.N. Singh[1], J.J. Lucas[1], G.R. Hayes[1], C.E. Costello[2], U. Sommer[2] and R.O. Gilbert[3]
[1]Department of Biochemistry & Molecular Biology, SUNY Upstate Medical University, Syracuse, NY, USA; [2]Boston University School of Medicine Mass Spectrometry Resource, Boston, MA, USA; [3]Department of Clinical Sciences, College of Veterinary Medicine, Cornell University, Ithaca, NY, USA

Tritrichomonas foetus is a serious veterinary pathogen, causing bovine trichomoniasis, a sexually transmitted disease leading to infertility and abortion. *T. foetus* infects the mucosal surfaces of the reproductive tract. Infection with *T. foetus* induces apoptosis in bovine vaginal and uterine epithelial cells (BVECs and BUECs) in culture. The objective of our study is to identify and characterize an element leading to host cell pathology.

Bacitracin affinity chromatography was used to purify a cysteine protease (CP) fraction from the soluble fraction elaborated by the parasites. It yielded a single band on SDS-PAGE having an apparent MW of 30 kDa (CP30). CP30 induced cytotoxicity/cell death in BVECs and BUECs in a species-specific manner. The specific cysteine protease inhibitor E-64 significantly inhibited the induction of cytotoxicity/cell death.

MALDI-TOF MS analysis of CP30 revealed a single peak with a MW of 23.7 kDa. Mass spectral peptide sequence analysis of proteolytically digested CP30 revealed homologies to a previously reported cDNA clone, CP8. Fluorescence microscopy along with the ELISA[PLUS] assay and flow cytometry analyses were used to detect apoptotic nuclear condensation, DNA fragmentation, and changes in plasma membrane asymmetry in host cells undergoing apoptosis in response to *T. foetus* infection or incubation with CP30. Furthermore, the activation of caspase-3 and inhibition of cell death by caspase inhibitors indicates that caspases are involved in BVEC and BUEC apoptosis.

These results imply that CP30 secreted by the parasites is involved in the pathogenesis of *T. foetus* infection. Developments of rational new treatments/vaccine for bovine trichomoniasis depend upon the understanding of the basic infection process and the role of specific parasite molecule such as CP30.

Acknowledgements
USDA support to B.N.S. and NIH/NCRR support to C.E.C.

Immunological approaches to enhanced production in food animals

R.J. Yancey, Jr.
Biologicals Development, Veterinary Medicine Research and Development, Pfizer Animal Health, Pfizer Inc, Kalamazoo. MI, 49001, USA

Abstract

Use of antimicrobial agents for disease prevention in livestock or as a method to enhance animal performance is under increasing scrutiny and alternatives to antimicrobial agents need to be developed. Vaccines and immunological approaches can be applied in the area of disease prevention and control and directly at the production/growth promotion level. Examples of vaccine use in livestock production occur for respiratory diseases in swine, enteric disease in swine, and bovine mastitis. Logically, the use of these vaccines to prevent disease and ensure healthier animals should result in increased productivity. The difficulties associated with vaccination include vaccine efficacy, timing of vaccination in relation to production management, duration of immunity, and the associated labor costs. Vaccines for *Mycoplasma hyopneumoniae* have been clearly shown to enhance growth and productivity of swine. Similarly, vaccines against coliform mastitis infections have be demonstrated to enhance the productivity of dairy cows.

Immune approaches to replace antibiotics for growth promotion have included mechanisms that affect the gut flora directly, reduce the level of inflammatory cytokines, use selected cytokines directly, or affect levels of endogenous hormones. A successful approach targeting the gut flora directly was suggested by early studies that indicated that specific components of the endogenous, "normal" gut flora were detrimental to optimal health and growth of domestic animals. Antibiotics promote a healthy, faster growing animal by targeting these endogenous components of the microflora. A specific mucosal immune response directed at these components also could reduce their numbers in much the same way as antimicrobials. Anti-inflammatory agents including cytokine receptor antagonists such as IL-1ra have been shown to enhance growth performance of pigs as well as reduce the clinical manifestations of certain bacterial infections. Direct use of cytokines such as IFN-γ and IL-5 has been shown to enhance growth performance and disease resistance in poultry and swine. Anti-hormonal vaccines such as those vaccines targeting gonadotropin-releasing hormone (GnRH) increase growth performance of intact boars and gilts, as well as preventing sexual development in both sexes and boar taint. Similar enhancement of performance and anti-reproductive activity has been demonstrated with anti-GnRH vaccines in cattle. Additional anti-hormonal vaccine approaches include active immunization against somatostatin, inhibin, vasoactive intestinal peptide, and myostatin.

Keywords: Vaccines, immunological approaches, cytokines, anti-hormonal vaccines, growth performance

Introduction

Vaccines and immunological approaches to enhance food animal performance can be applied in the area of disease prevention and control and directly at the production level. Types of immunological approaches that have been shown to enhance animal production include: active and passive immunization against clinical or sub-clinical disease, active and passive immunization against hormones, use of cytokines receptors, or use of cytokines themselves.

The efficiency of these approaches depends on a number of factors including overall herd health, the status of herd immunity, and the individual immunity of the animals within the herd. Vaccines tend to be restricted in their spectrum and are specific for a given immunogen. The method of delivery, the timing of the dose, the number of doses, and the duration of immunity are important considerations in maximal efficacy of immunological approaches. If an animal is too young when vaccinated or given immunotherapy, its immune system may be immature or the presence of maternal antibody may interfere with the immune response.

The benefit of an immunological approach to production enhancement must be weighed against the convenience and cost of its use. The benefit should also be compared to concerns about residues when antibiotics, steroids, or other chemicals are used to enhance growth. Handling of the animal is an important factor in that stress may also interfere with desired immune responses. On the other hand, labor considerations may favor vaccination or treatment at times of animal movement, sometimes high stress periods. While immunotherapy or vaccination during preconditioning may be more efficient, monetary considerations may preclude the practice.

Disease prevention/control and production enhancement

Examples of the use of vaccination to control disease, and at the same time enhance livestock production, include vaccination for respiratory diseases, enteric disease, and bovine mastitis. Logically, the use of vaccines to reduce disease and ensure healthier animals should result in increased productivity. However, proved returns on investment have only been realized in a few cases.

Vaccination for *Mycoplasma hyopneumoniae* has been associated with enhanced growth and productivity of pigs (Maes *et al.*, 1999; Moreau *et al.*, 2004). In a large multi-herd study, researchers found that vaccination against *M. hyopneumoniae* significantly reduced the incidence of pneumonia and enhanced average daily gain (ADG), feed conversion (FCR), and economic return for the producers (Maes *et al.*, 1999; Table 1). Productivity was enhanced even in the presence of concurrent infection with Porcine Reproductive and Respiratory Syndrome (PRRS) virus (Moreau *et al.*, 2004).

The use of spray-dried plasma (SDP) is a method for reducing enteric disease and enhancing productivity of pigs that has a probable immunologic basis (van Dijk *et al.*, 2001; Owusu-Asiedu *et al.*, 2002; Bosi *et al.*, 2004). Spray-dried plasma significantly increased ADG and reduced FCR, while also reducing disease due to enterotoxigenic *Escherichia coli*. Porcine plasma has a greater beneficial effect than bovine plasma. While the mechanism of SDP activity is not entirely clear, its activity has been attributed to the immunoglobulin fraction that has been shown to contain anti-*E. coli* antibodies. These antibodies may prevent attachment of the bacteria (Owusu-

Table 1. Effect of vaccination of pigs against M. hyopneumoniae (from Maes et al., 1999.).

Parameter	Controls	Vaccinates
Number of Pigs	3848	3730
% pigs with pneumonia	21.76	7.61[a]
ADG (g)	626	649[a]
FCR	2.90	2.83[a]
Days to 100 kg	194	189[a]
Additional net return/pig sold	NA	$1.34

[a]$P < 0.05$ compared to control value.

Asiedu et al., 2002; Bosi et al., 2004). It has also been found that feeding SDP reduces the level of proinflammatory cytokines such as TNF-α and IL-8 in the jejunal tissue (Bosi et al., 2004). Whether SDP reduces these cytokine levels by affecting the gut flora directly is not yet clear.

Bovine mastitis is the most costly disease for the US dairy industry with an annual estimated loss of $2 billion, primarily due to lower milk yields, reduced milk quality, and higher productions costs. There are a number of problems uniquely associated with vaccinating for bovine mastitis and effective vaccines are rare (Yancey, 1999). Nonetheless, vaccines that have been demonstrated to reduce mastitis due to staphylococci and coliforms have been shown to enhance the productivity of dairy cows. Using an experimental *Staphylococcus aureus* vaccine, Leitner and collaborators (Leitner et al., 2003) found that somatic cell counts were significantly reduced and milk yields were significantly enhanced (0.5 kg/cow/day increase) in vaccinated compared to control cows. This study was conducted in seven herds and included a total of 473 Holstein heifers. This benefit was observed even though there was little *S. aureus*-induced mastitis in the herds. Similarly, a partial budget analysis of vaccinating cows against clinical coliform mastitis with J5 vaccine concluded that when >1% of cow lactations resulted in clinical coliform mastitis, increased profits of $57/cow lactation could be obtained. As used in this model, this resulted in a 1700% return on investment from the J5 vaccination program (DeGraves and Fetrow, 1991).

Immune approaches directly for production/growth enhancement

Immune approaches to replace antibiotics for growth promotion have included mechanisms that affect the gut flora directly, reduce the level of inflammatory cytokines, use selected cytokines directly, or affect levels of endogenous hormones.

A successful approach targeting the gut flora directly was suggested by studies that indicated that specific components of the endogenous, "normal" gut flora were detrimental to optimal health and growth of domestic animals. Antibiotics promote a healthy, faster growing animal by targeting these endogenous components of the microflora (Visek, 1978). A specific mucosal immune response directed at these bacterial components potentially could reduce their numbers in much the same way as antimicrobials. Additional studies have suggested that the gut microflora alter metabolism indirectly by inducing cells of the gastrointestinal tract immune system to produce pro-inflammatory cytokines such as IL-1, TNF-α, and IL-6 (Roura et al., 1992, Johnson,

1997). A balanced, directed immune approach aimed at one or more of these cytokines might have an antibiotic-like, growth-permitting effect.

Along this same theme, anti-inflammatory agents including cytokine receptor antagonists such as IL-1ra were suggested as replacements for growth permitting antibiotics. IL-1ra was shown to enhance growth and performance of pigs as well as reduce the clinical manifestations of certain bacterial infections (Strom *et al.*, 2002a). Intramuscularly administered recombinant IL-1ra, at concentrations of 100-200 µg, enhanced weight gains of pigs from day 50 to market. These treatments also reduced production losses at weaning, and increased resistance of pigs to experimental challenge with hemorrhagic *E. coli*, *Brachyspira hyodysenteriae*, or *Actinobacillus pleuropneumoniae* (Strom *et al.*, 2002a). The authors claim but did not show that soluble, pro-inflammatory cytokine receptors such as TNF-α receptor and IL-6 receptor also enhance growth rates of animals.

Direct use of the cytokines IFN-α, IFN-γ, or IL-5 has been shown to enhance disease resistance and growth performance in poultry, cattle and swine. Cummins and others have published extensively on the preventive or therapeutic use of orally or intranasally administered INF-α in bovine respiratory disease complex in cattle (Babiuk *et al.*, 1991; Cummins *et al.*, 1993, 1999), rotavirus and TGE in pigs (Lecce *et al.*, 1990; Cummins *et al.*, 1995), and Newcastle disease in chickens (Marcus *et al.*, 1999). Likely due to its positive effect on health of animals affected by BRD complex, very low doses IFN-α significantly enhanced weight gain of cattle (Cummins *et al.*, 1999).

Prophylactic administration of interferon-γ was shown to enhance disease resistance of dairy cows to clinical coliform bovine mastitis (Babiuk *et al.*, 1991), and to enhance milk production of infected dairy cows. This cytokine has been shown to enhance activity of mammary gland neutrophiles and to down-regulate the production of pro-inflammatory cytokines in response to *E. coli* infection of the mammary gland (Campos *et al.*, 1993). Chicken INF-γ (ChIFN-γ) has been examined in poultry as an adjuvant, a disease control agent and as a replacement for growth promoting antibiotics (Lowenthal *et al.*, 2000; Hilton *et al.*, 2002). The cytokine was effective in all three cases. When 5000 U of recombinant ChIFN-γ was administered IP to chicks at day 0 and day one post-hatch and the chicks were challenged with *Eimeria acervulina* oocysts at day 1 of age, the chicks showed enhanced resistance to disease induced by this parasite. The treated group of chicks showed reduced weight loss, enhanced weight gain on recovery, and reduced oocyst production compared to saline treated controls (Lowenthal *et al.*, 1998). Birds treated

Table 2. Growth enhancement of chicks with rChIFN-γ (from Lowenthal et al., 1998).

Day Post-Hatch	Mean Body Weight (g)		Percentage Increase
	INF-γ (n=10)	Saline (n=10)	
1	138 ± 17	137 ± 16	0.7
6	300 ± 43	290 ± 42	3.4
11	469 ± 66	458 ± 62	2.4

with rChIFN-γ in the same manner, but not challenged with *Eimeria* were monitored for growth enhancement. Those data are shown in Table 2.

While the weight differences were not significant between groups, the authors attribute the lack of significant differences to the small group sizes. These authors have reported body weight differences between IFN-γ- and placebo-treated chicks of up to 8% in studies with older birds (Lowenthal *et al.*, 2000). When ChIFN-γ was delivered intraocular via a modified live viral vector, fowl adenovirus serotype 8, significantly enhanced weight gains were obtained compared to controls (Johnson *et al.*, 2000).

The cytokine IL-5 has recently been examined as a replacement for growth-promoting antibiotics and as a disease control agent in pigs (Strom *et al.*, 2002b). Recombinant IL-5 was used in groups of pigs (~20 per group) from a commercial piggery. The IL-5 was administered intramuscularly at a concentration of 100 μg/pig, twice a week for 6 weeks. Pigs treated with IL-5 had significantly increased growth rates and final growth weights compared to saline injected pigs (Strom *et al.*, 2002b). The authors also noted a reduced incidence in death from infectious diseases and reduced production loss at weaning in the treated versus the control pigs.

Anti-hormonal approaches that have been studied include those methods that enhance hormonal activity and those which interfere with the activity of endogenous hormone (Pell and Aston, 1995; Babiuk, 2002; Sillence, 2004).

Enhancement of growth promoting properties of growth hormone (GH) was demonstrated almost 30 years ago by using specific monoclonal antibodies to whole GH protein or to peptides of GH (Holder *et al.*, 1985). Passively administered anti-GH antibody has been shown effective in promoting growth rates of sheep. Active vaccination to GH from pigs was successful in enhancing lean tissue deposition in lambs (Pell and Aston, 1995), although passive immunization has generally been more effective than active immunization. However, others have failed to elicit lactogenic responses in dairy cows when GH vaccines or GH antibodies were administered. (Sillence, 2004).

The activities of other hormones such as growth hormone releasing hormone and insulin-like growth factor-1 also were enhanced by antibody-hormone complexes (Pell and Aston, 1995). The mechanism(s) for this enhancement are not completely clear. Protection of the hormone from degradation or soluble inhibitors, alteration of pharmacokinetics and biodistribution of the hormone, or enhanced interaction of the hormone with cell receptors have been proposed as the mechanisms for immune enhanced activity (Pell and Aston, 1995).

Interference with hormonal activity though anti-hormonal vaccines is another mechanism of growth enhancement. The most studied of the anti-hormonal vaccines are those targeting gonadotropin-releasing hormone (GnRH). GnRH is a 10 amino-acid peptide that is generally conjugated to a protein carrier such as diphtheria toxoid, tetanus toxoid, ovalbumin, serum albumin, *Mannheimia haemolytica* leukotoxoid, or keyhole limpet hemocyanin (KLH). The GnRH-conjugate is then formulated with an adjuvant. The efficacy of the vaccine depends both upon the carrier and the adjuvant formulation. There are two commercial anti-GnRH products that have been marketed in Australia. Vaxstrate® was licensed as an anti-reproductive vaccine for cattle. Improvac® was registered as a vaccine for immunosterilization, prevention of boar taint, and as a performance enhancer.

Anti-GnRH vaccines induced immunosterilization in both bulls and heifers (Adams and Adams, 1990; Adams and Adams, 1992; Hoskinson *et al.*, 1990). No production benefit has been shown in heifers and a depression in weight gain is observed that was overcome with implants containing anabolic steroid (Adams and Adams, 1990; Prendiville *et al.*, 1995). Enhanced performance has been demonstrated with anti-GnRH vaccines in bulls compared to steers (Aissat *et al.*, 2002). In this study small numbers of immunized bulls were comparable to steroid implanted steers in ADG, final carcass weight, and yield grade (Table 3). With a vaccine of GnRH conjugated to *M. haemolytica* leukotoxoid, however, performance was not enhanced in GnRH-vaccinated bulls compared to untreated bulls (Cook *et al.*, 2000).

Improvac® is an adjuvanted conjugate of diphtheria toxoid and the GnRH peptide. This vaccine increased growth and performance of intact boars and gilts, as well as preventing sexual development in both sexes and boar taint (Dunshea *et al.*, 2001; Oliver *et al.*, 2003). In a study at a commercial piggery, Improvac-treated boars grew more rapidly than control boars. This unexpected difference was attributed to a reduction in the aggressive behavior of vaccinates compared to the control intact boars (Dunshea *et al.*, 2001). Compared to barrows, these vaccinated pigs had better ADG and FCR, and were leaner, as measured by P2 backfat, than control barrows (Dunshea *et al.*, 2001; Table 4).

Table 3. Vaccination of bulls against GnRH (from Aïssat et al., 2002).

Parameter	Treatment		
	Steers (N=9)	Steers + Implant (N = 10)	Immunized Bulls (N = 10)
ADG (kg)	1.61	1.78	1.75
Carcass wt (kg)	310[a]	331[b]	329[b]
Yield grade	3.2	3.0	3.3

[a,b] Numbers in a row with differing superscripts differ significantly ($P < 0.05$).

Table 4. Vaccination of boars with Improvac (from Dunshea et al., 2001).

Parameter	Treatment		
	None (Boars)	Improvac (Boars)	None (Barrows)
Slaughter Wt. (kg)	113.3[a]	120.7[b]	117.1[b]
ADG (g)	858[a]	1119[b]	847[a]
FCR	3.30[a]	3.10[a]	3.73[b]
P2 backfat (mm)	12.6[a]	15.1[b]	17.1[c]

[a,b,c] Numbers in a row with differing superscripts differ significantly ($P < 0.05$).

Additional studies that have shown enhanced productivity of food animals include active or passive immunization against somatostatin, inhibin, vasoactive intestinal peptide, and myostatin. Somatostatin results have been varied depending upon the species. Results in sheep have generally been positive, while results in pigs have been negative (Elsaesser and Drath, 1995). Immunization against inhibin has resulted in increased ovulation rates in gilts, ewes, and mares and enhanced sperm production in rams (Geary and Reeves, 1996). Vasoactive intestinal peptide was reported to significantly increase egg production in turkeys (Caldwell *et al.*, 1999). Also, vaccinating against myostatin, leptin, or other hormones that affect lean to fat ratios has been suggested as a method to enhance growth of food-producing animals (Sillence, 2004).

Myostatin is a 15-kDa protein of the TGF-β family that has been demonstrated to have a role in muscle mass formation. Mutations in the myostatin gene have been linked to double muscling in such cattle breeds as Belgian Blues, Piedmontese, and Asturiana de los Valle (Kocamis and Killefer, 2002). Recently, immunization of turkey hens with a KLH-myostatin conjugate was shown to significantly enhance muscle mass of their male turkey progeny (El Halawani and You, 2002). These toms from vaccinated hens had significantly increased body, heart, breast, and thigh muscle weight at 17 weeks post-hatch (Table 5).

Summary

The use of vaccines that provide specific protection from disease and concurrently provide production gains occurs with vaccination for respiratory diseases in swine and bovine mastitis. Immune approaches to replace antibiotics for growth promotion have included mechanisms that affect the gut flora directly, reduce the level of inflammatory cytokines, use selected cytokines directly, or effect levels of endogenous hormones. Anti-inflammatory agents including cytokine receptor antagonists such as IL-1ra have been shown to enhance growth performance of pigs. Direct use of cytokines such as IFN-α, IFN-γ and IL-5 has been shown to enhance growth performance and disease resistance in cattle, poultry and swine. Anti-hormonal vaccines such as those vaccines targeting gonadotropin-releasing hormone (GnRH) increase growth performance of intact boars, gilts, and bulls. Additional anti-hormonal vaccine approaches include active immunization against somatostatin, inhibin, vasoactive intestinal peptide, and myostatin. While few of the cited immunological approaches to production enhancement have yet to be commercialized, the possibilities are exciting.

Table 5. Effect of passive vaccination on male turkeys at 17 weeks of age (El Halawani and You, 2002).

Parameter	Treatment Control (N = 9)	% Increase Immunized (N = 7)	
Weight (kg)	12.5 ± 0.2	15.1 ± 0.4	21[a]
Breast Muscle (kg)	2.5 ± 0.1	3.3 ± 0.1	33[a]
Thigh Muscle (kg)	2.6 ± 0.1	3.5 ± 0.1	35[a]

[a]$P < 0.01$ compared to control value.

References

Adams, T.E. and B.M. Adams, 1990. Reproductive function and feedlot performance of beef heifers actively immunized against GnRH. J. Anim. Sci. 68: 2793-2802.

Adams, T.E. and B.M. Adams, 1992. Feedlot performance of steers and bulls actively immunized against gonadotropin-releasing hormone. J. Anim. Sci. 70, 1691-1698.

Aïssat, D., J.M. Sosa, D.M. de Avila, K P. Bertrand and J.J. Reeves, 2002. Endocrine, growth, and carcass characteristics of bulls immunized against luteinizing hormone-releasing hormone fusion proteins. J. Anim. Sci. 80, 2209-2213.

Babiuk, L.A., 2002. Vaccination: a management tool in veterinary medicine. Vet. J. 164, 188-201.

Babiuk, L.A., L.M. Sordillo, M. Campos, H.P.A. Hughes, A. Rossi-Campos and R. Harland,. 1991. Application of interferons in the control of infectious diseases of cattle. J. Dairy Sci. 74, 4385-4398.

Bosi, P., L. Casini, A. Finamore, C. Cremokolini, G. Merialdi, P. Trevisi, F. Nobili and E. Menghrei, 2004. Spray-dried plasma improves growth performance and reduces inflammatory status of weaned pigs challenged with enterotoxigenic *Escherichia coli*. J. Anim. Sci. 82, 1764-1772.

Caldwell, S.R., A.F. Johnson, T.D. Yule, J.L. Grimes, M. Ficken and V.L. Christensen, 1999. Increased egg production in juvenile turkey hens after active immunization with vasoactive intestinal peptide. Poultry Sci. 78, 899-901.

Campos, M., D. Godson, H. Hughes, L. Babiuk and L.M. Sordillo, 1993. The role of biological response modifiers in disease control. J. Dairy Sci. 76, 2407-2417.

Cook, R.B., J.D. Popp, J.P. Kastelic, S. Robbins and R. Harland. 2000.The effects of active immunization against GnRH on testicular development, feedlot performance, and carcass characteristics of beef bulls. J. Anim. Sci. 78, 2778-2783.

Cummins, J.M., D.P. Hutcheson, M.J. Cummins, J.A. Georgiades and A.B. Richards, 1993. Oral therapy with human interferon alpha in calves experimentally infected with infectious bovine rhinotracheitis virus. Archivum Immunologiae et Therapiae Experimentalis 41, 193-197

Cummins, J.M., D. Guthrie, D.P. Hutcheson, S, Krakowka and B.D. Rosenquist, 1999. Natural human interferon-α administered orally as a treatment of bovine respiratory disease complex. J. Interferon Cytokine Res. 19, 907-910.

Cummins, J.M., R.E. Mock, B.W. Shive, S, Krakowka, A.B. Richards and D.P. Hutcheson, 1995. Oral treatment of transmissible gastroenteritis with natural human interferon-α: a field study. Vet. Immunol. Immunopathology. 45, 355-360.

DeGraves, F.J. and J. Fetrow, 1991. Partial budget analysis of vaccinating dairy cattle against coliform mastitis with *Escherichia coli* J5 vaccine. J. Amer. Vet. Med. Assoc. 199, 451-455.

Dunshea, F.R., C. Colantoni, K. Howard, I. McCauley, P. Jackson, K.A. Long, S. Lopaticki, E.A. Nugent, J.A. Simons, J. Walker and D.P. Hennessy, 2001. Vaccination of boars with a GnRH vaccine (Improvac) eliminates boar taint and increases growth performance. J. Anim. Sci. 79, 2524-2535.

El Halawani, M.E. and S. You, 2002.Use of passive myostatin immunization. US Patent Application Publication US 2002/0127234 A1.

Elsaesser, F. and S. Drath, 1995. The potential of immuno-neutralization of somatostatin for improving pig performance. Livestock Prod. Sci. 42, 255-263.

Geary, T.W. and J.J. Reeves, 1996. Production of a genetically engineered inhibin vaccine. Vaccine 14, 1273-1279.

Holder, A.T., R. Aston, M. Preece and J. Ivanyi, 1985. Monoclonal antibody mediated enhancement of growth hormone activity *in vivo*. J. Endocrinol. 107, R9-R12.

Hoskinson, R. M., R. D. G. Rigby, P. E. Matner, V. L. Huynh, M. D'Occhio, A. Neish, T.E. Trigg, B.A. Moss, M.J. Lindsey, G.D. Coleman and C.L. Schwartzkoff, 1990. Vaxstrate®:an anti-reproductive vaccine for cattle. Australian J. Biotechnology 4, 166-170.

Hilton, L.S., A.G.D. Bean and J.W. Lowenthal, 2002. The emerging role of avian cytokines as immunotherapeutics and vaccine adjuvants. Vet. Immunol. Immunopathology 85, 119-128

Johnson, M.A., C. Pooley and J.W. Lowenthal, 2000. Delivery of avian cytokines by adenovirus vectors. Developmental Comparative Immunology 24, 343-354.

Johnson, R.W. 1997. Inhibition of growth by pro-inflammatory cytokines, an integrated view. J. Anim. Sci. 75, 1244-1255

Kocamis, H. and J. Killefer, 2002. Myostatin expression and possible functions in animal muscle growth. Domestic Anim. Endocrinol. 23, 447-454.

Lecce, J.G., J.M. Cummins and A.B. Richards, 1990. Treatment of rotavirus infection in neonate and weanling pigs using natural human interferon-alpha. Mol. Biother. 2, 211-216.

Leitner, G., N. Yadlin, E. Lubashevsy, E. Ezra, A. Glickman, M. Chaffer, M. Winkler, A. Saran and Z. Trainin, 2003. Development of a *Staphylococcus aureus* vaccine against mastitis in dairy cows. II. Field trial. Vet. Immunol. Immunopathology 93, 153-158.

Lowenthal J.W., B. Lambrecht, T.P. van den Berg, M.E. Andrew, A.D.G. Strom and A.G.D. Bean, 2000. Avian cytokines – the natural approach to therapeutics. Developmental Comparative Immunology 24, 355-365.

Lowenthal, J.W., J.J. York, T.E. O'Neil, R.A. Steven, D.G. Strom and M.R. Digby. 1998. Potential use of cytokine therapy in poultry. Vet. Immunol. Immunopathology 63, 191-198

Maes, D., H. Deluyker, M. Verdonck, F. Castryck, C. Miry, B. Vrijens, W. Verbeke, J. Viaene and A. de Kruif. 1999. Effect of vaccination against *Mycoplasma hyopneumoniae* in pig herds with an all-in/all out production system. Vaccine 17, 1024-1034.

Marcus, P.I., L. van der Heide and M.J. Sekellick, 1999. Interferon action on avian viruses. I. Oral administration of chicken interferon-α ameliorates Newcastle disease. J. Interferon Cytokine Res. 19, 881-885.

Moreau, I.A., G.Y. Miller and P B. Bahnson, 2004. Effects of *Mycoplasma hyopneumoniae* vaccine on pigs naturally infected with *M. hyopneumoniae* and porcine reproductive and respiratory syndrome virus. Vaccine 22, 2328-2333.

Oliver, W.T., I. McCauley, R.J. Harrell, D. Suster, D.J. Kerton and F.R. Dunshea, 2003. A gonadotropin-releasing factor vaccine (Improvac) and porcine somatotropin have synergistic and additive effects on growth performance in group-house boars and gilts. J. Anim. Sci. 81, 1959-1966.

Owusu-Asiedu, A., S.K. Baidoo, C.M. Nyachoti and R.R. Marquardt, 2002. Response of early-weaned pigs to spray-dried porcine or animal plasma-based diets supplemented with egg-yolk antibodies against enterotoxigenic *Escherichia coli*. J. Anim. Sci. 80, 2895-2903.

Pell, J. M. and R. Aston, 1995. Principles of immunomodulation. Livestock Prod. Sci. 42, 123-133.

Prendiville, D.J., W.J. Enright, M.A. Crowe, M. Finnerty, N. Hynes and J.F. Roche, 1995. Immunization of heifers against gonadotropin-releasing hormone, antibody titers, ovarian function, body growth and carcass characteristics. J. Anim. Sci. 73, 2382-2389.

Roura, E., J. Homedes and K.C. Klasing, 1992. Prevention of immunologic stress contributes to the growth-permitting ability of dietary antibiotics in chicks. J. Nutr. 122, 2383-2390.

Sillence, M.N., 2004. Technologies for the control of fat and lean deposition in livestock. Vet. J. 167, 242-257.

Strom, A.D.G., A.G. Knowles and A. Husband, 2002a. A method of improving the growth performance of an animal. PCT, WO02/096216 A1

Strom, A.D.G., A.G. Knowles and M.E. Andrew. 2002b. Method of improving the growth performance of an animal. PCT, WO02/067979 A1

van Dijk, A.J., H. Everts, M.J.A. Nabuurs, R.J.C.F. Margry and A.C. Beynen, 2001. Growth performance of weanling pigs fed spray-dried animal plasma: a review. Livestock Prod. Sci. 68, 263-274.

Visek, W. J., 1978. The mode of growth promotion by antibiotics. J. Anim. Sci. 46, 1447-1469.

Yancey, R.J., 1999. Vaccines and diagnostic methods for bovine mastitis: fact and fiction. Advances Vet. Med. 41, 257-273.

Colostrum management in calves: effects of drenching versus bottle feeding

M. Kaske[1], A. Werner[1], H.-J. Schubert[2], H.W. Kehler[1] and J. Rehage[1]
[1]Clinic for Cattle; [2]Immunology Unit, School of Veterinary Medicine, Bischofsholer Damm 15, D–30173 Hannover, Germany

The objective of the study was to examine whether the administration of colostrum by use of a drencher (a) represents a safe method and (b) results in satisfying levels of immunoglobulins in the serum of newborn calves compared to bottle-fed calves.

Newborn HF-calves (N = 46; 43.7 + 6.1 kg birth weight; mean + SD) were used. Twenty-one calves received 1h post natum (p. n.) 2 L of fresh colostrum from the dam via a nipple bottle (group I). In 15 calves, 4 L colostrum of the dam were drenched 1h p. n. (group II). Thereafter, all calves were fed milk replacer exclusively. The concentration of immunoglobulins was analysed in a subsample of the colostrum after centrifugation (Sandwich-ELISA). Venous blood samples were taken from each calf prior to colostral administration and 12, 24, 48, 72, 96, 120, 144, 168 and 336h p. n.; total protein was analysed from plasma (Cobas Mira®). Serum immunoglobulin concentrations were determined from the sample taken 24h p. n.. To characterize the kinetics of immunoglobulin absorption, a catheter was introduced into the jugular vein of 5 calves (2 L colostrum 1h p. n. by nipple bottle; group III) and 5 calves (4 L colostrum 1h p. n. by drencher; group IV); 17 blood samples were taken within 72h p. n. and analysed for immunoglobulins and total protein.

Especially calves with a low birth weight appeared depressed for 12–24h after being drenched; adaptation to further feeding with a nipple bucket was more difficult compared to calves of group II. Concentration of immunoglobulins in colostral serum was 53.7 + 14.8 g/l. Calves total plasma protein concentration (prior to colostral supply: 43.7 + 3.6 g/l) rose within 24h p. n. by 11.1 + 4.6 g/l (group I) and significantly higher in drenched calves (group II; 16.5 + 7.3 g/l). Serum concentration of total immunoglobulins 24h p. n. was higher in drenched calves (group II: 25.2 g/l; 18.6/36.6; median and 25/75 percentiles) compared to bottle-fed calves (group I: 14.1 g/l; 11.7/19.1). Maximal immunoglobulin concentrations were found 12h p. n. in group III and IV. A slightly delayed increase of serum immunoglobulin concentrations (ca. 3h) was observed in group IV compared to group III. Again the drenched calves reached significantly higher immunoglobulin compared to the bottle-fed calves.

The proper application of 4 L of colostrum by a drencher was a safe and useful method to achieve an adequate transfer of immunoglobulins; the transfer of colostrum into the rumen seems to be without biological significance.

Immune system activation increases the tryptophan requirement in post weaning pigs

Nathalie Le Floc'h and Bernard Sève
Unité Mixte de Recherche sur le Veau et le Porc, INRA Saint Gilles, France 35590

Immune system activation prevents the pigs from reaching their growth potential by decreasing voluntary food intake but also by modifying nutrient metabolism and utilization for body protein accretion. Preliminary results obtained in our laboratory showed that pigs suffering from a chronic lung inflammation had lower plasma tryptophan (Trp) concentrations than healthy pair-fed pigs[1] suggesting an increase in tryptophan utilization. Moreover, inflammation caused an increase in indoleamine 2,3 dioxygenase (IDO) activity, an enzyme involved in Trp catabolism. The objective of the present experiment was to determine if a moderate immune system activation obtained by deteriorating the quality of environment where pigs were housed modified Trp requirement of post weaning growth in pigs.

Twenty blocks of four littermate piglets were weaned at 28 days of age and selected based on their body weight. Within a block, each piglet was affected to an experimental group according to a 2 x 2 factorial design: two levels of dietary Trp (14% and 20% of dietary lysine supply) and two level of sanitary status (clean environment vs unclean environment). Within a block, all pigs received the same amount of food. Pigs were weighed weekly and blood was taken 12, 33 and 47 days post weaning after an overnight fast for analysis of plasma amino acid and haptoglobin concentrations.

Pigs kept in unclean conditions had a lower growth rate and higher feed conversion ratio that control pigs. They also had lower plasma Trp concentrations and higher haptoglobin concentrations. Pigs fed the Trp-deficient diet had lower growth performance and plasma Trp concentration. Haptoglobin concentration was not affected by dietary Trp.

This study shows that a moderate activation of immune response obtained by deteriorating the sanitary quality of environment led to a reduction of growth performance and to a decrease in plasma Trp concentrations. Growth depression caused by immune system activation is limited by increasing the dietary Trp slightly above the recommended minimal level for maximum growth performance.

Acknowledgements
We acknowledge Ajinomoto-Eurolysine for their financial contribution.

References
[1] Melchior D., B. Sève, N. Le Floc'h, 2004. Chronic lung inflammation affects plasma amino acid concentrations in pigs. Journal of Animal Science 82, 1091-1099.

Periparturient negative energy balance and neutrophil function suppression are associated with uterine health disorders and fever in Holstein cows

D.S. Hammon[1], I.M. Evjen[1], J.P. Goff[2], T.R. Dhiman[1]
[1]Utah State University, Logan, UT, USA; [2]USDA, National Animal Disease Center, Ames, IA, USA

Eighty-three multiparous Holstein cows were used to investigate the association between periparturient energy balance, periparturient neutrophil functions, and uterine health disorders and fever.

Blood samples were collected at 1 wk prepartum, the week of calving, and at 1 wk, 2 wk, 3 wk, 4 wk, 5 wk and 8 wk postpartum for neutrophil function determination. Blood samples were collected weekly for plasma nonesterified fatty acids (NEFA) and Beta-hydroxybutyrate (BHB). Dry matter intake (DMI) was determined daily from wk-2 to wk-5. Neutrophil killing ability was evaluated by determining myeloperoxidase activity and Cytochrome C reduction activity in isolated neutrophils. Cows were examined at wk-3 for clinical endometritis (purulent cervical discharge on vaginal examination) and at wk-4 for subclinical (SC) endometritis (presence of neutrophils on endometrial cytological exam). Retained placentae (RP) were determined by visual and vaginal speculum examination on d-1 postpartum. Rectal temperatures were recorded from d-1 to d-10 postpartum. Fever was defined as a rectal temperature =103°F for= 2d. Differences in neutrophil myeloperoxidase activity and Cytochrome C reduction, DMI, NEFA, and BHB for cows with RP, fever, clinical endometritis, and SC endometritis were determined using a repeated measures of ANOVA.

Of 83 cows, 14 developed RP, 13 developed clinical endometritis, 61 developed SC endometritis, and 18 developed fever. Cows with RP had significantly (P<0.001) lower DMI beginning 1 wk before calving and for the first 3 wk of lactation compared to cows without RP. Neutrophil myeloperoxidase activity (P=0.23) and cytochrome C reduction (P=0.12) were not different for cows with retained placenta compared to cows without retained placenta. For cows with SC endometritis, neutrophil myeloperoxidase activity tended (P=0.06) to be suppressed beginning prior to calving and lasting until 2 wk postpartum. SC endometritis cows had significantly (P=0.01) lower DMI from wk-1 to wk-5, significantly (P=0.01) higher NEFA from wk-2 to wk-4, and significantly higher (P<0.04) BHBA from wk-1 to wk-4, compared to cows without subclinical endometritis. For cows with clinical endometritis, neutrophil myeloperoxidase activity was significantly (P<0.01) suppressed and neutrophil Cytochrome C reduction tended (P=0.09) to be suppressed beginning prior to calving and lasting until 4 wk postpartum, compared to cows without clinical endometritis. DMI, NEFA, and BHBA were similar for cows with or without clinical endometritis. Cows with fever, had significantly suppressed neutrophil myeloperoxidase activity (P<0.01) and neutrophil Cytochrome C reduction (P=0.03) beginning 1 wk prepartum and lasting until 1 wk postpartum and had significantly (P=0.05) lower DMI from wk-1 to wk-4, significantly (P=0.03) higher NEFA from wk-1 to wk-4, and significantly (P<0.03) higher BHBA wk-1 to wk-4, compared to cows without fever.

These data demonstrate that endometritis and fever in dairy cows are associated with suppressed neutrophil function and that some uterine health disorders and fever are preceded by negative energy balance that begins prior to calving and extends into early lactation. Since neutrophil function and energy balance declined before the onset of these disorders, these data suggest suppressed neutrophil function and negative energy balance increase the risk of uterine health disorders and fever developing in the fresh cow.

Acknowledgements

This research was supported by Pharmacia Animal Health, Utah State University, and the NADC, USDA

Activities of enzymes in peripheral leukocytes reflect metabolic conditions in fattening steers

T. Arai, A. Inoue, A. Takeguchi, S. Urabe, T. Sako, I. Yoshimura and N. Kimura
Department of Veterinary Science, Nippon Veterinary and Animal Science, Japan

As the Japanese Black Wagyu, which is very popular as fattening beef cattle in Japan, is fed on different diets in the process of fattening, various changes in energy metabolism may be induced in the animal tissues. The activities of certain enzymes related to energy metabolism in the peripheral leukocytes are considered to reflect the metabolic conditions in animal tissues. To investigate the availability of some peripheral leukocytes enzymes as an indicator to evaluate metabolic conditions, glucose, triglyceride, cholesterol and immunoreactive insulin (IRI) concentrations, some enzymes activities in plasma and activities of enzymes related to energy metabolism in peripheral leukocytes were measured in fattening Japanese Black Wagyu x Holstein steers fed on different diets at 8, 12, 16, 20 and 24 months of age.

High concentrations of roughage in the diet of stage I (6-12 months) animals were replaced with barley and wheat bran in the diets of stage II (13-20 months) and stage III (21-24 months) animals. After 16 months of age, the plasma glucose concentrations of the steers were significantly lower than those at 8 months of age. The plasma IRI concentrations at 20 and 24 months of age were significantly higher than those at 8 months of age. Activities of hexokinase (HK), glucose-6-phosphate dehydrogenase (G6PD), aspartate aminotransferase (AST), and malate dehydrogenase (MDH) in cytosolic fractions, and glutamate dehydrogenase (GLDH), MDH and AST in mitochondrial fractions in peripheral leukocytes of the steers at 24 months of age were significantly higher than those at 8 months. Increasing plasma IRI concentrations was considered to induce acceleration of glucose utilization in leukocytes of the fattening steers. The cytosolic ratio of MDH/lactate dehydrogenase (LDH) activity (M/L ratio) in leukocytes increased significantly in the fattening steers.

It has been reported that the M/L ratio is a useful parameter to evaluate metabolic conditions in the race horses with training[1]. The MDH activities and M/L ratio in the peripheral leukocytes were considered to be a useful indicator for evaluating changes in energy metabolism in fattening steers with aging.

References

[1] Arai T., M. Hosoya, M. Nakamura, *et al.*, 2002. Cytosolic ratio of malate dehydrogenase/lactate dehydrogenase activity in peripheral leukocytes of race horses with training. Research in Veterinary Science 72, 241-244.

Effects of 0.3% dietary β-glucan on nonspecific/specific immunity, oxidative/antioxidative status and growth performance in weanling pigs

S. Schmitz, S. Hiss and H. Sauerwein
Institute of Physiology, Biochemistry and Animal Hygiene, University of Bonn, Germany

Immunomodulatory feed additives offer alternatives to anti-microbial growth promoters in swine production. β-1,3/1,6-glucans are branched carbohydrates derived from yeast cell walls. Based on reports about modulatory effects of these substances on the immune system and considering potential overdosing in practise, we examined the influence of a provocatively high dose of β-glucans on various health-related parameters in weaned pigs to evaluate the potential benefit and/or risks of β-glucan concentrations tenfold higher than intended use.

30 piglets were randomly allocated to either the control group (C; n = 10) or the treatment group (T; n = 20). After weaning at 29 to 31 days of age, the two groups were fed the same conventional feed with 0 or 0.3% β-glucans (Antaferm-M, Dr. Eckel GmbH, Niederzissen, Germany). After 3.5 weeks of treatment all animals were fed a diet without β-glucans. Oxidative stress was monitored in whole blood directly after sampling using the D-ROM® system (detection of reactive oxygen metabolites)[1]. Nonspecific immune parameters recorded were phagocytic activity and oxidative burst of neutrophile granulocytes in whole heparinised blood via flowcytometry and also haptoglobin serum concentrations (Hp) by enzyme immuno assay. The specific immune defense was investigated by means of Immunoglobulin G and A serum concentrations (IgG and IgA) by ELISA. The total antioxidant capacity was quantified using the TEAC assay[2]. Blood samples were taken 1 week before weaning and in weekly intervals thereafter. Measurement of oxidative stress and neutrophile activity was limited to the samples obtained 1 week before weaning, and 3 and 6 weeks after weaning. Individual body weights were recorded weekly and feed intake was quantified daily per pen containing 10 pigs each. All data were analyzed with the PROC MIXED procedure of SAS. The model included terms for the fixed effects of dietary treatment, time and the appropriate interactions.

A treatment x time interaction was found for oxidative burst and serum Hp concentrations (P<0.01 and P<0.05, respectively). Concentrations of Hp in controls and treated pigs were elevated after weaning and returned to baseline levels for both groups, whereas Hp in treated pigs increased again. Phagocytic activity was not influenced by β-glucan feeding. A treatment x time interaction also was observed for IgA and oxidative stress (P<0.01 and P<0.05, respectively). In addition serum IgA was increased over time. Oxidative stress for both groups was elevated after weaning but returned to initial values for the C group, whereas oxidative stress in treated pigs remained elevated. β-glucan feeding resulted in a reduction in IgG concentrations and antioxidant capacity (P<0.01 and P<0.001, respectively) as compared to controls. There was a β-glucan feeding x time interaction (P<0.001) for the average daily gains (ADG). Weaning resulted in a reduction in ADG for controls and treated pigs (P<0.0001). There were no differences in food intake between the groups.

The current study demonstrated contradictory effects of β-glucans on the various parameters assayed. The results indicate a suppressive effect on IgG, TEAC and growth performance and a marginally stimulative effect on serum Hp concentrations. The administration of the β-glucan preparation used herein at the dosage investigated is thus not recommendable.

References

[1] Cesarone M.R., G. Belcaro, M. Carratelli, *et al.*, 1999. A simple test to monitor oxidative stress. Int. Angiol. 2, 127-130.

[2] Miller N.J., C. Rice-Evans, M.J. Davies, *et al.*, 1993. A novel method for measuring antioxidant capacity and its application to monitoring the antioxidant status in premature neonates. Clin. Sci. 84, 407-412.

Evaluation of Liver Abscess Incidence in Feedlot Cattle Fed a Dietary Antioxidant (AGRADO®) across Four Studies

M. Vázquez-Añón[1], F. Scott[1], B. Miller[1] and T. Peters[2]
[1]Novus International, Inc., St. Louis, Missouri, USA; [2]Dekalb Feeds, Rock Falls, IL, USA

Liver abscesses commonly occur in feedlot cattle when fed a finishing diet and is associated with lower performance. Four studies were conducted using over 50,000 feedlot cattle to evaluate the effect of feeding the antioxidant 6-ethoxyl-1, 2, 2, 4-trimethylquinoline (AGRADO® feed antioxidant; a registered trademark) on the incidence of liver abscesses.

Study 1: At Oklahoma State University, 75 crossbred steers (326 kg BW) housed in 15 pens were fed a ground corn-based diet supplemented with 0 or 135 PPM of Agrado (DM basis) during the last 28d prior to harvest. At the end of study, cattle were slaughtered and livers were examined for the presence and severity of abscesses. The incidence of liver abscesses was low for all the cattle, but was 34% lower (4.4% *vs.* 6.7%) for those cattle supplemented with Agrado.

Study 2: At Colorado Research Feedlot, Lamar, CO, 128 crossbred yearling steers (350 kg BW) were fed a steam-flaked corn-based diet and supplemented with 0 or 150 PPM of Agrado (as-fed basis) for the last 123d of the finishing period. All cattle received 15-30 g/tn of Monensin and 2-10 g/tn of Tylosin. The incidence of liver abscesses was 25% lower for cattle supplemented with Agrado (14.06% *vs.* 18.75%).

Study 3: At Texas Research Feedlot Inc, 1991 beef steers (566 lb BW) were fed a steam-flaked corn-based diet supplemented with none, 125 IU vit E/day or 125vit E +150 PPM Agrado during the finishing phase. Incidence of liver abscess in cattle supplemented with 125 IU vit E was slightly reduced from 13.8 to 12.0%, but was further reduced to 10.2% when supplemented with 125 IU vit E+150 PPM Agrado.

Study 4: Liver abscess from 485 beef cattle raised in 9 feedlots supplemented daily with 500 IU vit E + 150 PPM Agrado during the finishing period were compared at PM Beef packing plant, Windon, MN, to 803 beef cattle raised in 12 feedlots supplemented daily with 1000 IU vit E. None of the cattle received Tylosin or Monensin. Incidence of liver abscess in groups of cattle fed Agrado averaged 4.9% and ranged from 0 to 7.5% and cattle not fed Agrado averaged 20.5% and ranged from 0 to 37%. Therefore, cattle with diets containing Agrado had 15.3% less liver abscesses than those fed diets not treated with Agrado.

From the results of the four studies, it appears that addition of antioxidants such as Agrado to feedlot cattle diets reduces the incidence of liver abscesses.

Pathogenic *Escherichia coli* in dairy cows held in farms with organic production (OP) and with integrated production (IP)

P. Kuhnert[1], C. Dubosson[1], M. Roesch[2], E. Homfeld[2], M.G. Doherr[3] and J.W. Blum[1, 2]
[1]Institute of Veterinary Bacteriology; [2]Division of Nutrition and Physiology; [3]Division of Clinical Research, Vetsuisse Faculty, University of Berne, Berne, Switzerland

The bacterium *Eschericia coli* is normally a non-problematic inhabitant of the intestine of animals and man. However, some types are pathogenic and might cause diarrhea, enteritis and can damage other organs and thus have to be considered as a health risk for animals and humans. Certain pathogenic types of *E. coli* (termed STEC) produce Shiga-like toxins (Stx1 and Stx2). A few of these *E. coli* (termed EHEC) produce additional toxins and are very pathogenic. Several severe disease outbreaks in humans have been described in the past due to ingestion of EHEC. Cattle feces are considered to be the most important source of STEC and EHEC and diseases are therefore primarily caused by contamination of foods and feeds with cattle feces. Recent studies indicate that the number EHEC in cattle may increase in association with a high intake of energy-rich concentrates. Since IP farms are expected to feed more energy-rich concentrates than OP farms because milk production is higher in IP than in OP farms, we have hypothesized that the prevalence of EHEC - and possibly of STEC as well - is higher in IP than OP farms.

Based on these premises we have studied the prevalence of STEC and EHEC in feces collected from 483 cows in 60 OP farms and from 482 cows in 60 IP farms between 21 and 43 days post partum, *i.e.* when the greatest differences in feeding of concentrates between OP and IP farms could be expected. Farms were located in the midland and prealpine zones and were a representative sample of the situation in the Canton of Berne, Switzerland. IP farms matched to OP farms were comparable in terms of community, agricultural zone and number of cows/farm. Depending on the number of cows per farm, 5, 8, or 13 cows were selected. *E. coli* were grown overnight in an enrichment medium, followed by DNA isolation and PCR analysis using two specific TaqMan systems for STEC and EHEC, respectively.

STEC were detected in 100% of OP farms and 100% of IP farms and EHEC were found in 23% of OP farms and 17% of IP farms. On a cow level STEC were found in 60% and EHEC were found in 4% of feces. Multivariate statistical analyses revealed that the presence of STEC was significantly enhanced if cows were held for more than 3 h/day on pastures [odds ratio (OR) = 4.0; P<0.001], if cows were held in paddocks for up to 1h/day (OR = 1.9; P<0.01), if total-mixed-rations were fed with a wagon (OR = 2.9; P<0.001), and if milk urea concentration was elevated (OR =1.03; P<0.001), suggesting that these situations enhanced the risk of cross-contamination of feeds and and cross-infection of cows. The OR were relatively low (OR = 0.65; P<0.02) if cows were held in OP farms. The low number of cows with EHEC-positive fecal samples (N=24/967 fecal samples) did not allow epidemiological analyses.

In conclusion, the prevalence of STEC and EHEC was similar in OP and IP farms, although the risk that cows were carriers of STEC was reduced if held in OP farms. The prevalences of STEC and EHEC were similar as reported in other countries. Because the incidence of diseases caused by STEC and EHEC in humans in Switzerland is low, the risk that humans get infected is low despite a relatively high prevalence in cattle.

Section E.
Reproductive health

Rearing conditions and disease influencing the reproductive performance of swedish dairy heifers

J. Hultgren
Dept. of Animal Environment and Health, Swedish University of Agricultural Sciences, Sweden

To investigate effects of calfhood and rearing on heifer fertility, a longitudinal study of 122 dairy herds in soutwestern Sweden was performed. Totally 3081 heifers born in 1998 were followed from birth to first calving. Information about housing, management, breeds and dates of birth, breeding and calving was obtained from farmers. Diseases were recorded by farmers and veterinarians.

About a hundred potential predictors were recorded and considered for analysis. The outcomes were age at first service (days; log-transformed), being bred at 18 mo of age (binary), first-service conception risk (binary), age at first calving (days; log-transformed), and having calved at 28 mo of age (binary). Animals mistakenly covered by a bull too early in life (as stated by the farmer) were excluded. In the analysis of conception risk, animals that calved between 260 and 290d after first service were coded as 1, those that calved after this period were coded as 0, and those that either were not bred by AI or reported to calve <260 d after first service were excluded. For each outcome, a 2-level (heifer; herd) variance-component model was built in MLwiN[1], accounting for herd clustering. A binomial distribution and the logit link was applied for binary outcomes. Between 1844 and 2356 heifer records were used.

Median time to first service was 533d, 64% of all heifers bred by AI conceived at first service, and median time to calving was 841d. Approximately 42% of the heifers developed disease in the first 24 mo of life but before conception, and 23% during the first 3 mo. Heifers born in March to June had an approximately 2-3% longer time to first service and calving than those born in July to February. Heifers kept in individual pens during the first month had a 4% shorter time to first service than those kept in group pens. Heifers grazed before start of breeding had a 6% longer time to calving than heifers that were not grazed. If grazed before start of breeding, heifers kept in group pens during the first month and thereafter in boxes with concrete slatted flooring had a 3% longer time to calving than those kept in individual pens during the first month. Heifers with a history of disease other than respiratory illness, diarrhea or ring-worm later than 6 mo of age (mainly unspecified infection, traumatic lesion, interdigital necrobacillosis or abortion) had an approx. 5% longer time to calving and a 3 to 5-fold lower odds of calving before 28 mo than heifers without such disease. Heifers in herds where the start of breeding was decided based on heart-girth measurements had a 10% longer time to first service and a 6% longer time to calving than those in herds with other breeding routines. For conception risk at first service, only 3.2% of the unexplained variation resided at the herd level; for the remaining outcomes, this portion ranged between 40 and 57%.

I conclude that the reproductive performance of dairy heifers is highly influenced by herd-level factors, *e.g.* routines for grazing and breeding, which are probably also related to animal health. However, this study does not provide good support for earlier findings that infectious disease early in life is associated with delayed maturing and impaired reproductivity[2].

Acknowledgements

Professor Catarina Svensson planned the study and Karin Lundborg collected data. Professor Pascal A. Oltenacu and Daniel Maizon assisted in the statistical analysis.

References

[1] Rasbach J., W. Browne, H. Goldstein, *et al.*, 2002. A user's guide to MLwiN, version 2.1d for use with MLwiN 1.10. Centre for Multilevel Modelling, Institute of Education, London.

[2] Correa M.T., C.R. Curtis, H.N. Erb, *et al.*, 1988. Effect of calfhood morbidity on age at first calving in New York Holstein herds. Preventive Veterinary Medicine 6, 253-262.

Factors influencing conception rate after synchronization of ovulation and timed artificial insemination

B.-A. Tenhagen, C. Ruebesam and W. Heuwieser
FU Berlin, Clinic for Reproduction, Germany, www.bestandsbetreuung.de

The objective of the study was to investigate factors that may influence conception rate (CR) after synchronization of ovulation and timed artificial insemination (TAI). A total of 172 lactating Holstein dairy cows were submitted 63 ± 3 days postpartum to a synchronization protocol consisting of two treatments with GnRH (0.02 mg buserelin) 9 days apart and one treatment with PGF_{2a} (0.15 mg D-cloprostenol) 7 days after the first administration of GnRH. TAI was carried out either 6 or 16 hours after the second treatment with GnRH. Primiparous cows were kept in the same barn but in a group separated from multiparous cows. Blood samples were collected from all cows 7 days after TAI to determine the presence of a functional corpus luteum and to assess the metabolic situation at arrival of the conceptus in the uterus. Furthermore, body condition, milk production during the week prior to PGF_{2a} and apparent clinical diseases were recorded. Data were compared using Chi-square test and Mann-Whitney-U-test.

In accordance with other reports[1], timing of AI 6 or 16 hours after the second GnRH treatment had no significant impact on CR (23.3% *vs.* 27.9%) and was higher in primiparous than in older cows[2] (45.1 *vs.* 17.4%, P<0.01). A body condition < 2.75 was associated with poorer conception rates compared to cows in better condition (17.0 *vs.* 35.9%, P<0.01). Cows with at least one other health disorder had lower conception rates than those without disorders (17.8 *vs.* 37.3%, P<0.05). Both parameters were significantly worse in multiparous than in primiparous cows. Therefore, low BCS and presence of health disorders may have contributed to the lower CR in multiparous cows. Differences between age groups may have been biased by separate housing of primiparous cows. However, both age groups were housed opposite of each other in identical facilities and received the same total mixed ration in the same feeding trough.

Activity of AST and GLDH and serum levels of bilirubin did not differ significantly between cows that conceived and those that did not. This indicates that the liver metabolism one week after TAI is of limited importance for the risk of conception. Serum urea, on the other hand, was higher in cows that conceived (338 *vs.* 313 mg/l, P<0.05). This is in contrast to reports on the negative effect of high serum urea on fertility[3]. The reason for this finding remains unclear. There was no association of the metabolic parameters with the two parity classes. Milk production level did not affect conception rates in cows of second and greater lactation (39.7 *vs.* 39.3 kg/day). In primiparous cows, production was lower in cows that conceived than in those that did not conceive (28.4 ± 4.1 *vs.* 31.0 ± 4.3, P<0.05). In a recent study, milk production level had no significant effect on CR after TAI using the same synchronization protocol[4]. It has been proposed, that insufficient luteal activity may contribute to conception failure in high yielding cows. In our study, progesterone levels 7 days after TAI did not differ significantly between cows that conceived and non pregnant cows.

Overall, results underline the importance of the general condition of cows for the success of reproductive management programs. More research is needed to analyze the differences between primiparous and older cows in this and in other studies and their possible relations to separate housing of primiparous cows.

Acknowledgements

The authors gratefully acknowledge the support of Intervet Deutschland GmbH and the farm staff.

References

[3] Butler, W.R., J.J. Calaman, *et al.*, 1996. Plasma and milk urea nitrogen in relation to pregnancy rate in lactating dairy cattle. J. Anim. Sci. 74, 858-865.

[1] Pursley, J.R., R.W. Silcox, *et al.*, 1998. Effect of Time of Artificial Insemination on Pregnancy Rates, Calving Rates, Pregnancy Loss, and Gender Ratio After Synchronization of Ovulation in Lactating Dairy Cows. J. Dairy Sci. 81, 2139-2144.

[4] Tenhagen, B.A., C. Vogel, *et al.*, 2003. Influence of stage of lactation and milk production on conception rates after timed artificial insemination following Ovsynch. Theriogenology 60, 1527-1537.

[2] Tenhagen, B.A., R. Surholt, *et al.*, 2004. Use of Ovsynch in dairy herds—differences between primiparous and multiparous cows. Anim. Reprod. Sci. 81, 1-11.

The incidence of endometritis and its effect on reproductive performance of dairy cows

R.O. Gilbert[1], S.T. Shin[1], M. Frajblat[1], C.L. Guard[2], H.N. Erb[2] and H. Roman[1]
[1]Department of Clinical Sciences, College of Veterinary Medicine, Cornell University;
[2]Department of Population Medicine and Diagnostic Sciences, College of Veterinary Medicine, Cornell University, Ithica NY, USA

In two separate studies the prevalence and reproductive consequences of subclinical endometritis were ascertained. Endometritis was diagnosed by endometrial cytology and compared with subsequent reproductive performance. In the first trial, 141 cows in 5 commercial dairy herds were examined once between 40 and 60 days postpartum. A low volume uterine lavage was obtained by infusing 20 ml of sterile saline into the uterus, agitating it briefly, and aspirating part of the volume. The aspirate was centrifuged onto a glass slide, air dried and stained with a modified Wright-Giemsa stain (Diff-Quik®). Endometritis was diagnosed in 53% of cows. Concordance between examiners was excellent (kappa = 0.864; P<0.0001). Median days open was longer for cows with endometritis than without (203 days *vs.* 118; P<0.0001). Fewer endometritis positive cows were pregnant by 100 days postpartum (16 *vs.* 43%; P=0.002). Ultimately 84% of cows became pregnant during the study period (74% of endometritis-positive cows and 94% of endometritis-negative cows; P=0.003).

In the second trial samples were obtained from 529 cows in 6 commercial dairy herds at 3, 5 and 7 weeks postpartum. Although prevalence of inflammation was high at 3 and 5 weeks (92 and 67%), this had no significant effect on subsequent reproduction. At 7 weeks postpartum 51% of cows had endometritis. Endometritis at this stage impaired overall conception rate determined by survival analysis (P<0.002). Median days open was 158 for cows with endometritis (*vs.* 122 d); for this group 30% of cows left the herd or failed to become pregnant by 300 days (*vs.* 18%; P=0.003). Endometritis did not affect the days to first service, but first service conception rate was lower for cows with endometritis (26% *vs.* 34%; P=0.008) and endometritis-positive cows required more services per conception (median 2 *vs.* 3; P=0.004). In both trials herd had a pronounced influence on prevalence of endometritis and on reproductive performance.

These results show that endometritis near the end of the voluntary waiting period in high producing dairy cows is prevalent (approximately 50%). Furthermore, it exerts a profound influence on reproductive performance, lowering first service conception rate, requiring more service per conception, resulting in a lower overall pregnancy rate (determined by survival analysis), more days open, and more cows failing to become pregnant again by 300 days postpartum. These findings suggest that subclinical endometritis, a hitherto under-recognized condition, is a major contributor to impaired reproductive performance in dairy cows. Based on these findings, subclinical endometritis is likely to be extremely costly to the industry in additional days open and failure to become pregnant within an economical period.

Acknowledgements
Funded, in part, by USDA Animal Health and Disease Funds, Cornell University College of Veterinary Medicine and Pharmacia, Inc.

Use of *Neospora caninum* vaccine in a dairy herd undergoing an abortion outbreak

R.W. Meiring and P.J. Rajala-Schultz
Dept. of Veterinary Preventive Medicine, The Ohio State University, Columbus OH, USA

Neospora caninum is a major cause of abortion in dairy cattle in the U.S. and around the world. A major route of transmission for *N. caninum* is congenital, and no horizontal cow-to-cow transmission has been demonstrated. Abortion is the only recognized clinical sign in adult cattle. Infected cows may give birth to calves with neurological problems and to persistently infected, normal appearing calves. The latter serve to maintain the infection in the herd [1-5]. The objective of this paper is to report the use of a recently introduced commercial Neospora vaccine (NeoGuard™, Intervet) in a 400-cow dairy undergoing an abortion storm where *N. canimum* was diagnosed as a causal agent.

The herd experienced about 14% abortion rate between autumn 2001 and spring 2002 (normally < 5%/year) and *N. caninum* was demonstrated in several aborted fetuses during the outbreak. Between June 2002 and May 2003, all cows and replacement heifers in this herd were enrolled in a vaccine trial if they were diagnosed pregnant 35-60 days after breeding. They were randomly assigned either to receive or not to receive a commercial Neospora vaccine. Administration of the vaccine was done according to the manufacturer's label instructions. Serum samples were collected at this point to evaluate whether the animals were seropositive for *N. caninum* prior to the vaccination. Cows were rechecked between 90 and 110 days of pregnancy and all pregnancy losses as well as later abortions were recorded. Serum samples for *N. caninum* antibody detection were collected from all animals with fetal loss during the study period.

Twenty-two (22) out of the 411 cows and heifers enrolled in the study (5.4%) were seropositive for *N. caninum* at the first pregnancy check. 15% of the seropositive cows and 7.3% of the seronegative animals were diagnosed open at the second pregnancy check at approximately 100 days. Twenty-nine cows had fetal loss from the second pregnancy check through 250 days of gestation, 13 vaccinates and 16 non-vaccinates. Overall, 11% of the vaccinates and 13.9% of the non-vaccinates experienced fetal loss during the study period. Twenty-seven of the vaccinates and 25 of the non-vaccinates were sold during the study, for reasons unrelated to their reproductive status.

The abortion rate between the vaccinates and non-vaccinates did not differ significantly. By the time the herd started using the new Neospora vaccine, the abortion outbreak was already tapering off and the vaccine did not appear to provide any additional protection from abortions in this herd.

References

[1] Anderson M.L., A.G. Andrianarivo and P.A. Condrad, 2000. Neosporosis in cattle. Anim. Reprod. Sci. 60, 417-431.

[2] Dubey J.P., 1999. Recent advances in Neospora and neosporosis. Vet. Parasitol. 84, 349-367.

[3] Hernandez J., C. Risco and A. Donovan, 2002. Risk of abortion associated with *Neospora caninum* during different lactations and evidence of congenital transmission in dairy cows. J. Am. Vet. Med. Assoc. 221, 1742-1746.

[4] Pfeiffer D.U., N.B. Williamson, M.P. Reichel, *et al.*, 2002. A longitudinal study of Neospora infection on a dairy farm in New Zealand. Prev. Vet. Med. 54, 11-24.

[5] Sager H., I. Fischer, K. Furrer, *et al.*, 2001. A Swiss case-control study to assess *Neospora caninum* associated bovine abortions by PCR, histopathology and serology. Vet. Parasitol. 102, 1-15.

Ovulatory cycles, metabolic profiles, body condition scores and their relation to fertility of multiparous Holstein dairy cows

J. Samarütel[1], K. Ling[1], A. Waldmann[2], H. Jaakson[1], A. Leesmäe[3] and T. Kaart[1,3]
[1]Department of Animal Nutrition; [2]Department of Reproductive Biology; [3]Department of Animal Breeding, Estonian Agricultural University, Tartu 51014, Estonia

Research has shown that fertility of lactating cows is highly dependent on the start of luteal activity post partum. Conception rate has been related to the number of ovulatory cycles preceding first insemination. The aim of our study was to investigate the influence of onset of postpartum ovarian activity on subsequent fertility in multiparous Holstein dairy cows.

The study was carried out on a commercial dairy farm of about 250 Holstein dairy cows during the four-year period from 1999 to 2002. Average 305-day milk production a farm was 7922 kg. Cows, involved in the study (n=71), calved during the indoor period, from the end of December to the beginning of March. In order to study the effect of some metabolites associated with cows energy and/or protein status, blood samples were taken 1-14 days before calving, 1-14, 28-42 and 63-77 days after calving and were analysed for aspartate aminotransferase, glucose, ketone bodies, total cholesterol, urea, total lipids and triglycerides concentrations. Body condition scoring (BCS), using a 5-point scale with quarter-point divisions, was started about three weeks before calving and thereafter was performed every ten days. Milk samples for progesterone (P_4) concentration measurement were taken twice a week; P_4 levels measured in whole milk by enzyme immunoassay. In data analyses cows were categorized into four groups according to their milk P_4 profiles. The general and generalized linear models procedures with SAS system were used to compare reproduction traits, blood metabolites and body condition scores of different groups. The effect of lactation number and the effect of repeated measurements were taken into account. For non-normal traits the logarithm or logit transformation were used. For service period having many zero values the median test was implemented

More than half of the cows (54%) had normal progesterone profiles (interval from calving to first luteal response, P_4>5ng/ml, up to 50 days followed by regular cyclicity). Others (25%), whose interval from calving to first luteal response (P_4>5ng/ml) was more than 50 days, were categorized as delayed resumption of ovarian cyclicity (DOV 1); 4% of cows had cessation of cyclicity (DOV2) and 16% had prolonged luteal phase (PCL; progesterone levels remained elevated for > 20 days without preceding insemination). Fertility was best in the group DOV1 - days open 96 ± 35.4, services per conception 1.6 ± 1.0, first service conception rate 64.7%. Cows with normal progesterone profiles, DOV2 and PCL cows had similar reproductive performance. BCS decline during the 40-day postpartum period of normal, DOV1 and PCL groups was moderate (0.5-0.6 BCS units), DOV2 cows lost 0.81 BCS units. DOV1 cows had 1-2 week postpartum higher AST concentrations and slightly higher ketone body values compared with other three groups.

In conclusion, delayed resumption of ovarian cyclicity (DOV1 first P_4 rise 71 ± 16 days post partum) did not affect negatively fertility of cows in moderate negative energy balance during the postparturient period.

Field trial on blood metabolites, body condition score (BCS) and their relation to the recurrence of ovarian cyclicity in Estonian Holstein cows

Katri Ling[1], Andres Waldmann[2], Jaak Samarütel[1], Hanno Jaakson[1], Tanel Kaart[1], Andres Leesmäe[3]
[1]*Institute of Animal Science, Estonian Agricultural University, Kreutzwaldi 1, 51014 Tartu, Estonia,, kling@eau.ee;* [2]*Faculty of Veterinary Medicine, Estonian Agricultural University, Kreutzwaldi 62, 51014, Tartu, Estonia, waldmann@ut.ee;* [3]*Piistaoja Experimental Station, 86801 Piistaoja, Pärnu County, Estonia,*

Coping with metabolic adaptations needed for the transition from pregnancy to lactation is a prerequisite for the success of the next reproduction cycle. The present study was undertaken to assess the variations in metabolism and BCS in Estonian Holstein dairy cows *pre-* and *post-partum* and to relate these parameters to the resumption of ovarian activity.

The investigation on 60 dairy cows (11 animals in consecutive years, 2nd to 8th lactation) was carried out during a 4-year period on a 250-head commercial dairy farm. The cows were milked twice a day and the farm average milk yield was 6791, 7467, 8665 and 8984 kg in years 1999, 2000, 2001 and 2002, respectively. Blood samples taken 1-14 days before calving, 1-14, 28-42, and 63-77 days after calving were analysed for aspartate aminotransferase, glucose, ketone bodies, triglycerides, non-esterified fatty acids and cholesterol. Milk progesterone (P_4) profiles (collected twice a week, P_4 levels measured in whole milk by enzyme immunoassay) were used to evaluate the interval from calving to first luteal response, $P_4 > 5$ng/ml, and the interval from calving to first normal cycle. Cows with clinical signs of puerperal disorders were excluded from the study. The repeated measures general linear models analyses using SAS system were performed to estimate the influence of the measured parameters on the interval from calving to first luteal response and the interval from calving to first normal cycle, the regression procedures with stepwise and R^2 model selection methods were implemented to find the most relevant combinations of parameters. For non-normal traits the logarithm or logit transformation were used.

Higher BCS at calving as well as greater BCS decrease during 40 *post-parturient* days increased the interval from calving to first luteal response and the interval from calving to first normal cycle. Higher concentration of ketone bodies and lower concentration of cholesterol before calving had a positive effect on the first luteal response and the recurrence of ovarian cycles. There was a positive correlation between triglycerides concentration *pre-partum* and the interval from calving to first luteal response. Higher *periparturient* aminotransferase showed a tendency to increase the interval from calving to first cycle. In the cows not having had the first cycle before 28 days *post-partum,* lower urea during that period tended to favour the first luteal response and the recurrence of the cycles; at the same time higher glucose during the periods of 28-42 and 63-77 days after calving tended to favour the recurrence of the cycles in the cows not having had them before.

In conclusion, our results indicate that metabolic adaptations already *pre-partum* were essential for the resumption of postpartum ovarian activity. High BCS before calving and extensive BCS loss during 40 *postparturient* days increased the risk of delayed recurrence of ovarian activity.

Acknowledgements

The study was supported by Estonian Science Foundation (Grants 5422 and 4822)

In vitro embryo production: growth performance, feed efficiency, health status, and hematological, metabolic and endocrine traits in veal calves

M. Rérat[1], Y. Zbinden[1], R. Saner[2], H. Hammon[1], and J.W. Blum[1]
[1]Division of Nutrition and Physiology, Vetsuisse Faculty, University of Berne, Berne, Switzerland;
[2]Swiss Association for Artificial Insemination, Mülligen, Switzerland

The large-offspring syndrome and other problems have been described for *in vitro* produced bovine embryos (IVP). Growth performance, feed efficiency as well as hematological, metabolic, and endocrine traits in calves derived from *in vitro*-produced embryos (IVP; n = 11) were compared with calves derived from artificial insemination (AI; n = 8). Provided feeds were identically composed for both groups and daily amounts fed dependend on body weight. Blood samples were taken preprandially on d 1, 2, 3, 4, 7, 14, 28, 56, and 112 of life and every 20 min between 08:30 and 16:30 on d 7 and d 112 for the evaluation of growth hormone secretory patterns.

Gestation length was longer (P<0.05) in IVP than in AI calves, but birth weights were similar in both groups. Feed intake, average daily gain and body length during the whole experimental period, body weight from wk-8 to wk-16, gain/feed ratio during the first month of life were higher (P<0.05) in IVP than AI calves. At birth potassium, 3.5.3'-triiodothyronine and thyroxine concentrations were lower (P<0.05) in IVP than in AI calves. Concentrations of sodium and potassium on d 7, of triglycerides on d 28, and of albumin on d 56 were higher (P<0.05) in IVP than in AI calves. There were no differences between AI and IVP calves of blood acid-gas status and pH, hematological traits (hemoglobin concentration, hematokrit, erythrocyte and leucocyte numbers), concentrations of other metabolites (total protein, urea, creatinine, glucose, non-esterified fatty acids, triglycerides, cholesterol) aand hormones (growth hormone, insulin, glucagon, insulin-like growth factors-1 and –2, leptin).

In conclusion, IVP calves had a higher feed intake and growth rate and (in the first month of life) a better feed efficiency than AI calves, but this was not mirrored by consistent hematologic, metabolic or endocrine changes.

Dairy farms with organic production (OP) and with integrated production (IP): comparison of management, feeding, production, reproduction and udder health

M. Roesch[1], E. Homfeld[1], M.G. Doherr[2] and J.W. Blum[1]
[1]Division of Nutrition and Physiology; [2]Division of Clinical Research, Vetsuisse Faculty, University of Berne, Berne, Switzerland

In a case-control study differences between 60 organic farms (OP) and 60 farms with integrated production (IP). They were located in the midland and prealpine zones and were a representative sample of the situation in the Canton of Berne, Switzerland. IP farms matched to OP farms were comparable in terms of community, agricultural zone and number of cows/farm. Depending on the number of cows/farm, 5, 8, or 13 cows were selected, resulting in 970 multiparous cows that were studied at about 30 days pre partum, and 30 and 100 days post partum (visits 1, 2 or 3). At each visit, body weight, body condition score, type and amounts of provided feeds and farm management were documented. Blood and fecal samples were collected at the second visit. Milk samples were obtained from every quarter with California mastitis test 2 for bacteriological analyses.

Characteristics of OP and IP farms were as follows: mean size = 17.7 and 16.9 ha; mean dairy cow number = 14.0 and 15.5; mean annual milk quota = 65,900 and 70,000 kg; loose housing systems = 18 and 7% (P=0.053); outside paddocks = 98 and in 75% (P<0.05); Simmental x Red Holstein cows = 87 and 75%; Holstein cows = 42 and 60%. In winter rapeseed or soya protein and succulent feeds were more often fed (P>0.1) in IP than OP farms. Main causes for cow replacements in OP and IP farms were fertility disorders (45 and 45%), age (40 and 42%), sale (30 and 37%) and udder health (35% and 13%; P<0.01). Regular teat dipping after milking was performed in 25% and 43% (P<0.05) in OP and in IP farms. Blanket dry cow treatments were performed in 45% of IP, but not in OP farms (because not allowed). In the lactation preceding the actual study in OP and IP farms the 305-day energy-corrected milk yield was 5,767 and 6,331 kg (P<0.005) and somatic cell counts (SCC) in milk were 119,000 and 115,000/mL. In the actual lactation, at the second and the third visit SCC were 104,000 and 92,000/mL in OP farms and 73,000 and 80,000/mL in IP farms. Reproduction data during the preceding lactation were similar and on a high level in both OP and IP farms. Alternative veterinary methods in OP and IP farms were used in 55 and 17%. Thus, OP and IP farms differed in several respects.

Milk production, nutritional status, and fertility in dairy cows held in organic farms and in farms with integrated production

E. Homfeld[1], M. Roesch[1], M.G. Doherr[2] and J.W. Blum[1]
[1]Division of Nutrition and Physiology; [2]Division of Clinical Research, Vetsuisse Faculty, University of Berne, Berne, Switzerland

Milk production of cows in farms with organic production (OP) tends to be lower than in farms with conventional or integrated production (IP). We have performed a longitudinal case-control study to investigate management, nutritional, metabolic and endocrine risk factors that may be associated with low milk production in OP farms. In 60 OP and 60 IP farms, matched in size, location and altitude abve sea level, 970 cows were selected. Cows were visited once pre and twice post partum (p.p.). General farm data were collected at first visit. Body condition scores (BCS) and body weights (BW) were determined three times and blood was sampled once at 21 to 43 days p.p. to determine plasma concentrations of glucose, non-esterified-fatty-acids, β-(OH)-butyrate, albumin, urea, insulin-like growth factor-1 (IGF-1) and 3,5,3'-trijodothyronine. A multivariate model was used based on stepwise forward selection and stepwise backward deletion analysis to determine factors that may have an impact on low milk yield.

Amounts of concentrates and protein fed were lower ($P<0.05$) in OP than in IP cows. Energy-corrected milk yields at the second and third visit, and milk fat, protein and urea concentrations at 100 days p.p. were lower ($P<0.01$) in OP than in IP cows. The BW was lower in OP than IP cows (mean 650 and 673 kg, respectively; $P<0.001$). No significant differences were found in fertility traits between OP and IP cows. Plasma albumin, urea and IGF-1 concentrations were lower ($P<0.05$) in OP than IP cows. Based on the determination of odds ratios (OR) the following factors were positively and significantly ($P<0.05$) associated with low milk yield: Simmental breed (OR 2.3 *vs.* Simmental x Red Holstein crossbreed), teat dipping (OR 1.8), California mastitis test (CMT) positive on both hind quarters (OR 1.8), and a regular CMT (OR 2.0). There were significant negative OR of low milk yield with Holstein breed (OR 0.5), age (OR 0.75), calving interval (OR 0.99), hours spent outside (OR 0.7), amounts of supplemented concentrates and protein (OR 0.8 and 0.6, respectively), BW (OR 0.99), BCS<3 (OR 0.6), milk fever (OR 0.3), plasma albumin concentration (OR 0.9), milk fat content in preceding lactation (OR 0.4), milk somatic cell count (OR 0.998), and udder suspension (OR 0.5).

In conclusion, low milk yields were associated with farm type and due to interactions between genetics, nutrition, management, udder health, and agricultural zone.

Oviductal prolapse led to more than eleven percent of hens dead in a highly inbred line of white leghorn chickens

H.M. Zhang, A.R. Pandiri and G.B. Kulkarni
USDA-ARS Avian Disease and Oncology Laboratory, East Lansing, MI, USA

Oviduct prolapse is a serious and often deadly condition where the oviduct of a hen protrudes through the vent[1], which results in reduced egg production and higher disease susceptibility. Oviduct prolapse was observed in three consecutive generations of a highly inbred white leghorn line where the mortality ranged from 3.7 to 11.3%. It is reported that oviductal prolapse may be caused by nutritional imbalance[2], escalated by lower plasma concentration of 17 beta-estradiol[3], or over-sized eggs[4].

Our data showed that the frequency of incidence was different between some of the maternal families. Furthermore, selection for non-prolapsed hens as breeders reduced prolapse incidence in subsequent generations. These preliminary results suggest that genetics might play a role in governing the occurrence of the condition.

References

[1] http://edis.ifas.ufl.edu/BODY_PS029
[2] http://edis.ifas.ufl.edu/BODY_PS029
[3] Shemesh M., L. Shore, *et al.*, 1984. The role of 17 beta-estradiol in the recovery from oviductal prolapse in layers. Poul. Sci. 63, 1638-43.
[4] http://www.geocities.com/farmlinks/feedinghens.html#3

Section F.
Epidemiology of production diseases

Epidemiology of subclinical production diseases in dairy cows with an emphasis on ketosis

T.F. Duffield
Department of Population Medicine, Ontario Veterinary College, University of Guelph, Guelph, Ontario, Canada N1G 2W1

Abstract

Subclinical metabolic disease frequently goes unnoticed and may be associated with clinical disease risks, impaired production, reduced reproductive performance, and increased risk of culling. Of the major subclinical metabolic diseases, the most published data currently exists for subclinical ketosis. Subclinical ketosis is common in lactating dairy cows with incidence rates above 40% in many herds. Risk of subclinical ketosis is influenced by many factors including parity, body condition score, season of calving, and herd management. Risk increases with increasing parity and higher body conditions precalving. There is a wide variation in herd incidence. Prevention is achieved largely through effective dry cow programs that encompass both good nutrition and excellent cow management.

Keywords: subclinical ketosis, epidemiology, risk factors, prevention

Introduction

Most periparturient abnormalities have some metabolic element as a component of the sufficient cause of clinical disease. The metabolic disturbance of milk fever can be measured through low serum calcium concentrations. Negative energy balance, fat mobilization and subsequent elevations in ketone body concentrations play a contributing role in the expression of fatty liver syndrome, clinical ketosis, and abomasal displacement. A negative energy balance may also increase the risk of retained placenta, metritis, and mastitis through impaired immune function. A third category of metabolic disease in early lactation might include rumen acidosis, which is marked by low rumen pH. Thus, calcium homeostasis, energy balance and rumen pH are important considerations for disease prevention in transition dairy cows (Goff and Horst, 1997).

In general, subclinical disease incidence is far more common than clinical disease, frequently goes unnoticed and may be associated with significant clinical disease risks, impaired production and reduced reproductive performance. Subclinical ketosis is associated with both losses in milk production and increased risk of periparturient disease. Prevention depends on several factors such as proper transition cow nutrition and cow management, control of body condition, and may be helped through the use of certain feed additives such as niacin, propylene glycol, rumen protected choline and ionophores. It is commonly accepted that subclinical hypocalcemia is an important disease but very little is published on the impact of this problem on subsequent risk of disease or production loss. Subclinical rumen acidosis is also thought to be a major problem on many dairies but is difficult to measure and very little controlled research exists for this syndrome. This article will focus on the epidemiology and impact of subclinical ketosis.

Incidence/prevalence

The reported prevalence of subclinical ketosis ranges widely. Reported prevalences for hyperketonemia in the first two months of lactation have ranged from 8.9 to 34% in various studies (Andersson and Emanuelson, 1985; Dohoo and Martin, 1984a; Duffield *et al.*, 1997; Kauppinen, 1983a; Nielen *et al.*, 1994). Studies reporting incidence often find higher rates. Emery *et al.* used weekly milk ketone testing and found a lactational incidence of ketolactia of 29% in one herd (Emery *et al.*, 1964). Simenson *et al.* (1990) measured acetoacetate in weekly milk samples from cows within five herds that had a high clinical ketosis incidence rate and discovered 46% of cows in the first six weeks of lactation had milk acetoacetate values above 0.1 mmol/l. Emery *et al.* (1964) suggested that 50% of all lactating cows go through a stage of subclinical ketosis in early lactation. Duffield *et al.* (1998a) reported the cumulative incidence of subclinical ketosis over the first 9 weeks of lactation in 507 untreated cows from 25 Holstein dairy farms was 59% and 43% using cutoff threshold beta-hydroxybutyrate (BHBA) concentrations of 1200 and 1400 µmol/l respectively. It is difficult to compare these numbers across studies since numerous factors beyond cow and herd level risk influence the rates. The test characteristics (sensitivity and specificity) influence prevalence, as does the timing and intensity of sampling. Generally studies that measured rates with less sensitive milk tests report much lower prevalences than those using serum or urine. For continuous measures of ketones such as serum BHBA, the threshold used for defining subclinical ketosis will influence rates of disease with lower thresholds giving higher rates. There is a need for the use of objective reasons for setting cutpoints or threshold values for subclinical metabolic diseases.

Subclinical ketosis is commonly reported for the first two months of lactation since this is the primary risk period (Baird, 1981). Dohoo and Martin (1984a) found more positive milk ketone tests in the first versus the second month of lactation, and observed that the peak prevalence of hyperketonemia occurred in the third and fourth week of lactation. Other researchers also found the peak prevalence of hyperketonemia occurred during the third and fourth week postcalving (Andersson and Emanuelson, 1985; Kauppinen, 1983a; Simenson *et al.*, 1990). However, recent work suggests that there is an earlier peak that occurs in the first two weeks postcalving (Duffield *et al.*, 1997, 1998a). It may be that advances in genetics and feeding management have pushed the metabolic challenge closer to calving. Alternatively, differences in peak occurrence for subclinical ketosis may reflect differences in etiology; with early lactation occurrence reflecting suboptimal dry cow management and expression of fatty liver, while later occurrences may indicate deficiencies in lactating cow management (Cook *et al.*, 2001). Very little if any information exists on the average duration of subclinical ketosis. Dohoo and Martin (1984b) estimated the duration to be about 16 days in their study, assuming that the prevalence was 12% and the lactational incidence was 50%.

Cow level risk factors

Important cow-level risk factors that may influence the occurrence of subclinical ketosis include parity, body condition, genetics and season.

Several authors have reported a higher prevalence of hyperketonemia with increasing lactation number (Andersson and Emanuelson, 1985; Dohoo and Martin, 1984b; Duffield *et al.*, 1997). Body condition score (BCS) prior to calving is also an important risk factor for subsequent development of subclinical ketosis during lactation. Dohoo and Martin reported that a prolonged previous calving interval increased the risk of subclinical ketosis in the subsequent lactation (Dohoo and Martin, 1984b). Duffield *et al.* (1998b) reported that fat (BCS ≥ 4.0) cows had both highest BHBA concentrations postcalving and were at highest risk of developing subclinical ketosis compared to cows in moderate and thin body condition prior to calving (Duffield *et al.*, 1998a).

Genetics and breed are other potential sources of herd variation. Andersson and Emanuelson (1985) found that Swedish Red and White cows had significantly higher milk acetone levels than Swedish Friesians. Later studies observed the same breed differences with respect to the incidence of clinical ketosis, and to the risk of hyperketonemia (Bendixen *et al.*, 1987; Gustafsson *et al.*, 1993). Estimates of the heritability of subclinical ketosis vary. Dohoo *et al.* (1984b) calculated the heritability of clinical ketosis to be 0.32 but subclinical ketosis was found to have no genetic component. Tveit *et al.* (1992) reported a heritability estimate for acetoacetate of 0.11 with a genetic correlation to milk yield of 0.87 in Norwegian cattle. Mantysaari *et al.* (1991) estimated the heritability of clinical ketosis to be 0.07 to 0.09 in Finnish Ayrshire cattle. The heritability of clinical ketosis in Canadian Holstein cows was recently estimated to be 0.09 (Uribe *et al.*, 1995). The general conclusion is that the heritability of both clinical and subclinical ketosis appears to be low to moderate.

Season has been reported to influence the degree of hyperketonemia in dairy cattle in several countries. A Norwegian study found increasing levels of plasma acetoacetate with each month, for calvings from August to December (Tveit *et al.*, 1992). The authors suggested this observation may have been due to a systematic increase in body weight at freshening during the calving season and therefore an increasing tendency for fat cows at parturition. In addition, the forage levels of butyric acid may have increased during the same period (Tveit *et al.*, 1992). Whitaker *et al.* (1983) reported lower values of BHBA in May compared to other months of the year for dairy cows in the United Kingdom. Andersson and Emanuelson (1985) found levels of milk acetone to be higher from October to February versus the remainder of the year in Swedish dairy cows. This is in agreement with Grohn *et al.* (1984) who reported the risk of clinical ketosis was highest during the indoor feeding season (September to May). Clinical ketosis was diagnosed most often in cows that calved near the end of January. However, no association between season and subclinical ketosis was found in a Canadian study (Dohoo and Martin, 1984b). The authors of the Canadian project suggested that there may be certain factors that cause the expression of clinical ketosis, which are not associated with the subclinical condition. However, it could also be argued that the interval between samples in this study underestimated the level of subclinical ketosis, and therefore reduced the opportunity to observe a seasonal effect. In a subsequent Canadian study, both highest BHBA concentrations and highest subclinical ketosis prevalence occurred in the summer months (Duffield *et al.*, 1998a). This may be the combined result of suppression of intake due to summer heat stress, changes in forages and reduced management intensity common to Ontario dairy farms from June to September.

From the above reports, with varying seasonal risk patterns, it is most likely that the seasonal effects are driven by either environmental, nutritional, or cow level risk factors that are influencing dry matter intake.

Herd level risk factors

There are many reports of herd differences in the prevalence of hyperketonemia (Andersson and Emanuelson, 1985; Dohoo and Martin, 1984a; Duffield et al., 1997; Nielen et al., 1994). Dohoo and Martin (1984a) discovered a range in the herd prevalence of ketolactia from 0 to 34%. Cow and herd accounted for 14 and 6 percent of the total variation in beta-hydroxybutyrate in a Canadian study (Duffield, 1997). Differences in the distribution of fat, moderate and thin cows between herds may also explain herd differences in the occurrence of subclinical ketosis. Herd differences found in these studies might also be partly attributed to variations in the average herd parity. However, parity and body condition explained only 11% of the variation in herd incidence in one 25 herd-study (Duffield, unpublished). Increasing the feeding frequency and increased volume of concentrates fed have been shown to reduce the incidence of hyperketonemia (Gustafsson et al., 1995). This could help explain some differences in the prevalence and incidence rates of subclinical ketosis among herds. Tveit et al. reported an association between year and level of acetoacetate and attributed this relationship to increased levels of butyric acid in the silage for certain years (Tveit et al., 1992). Herd factors such as cow management, environment, nutrition, and cow comfort likely play major roles in influencing dry matter intake and the risk of subclinical ketosis.

Association with periparturient disease

Several investigations have evaluated the relationship between clinical ketosis and other health parameters, but few studies have attempted to assess the associations between subclinical ketosis and periparturient disease. Although subclinical and clinical ketosis are both part of the same continuum, we can only assume the associations found for clinical ketosis would be the same for the subclinical condition. Subclinical ketosis as a risk factor for subsequent disease occurrence has been linked with clinical ketosis, abomasal displacement, metritis, mastitis, and cystic ovarian disease. The most consistent associations have been between subclinical ketosis, clinical ketosis and displaced abomasum. Impaired immune function either through reduced energy or by direct effects of ketones on white blood cells, are likely the reason for effects on infectious disease such as metritis and mastitis. Dohoo and Martin (1984a) found that cows with subclinical ketosis had an increased risk of metritis or clinical ketosis four days later. However, the authors argued that since metritis is a condition which normally develops at calving, subclinical ketosis is more likely a result rather than a cause of metritis.

The relationship between displaced abomasum and ketosis has been identified as bi-directional (Curtis et al., 1993; Grohn et al., 1989). That is, ketosis may be a cause of displacement and abomasal displacement may lead to ketosis. Correa et al. (1993) found that ketosis increased the risk of abomasal displacement, but not the reverse. Dohoo and Martin (1984b) could find no direct association between the two conditions. However, ketosis as an inciting or predisposing cause of abomasal displacement can be further supported by some subsequent research. Elevated BHBA concentrations above 1000 µmol/l increased the likelihood of abomasal displacement (Geishauser et al., 1997). Cows with concentrations of BHBA at or above 1400 µmol/l in the first two weeks post calving were three times more likely to subsequently develop either clinical ketosis or abomasal displacement (Duffield, 1997).

Impact on milk production and milk components

In general there is consensus that a negative association between hyperketonemia and milk production exists, however there are conflicting reports. In one study, the loss of production associated with a positive milk ketone test was 1.0 to 1.4 kg of milk per day (Dohoo and Martin, 1984a). This represented 4.4 to 6.0% of the mean test day milk production. In this study, both the milk samples and the milk weights were obtained on the same day, and thus ketolactia and milk production were evaluated concurrently. Test day milk production was negatively correlated with milk acetone levels in four separate Scandinavian projects (Andersson and Emanuelson, 1985; Gustafsson et al., 1993; Miettenen and Setala, 1994; Steen et al., 1996). By contrast, Kauppinen (1983b) showed a significant positive correlation between both the BHB and acetoacetate concentrations in blood and milk yield. Kauppinen (1984) subsequently reported that subclinically ketotic cows had significantly higher annual milk yields than nonketotic cows. Herdt et al. (1981) found higher levels of BHBA in higher producing cows; but individual milk tests preceded blood measurement for BHBA. It is possible that higher milk yields put cows at increased risk of developing subclinical ketosis. Increased levels of milk production may be associated with increased fat mobilization and a greater risk of hyperketonemia (Lean et al., 1991). In a review of several observational studies, Erb (1987) concluded that higher milk yields in the previous lactation did not increase the risk of ketosis in the subsequent lactation. However, without objective measures for ketosis, the definition might vary widely in observational studies such as this. One Finnish study determined increased previous milk yield to be a risk factor for ketosis (Grohn et al., 1989), but this association was not identified in two North American projects (Curtis et al., 1989; Dohoo and Martin, 1984b). Regardless of the threshold chosen for hyperketonemia, the incidence of subclinical ketosis should be considerably higher than that of clinical ketosis. Assuming that higher producing cows are more likely to be hyperketonemic, a larger proportion of subclinically ketotic cows combined with misclassification bias of clinical ketosis cases could mute any association between clinical ketosis and previous lactation milk yield. One study that evaluated test day milk production and clinical ketosis in over 60,000 Finnish Ayrshires demonstrated a lactation curve depression associated with ketosis and a loss of 44.3 kg of milk (Detilleux et al., 1994). However, cows with ketosis had 141.1 kg more 305 day milk yield than normal cows (Detilleux et al., 1994). Inverted milk curves in hyperketonemic cows were also noted in a large Swedish study (Gustafsson et al., 1995). Most of the loss in milk production occurred in the first 100 days of lactation and amounted to 328 kg loss in fat corrected milk yield over 200 days. Recent studies support both the subclinical ketosis, clinical ketosis temporal association and the negative impact on milk production. Milk yield was lower 2 to 4 weeks in one study (Rajala-Schultz et al., 1999) and 8 to 9 days in another study (Edwards and Tozer, 2004) prior to the diagnosis of clinical ketosis. Noticeable milk yield losses weeks prior to actual diagnosis is strong support for the negative and unrealised impact of subclinical ketosis on milk production.

Milk fat and milk protein are significantly altered in hyperketonemia. Milk fat percentage was increased in subclinically ketotic cows (Miettenen, 1994; Miettenen and Setala, 1994). Mean annual milk fat yield was significantly higher in both subclinically ketotic and clinically ketotic cows compared to normal cows (Kauppinen, 1984). The association between milk fat and hyperketonemia is, presumably, because of increased availability of BHBA and fatty acids for milk fat synthesis. It is unclear whether increased levels of circulating ketones cause increased milk fat, or if cows that are prone to higher milk fat yields are more susceptible to subclinical ketosis. Milk protein percent has been reported to be lower in cows with subclinical ketosis (Miettenen,

1994; Miettenen and Setala, 1994). This may be the result of a reduced energy supply, since milk protein percent is positively associated with net energy balance (Grieve *et al.*, 1986).

Impact on reproductive performance

There has been conflicting research findings on the impact of subclinical ketosis on reproductive performance. However recent work seems to be building a consensus that there is a negative impact. At least two studies have identified no effect of either subclinical or clinical ketosis on individual cow fertility (Andersson and Emanuelson, 1985; Kauppinen, 1984). Significant correlations between the herd prevalence of hyperketonemia and herd mean intervals from both calving to first service and calving to last service have been noted, but this is herd level data and doesn't necessarily imply a cow-level association (Andersson and Emanuelson, 1985). A link between subclinical ketosis and the increased incidence of cystic ovaries has been reported in at least two studies, suggesting an impact of subclinical ketosis on ovarian function (Andersson and Emanuelson, 1985; Dohoo and Martin, 1984a). A significant inverse relationship between milk fat percentage and first insemination pregnancy rates has also been identified (Kristula *et al.*, 1995). Miettenen and Setala (1994) found an increased interval from calving to conception in cows with high milk yield and high fat yield. The associations between fertility and increased fat and milk yield do not necessarily imply a relationship between impaired fertility and hyperketonemia. The duration of either clinical or subclinical ketosis may be too short to exert a negative effect on calving interval (Kauppinen, 1984). Whitaker *et al.* (1993) found cows with a better energy status at 14 days postpartum had a reduced interval from calving to the onset of cyclicity and fewer services per conception. No effect was observed when energy status was evaluated at 21 days postpartum or at first service. This study was only conducted on 24 cows within one herd. More recently, however, studies with larger numbers of cows have also identified some negative impacts of subclinical ketosis on fertility. Walsh *et al.* (2004) has reported that 1st service conception risk was decreased by 50% in cows with serum BHBA concentrations $\geq$ 1400 umol/l in the 2nd week post-calving, based on 869 cows having at least one breeding postcalving. In addition, the odds ratio decreased from 0.7 at 1100 umol/l BHBA to 0.31 at 1900 umol/l BHBA, implying in a sense, a dose effect of ketone concentration on conception risk. Ketotic cows measured through milk acetone had longer calving to conception intervals (139 versus 85 days) (Cook *et al.*, 2001). Koller *et al.* (2003) found increased concentrations of ketone bodies in the first 6 weeks post calving delayed conception.

Culling

The influence of subclinical ketosis on the risk of culling has not been well described. Culling may be more likely because of impaired health, reduced production and potentially impaired reproductive performance. Clinical ketosis increased the risk of culling both early in lactation and late in lactation (Grohn *et al.*, 1998). Presumably the early lactation risk was associated with reduced production and increased disease risk, while the later culling association might be related to reduced reproductive performance. Cook *et al.* (2001) reported an increased culling risk for ketotic cows measured through milk acetone in a 410 cow study. Both subclinical ketosis and subclinical hypocalcemia appear to increase the risk of culling. Cows with serum calcium concentrations less than 2.0 mmol/l and not diagnosed with milk fever were approximately 1.5 times and cows with serum BHBA concentrations above 1400 umol/l were 1.4 times more likely to be culled in the first 95 days of lactation, after controlling for the risk of clinical disease on culling and the random effect of herd (Duffield, unpublished).

Thresholds for defining subclinical ketosis

Most thresholds in the literature for defining subclinical ketosis are arbitrary. Since higher concentrations of ketone bodies are expected postcalving, arbitrarily assigning a threshold based on a best guess or data distribution, runs the risk of just being on the higher end of normal. Few studies have actually reported threshold concentrations based on either negative impacts on health, production, reproduction or culling. Reist *et al.* (2003) used milk acetone at 0.4 mmol/l and serum BHBA at 2.3 mmol/l and found significantly reduced milk yield at both cutpoints and increased risk of endometritis at the milk acetone threshold. Duffield (1997) and Geishauser *et al.* (1998) have both reported a three-fold increased risk of clinical ketosis and displaced abomasums at serum concentrations of BHBA at or above 1400 umol/l. Duffield (1997) also found a significant reduction in milk yield at first DHI test at 1800 umol/l BHBA, with a non-significant (P=0.13) reduction at 1600 umol/l. However, the increasing thresholds were associated with increasing reductions in milk yield. Walsh *et al.* (2004) has identified reductions in fertility starting at a cutpoint of 1100 umol/l and Duffield (unpublished) has identified increased culling risk at 1400 umol/l BHBA. Based on the above data, it would appear that a threshold for defining subclinical ketosis with serum BHBA should start at about 1400 umol/l.

Understanding subclinical ketosis through prevention research

If the negative associations between subclinical ketosis and health and production impairment are genuine, prevention studies should validate many of these associations. There are many randomised clinical studies of monensin that can be useful for this purpose. Monensin is a feed additive that provides the cow with increased propionate through alterations in rumen microflora. The objective of examining the effects of these data isn't to extol the virtues of monensin but rather to examine the impact of subclinical ketosis on cow health and performance.

The antiketogenic properties of monensin were first studied in a Canadian trial involving two levels of monensin and three groups of 12 Holstein cows (Sauer *et al.*, 1989). Monensin included at 30 grams per ton of total ration (high group), decreased the incidence of subclinical ketosis and significantly reduced blood BHB levels in the first three weeks postpartum (Sauer *et al.*, 1989). The incidence of subclinical ketosis, defined as total blood ketones > 9 mg/100 ml (900 µmol/l), was reduced by 50% and blood BHB levels were reduced by 40% for the high monensin group (Sauer *et al.*, 1989). A controlled release capsule that delivers 335 mg of monensin sodium per day for 95 days reduced the incidence of subclinical ketosis by 50% and also decreased the duration of the condition when it was administered 3 weeks prior to expected calving (Duffield *et al.*, 1998a). Administration of a monensin controlled release capsule precalving reduced the incidence clinical ketosis by 40%, abomasal displacement by 40%, and tended to reduce the incidence of retained placenta (by 24%) in the analysis of data pooled from two different studies (Duffield *et al.*, 2002). The milk production response in the first of those studies depended on body condition and was 0.85 kg/day at peak lactation in cows with a precalving BCS of 3.25 to 3.75, and was 1.2 kg/day for the first 90 days of lactation in fat cows (BCS ≥ 4.0) (Duffield *et al.*, 1999). No milk production response was noted in thin cows presumably because they had the lowest BHBA concentrations and were at decreased risk of subclinical ketosis. Few studies have found any positive or negative reproductive effects. However, in a small study, monensin treated cows had a significantly reduced days to first ovulation (Tallam *et al.*, 2003).

Thus, a reduction in subclinical ketosis, improved cow health for those diseases associated (at least in part) with energy impairment. Milk production was enhanced, but only for cows at higher risk of subclinical ketosis. Limited impacts of monensin on reproductive performance have been identified, but, the observed reduction in days to first ovulation is consistent with the expected effect when there is increased energy supply.

Conclusions

Subclinical ketosis is an important disease of dairy cattle that is associated with increased risk of clinical diseases, lost milk production, reduced reproductive performance and increased culling risk. Estimates indicate it costs at least $78 US (Geishauser *et al.*, 2001) per occurrence, which when examined at the herd level, is considerably more than most clinical diseases, since subclinical disease is far more frequent. Cow-level risk factors are parity, body condition score and season of calving. Herd variation for this disease is wide and herd level risk factors are poorly described. However, herd level risk factors most likely involve combinations of factors such as management, feed quality and nutritional programs, cow comfort, environment, and other variables that influence dry matter intake.

References

Andersson L and Emanuelson U., 1985. An epidemiological study of hyperketonaemia in Swedish dairy cows; determinants and the relation to fertility. Prev. Vet. Med. 3, 449-462.

Baird DG., 1982. Primary ketosis in the high-producing dairy cow: clinical and subclinical disorders, treatment, prevention and outlook. J. Dairy Sci. 65, 1-10.

Bendixen PH, Vilson B, Ekesbo BI, *et al.*, 1987. Disease frequencies in dairy cows in Sweden. IV. Ketosis. Prev. Vet. Med. 5, 99-109.

Correa MT, Erb H and Scarlett J., 1993. Path analysis for seven postpartum disorders of holstein cows. J. Dairy Sci. 76, 1305-1312.

Cook NB, Ward WR and Dobson H., 2001. Concentrations of ketones in milk in early lactation, and reproductive performance of dairy cows. Vet. Rec. 148, 769-772.

Curtis CR, Erb HN, Sniffen CJ, *et al.*, 1985. Path analysis of dry period nutrition, postpartum metabolic and reproductive disorders, and mastitis in Holstein cows. J. Dairy Sci. 68, 2347-2360.

Detilleux JC, Grohn YT and Quass RL., 1994. Effects of clinical ketosis on test day milk yields in Finnish Ayrshire cattle. J. Dairy Sci. 77, 3316-3323.

Dohoo IR and Martin SW., 1984b. Disease, production and culling in Holstein-Friesian cows III. Disease and production as determinants of disease. Prev. Vet. Med. 2, 671-690.

Dohoo IR and Martin SW.,1984a. Subclinical ketosis: Prevalence and associations with production and disease. Can. J. Comp. Med. 48,1-5.

Duffield TF., 1997. Effects of a monensin controlled release capsule on energy metabolism, health, and production in lactating dairy cattle. DVSc Thesis dissertation, Univ. of Guelph.

Duffield T, Bagg R, DesCoteaux L, Bouchard E, Brodeur M, DuTremblay D, Keefe G, LeBlanc S and Dick P., 2002. Prepartum monensin for the reduction of energy associated disease in postpartum dairy cows. J. Dairy Sci. 85, 397-405.

Duffield TF, Kelton DF, Leslie KE, *et al.*, 1997. Use of test day milk fat and milk protein to predict subclinical ketosis in Ontario dairy cattle. C. V. J. 38, 713-718.

Duffield TF, Sandals D, Leslie KE, *et al.*, 1998b. Effect of prepartum administration of a monensin controlled release capsule on postpartum energy indicators in lactating dairy cattle. J. Dairy Sci. 81, 2354-2361.

Duffield TF, Sandals D, Leslie KE, *et al*, 1998a. Efficacy of monensin for the prevention of subclinical ketosis in lactating dairy cows. J. Dairy Sci. 81, 2866-2873.

Duffield TF, Leslie KE, Sandals D, *et al.*, 1999. Effect of prepartum administration of a monensin controlled release capsule on milk production and milk components in early lactation. J. Dairy Sci. 82, 272-279.

Edwards JL and Tozer PR., 2004. Using activity and milk yield as predictors of fresh cow disorders. J. Dairy Sci. 87, 524-531.

Emery RS, Burg N, Brown LD, *et al.*, 1964. Detection, occurrence, and prophylactic treatment of borderline ketosis with propylene glycol feeding. J. Dairy Sci. 47, 1074-1079.

Erb HN., 1987. Interrelationships among production and clinical disease in dairy cattle: a review. Can. Vet. J. 28, 326-329.

Geishuaser T, Leslie K, Duffield T, *et al.*, 1997. Evaluation of aspartate aminotransferase activity and β-hydroxybutyrate Concentration in blood as tests for left displaced abomasum in dairy cows. Am. J. Vet. Res. 58, 1216-1220.

Geishauser, T., Leslie, K., Kelton, D. and Duffield, T., 2001. Monitoring for subclinical ketosis in dairy herds. Compendium of Continuing Education 23, s65-s71.

Grieve DG, Korver S, Rijpkema YS, *et al.*,1986. Relationship between milk composition and some nutritional parameters in early lactation. Lact. Prod. Sci. 14, 239-254.

Grohn Y, Thompson JR and Bruss ML., 1984. Epidemiology and genetic basis of ketosis in Finnish Ayrshire cattle. Prev. Vet. Med. 3, 65-77.

Goff JP and Horst R L., 1997. Physiological changes at parturition and their relationship to metabolic disorders. J. Dairy Sci. 80, 1260-1268.

Grohn YT, Eicker SW, Ducrocq V and Hertl JA., 1998. Effect of diseases on the culling of Holstein dairy cows in New York State. J. Dairy Sci. 81, 966-978.

Grohn YT, Erb HN, McCulloch CE, *et al.*, 1989. Epidemiology of metabolic disorders in dairy cattle: Association among host characteristics, disease, and production. J. Dairy Sci. 72, 1876-1885.

Gustafsson AH, Andersson L and Emanuelson U.,1993. Effect of hyperketonemia, feeding frequency and intake of concentrate and energy on milk yield in dairy cows. Anim. Prod. 56, 51-60.

Gustafsson AH, Andersson L and Emanuelson U., 1995. Influence of feeding management, concentrate intake and energy intake on the risk of hyperketonaemia in Swedish dairy herds. Prev. Vet. Med. 22, 237-248.

Herdt T. H, Stevens JB, Olson WG, *et al.*, 1981. Blood concentrations of β-hydroxybutyrate in clinically normal Holstein-Friesian herds and in those with a high prevalence of clinical ketosis. Am. J. Vet. Res.12(3), 503-506.

Kauppinen K., 1983b. Correlation of whole blood concentrations of acetoacetate, B-hydroxybutyrate, glucose and milk yield. Acta. Vet. Scand. 24, 337-348.

Kauppinen K., 1983a. Prevalence of bovine ketosis in relation to number and stage of lactation. Acta. Vet. Scand. 24, 349-361.

Kauppinen K., 1984. Annual milk yield and reproductive performance of ketotic and non-ketotic dairy cows. Zbl. Vet. Med. A. 31, 694-704.

Kristula MA, Reeves M, Redlus H, *et al.*, 1995.. A preliminary investigation of the association between the first postpartum milk fat test and first insemination pregnancy rates. Prev. Vet. Med. 23, 94-100.

Koller A, Reist M, Blum JW and Kupfer U., 2003. Time empty and ketone body status in the early postpartum period of dairy cows. Reprod. Dom. Anim. 38, 41-49.

Lean IJ, Bruss ML, Baldwin RL, *et al.*, 1992.. Bovine Ketosis: a review II. biochemistry and prevention. Vet. Bulletin. 62, 1-14.

Mantysaari EA, Grohn YT and Quaas RL., 1991. Clinical ketosis: phenotypic and genetic correlations between occurrences and with milk yield. J. Dairy Sci. 74, 3985-3993.

Miettenen PVA., 1994. Relationship between milk acetone and milk yield in individual cows. J. Vet. Med. A. 41, 102-109.

Miettinen PVA and Setala JJ., 1993. Relationships between subclinical ketosis, milk production and fertility in Finnish dairy cattle. Prev. Vet. Med. 17, 1-8.

Nielen M, Aarts MGA, Jonkers GM, *et al.*, 1994.. Evaluation of two cowside tests for the detection of subclinical ketosis in dairy cows. Can. Vet. J. 35, 229-232.

Rajala-Schultz PJ, Grohn YT and McCulloch CE., 1999. Effects of milk fever, ketosis, and lameness on milk yield in dairy cows. J. Dairy Sci. 82:288-294.

Reist M, Erdin DK, Von Euw D, Tschumperlin KM, Leuenberger H, Hammon HM, Kunzi N and Blum JW., 2003. Use of threshold serum and milk ketone concentrations to identify risk for ketosis and endometritis in high-yielding dairy cows. Amer. J. Vet. Res. 64, 188-194.

Sauer FD, Kramer JKG and Cantwell WJ., 1958. Antiketogenic effects of monensin in early lactation. J. Dairy Sci. 72, 436-442, 1989.

Schultz LH and Myers M., 1959.Milk test for ketosis in dairy cows. J. Dairy Sci. 42, 705-710.

Simenson E, Halse K, Gillund P, *et al.*, 1990. Ketosis treatment and milk yield in dairy cows related to milk acetoacetate levels. Acta. Vet. Scand. 31, 433-440.

Steen A, Osteras O and Gronstol H., 1996. Evaluation of additional acetone and urea analyses, and of the fat-lactose-quotient in cow milk samples in the herd recording system in Norway. J. Vet. Med. A. 43, 181-191.

Tallam, S.K., Duffield, T.F., Leslie, K.E., Bagg, R., Dick, P., Vessie, G. and Walton, J.S., 2003. Ovarian follicular activitiy in lactating Holstein cows supplemented with monensin. J. Dairy Sci. 86, 3498-3507.

Tveit B, Lingaas F, Svendsen M, *et al.*, 1992. Etiology of acetonemia in Norwegian cattle. 1. Effect of ketogenic silage, season, energy level, and genetic factors. J. Dairy Sci. 75, 2421-2432.

Uribe HA, Kennedy BW, Martin SW, *et al.*,1995. Genetic parameters for common health disorders of Holstein cows. J. Dairy Sci. 78, 421-430.

Walsh R, LeBlanc S, Duffield T and Leslie K., 2004. The association between subclinical keotsis and conception rate in Ontario dairy herds. Tri-State Dairy Symposium, Ft Wayne, Indiana.

Whitaker DA, Kelly JM and Smith EJ., 1983. Subclinical ketosis and serum beta-hydroxybutyrate levels in dairy cattle. Br. Vet. J. 139, 462-463.

Whitaker DA, Smith EJ, da Rosa GO, *et al.*, 1993. Some effects of nutrition and management on the fertility of dairy cattle. Vet. Rec. 133, 61-64.

Associations of cow and management factors with culture positive milk samples

R. van Dorp[1] and D. Kelton[2]
[1]Department of Animal Science, Michigan State University, East Lansing, MI, USA; [2]Department of Population Medicine, University of Guelph, Guelph, ON, Canada

Mastitis is a disease of dairy cattle that has significant economic consequences in both its clinical and sub-clinical forms. Dairy health management programs have focused on both aspects of the disease, but often without a clear appreciation of the component strategies that are most effective at each level. While clinical mastitis is predominantly caused by environmental intramammary pathogens, sub-clinical mastitis is often the result of infection with contagious pathogens. Given that the contagious and environmental pathogens differ in their epidemiology, there is a need to study specific factors of cows and herds in relation to the incidence of clinical and sub-clinical mastitis caused by contagious versus environmental pathogens.

The objective of this study was to identify associations between cow and herd level risk factors and the prevalence of sub-clinical infection (defined as a culture positive composite milk sample (CPMS)), caused by major environmental (ENV) and major contagious pathogens (CONT), separately.

The "Sentinel Herd Project" (Guelph, Canada) was initiated to describe the udder health and milk quality of dairy cows on farms[1]. Herds were visited 3 to 6 times from July 1997 to February 1999. During herd visits composite milk samples were taken from all milking cows. These samples were cultured to identify the presence and type of bacterial pathogen. CPMS caused by ENV and CONT were compared to samples identified with minor bacterial pathogens or no bacterial growth. The final data set consisted of 13,419 observations on 5,158 Holstein cows from 57 herds. Information about herd management was collected using an orally administered questionnaire. Test day milk production (TD-milk) and TD somatic cell score (TD-SCS) records were provided by the Ontario Dairy Herd Improvement Corporation (DHI). Data consisted of 2 levels of clustering, the cow and the herd. A cluster-specific statistical model was chosen to analyze the data and to account appropriately for the random effects of herd and cow[2]. Variances of herd, cow nested within herd, fixed cow level effects [days in milk (dim), lactation number, TD-milk and TD-SCS] and herd management factors were estimated simultaneously.

Results showed differences in association of cow and herd factors with ENV and CONT infections. Cows CP for both ENV and CONT showed a seasonal pattern, with more CP samples in the fall, winter and the spring months, compared to the summer. Higher prevalence of ENV infection was associated with lower dim, higher lactation number, and higher TD-SCS. Infections with CONT were associated with increased dim, higher lactation number, higher TD-SCS, higher TD-milk and a day in milk with TD-SCS interaction. Prevalence of ENV was increased when cows were housed in free stall or bedded-pack stalls compared to cows housed in tie-stalls facilities.

The cow factors, lactation number, dim, and TD-SCS, were significantly associated with the prevalence of ENV and CONT infections.

References

[1] Kelton, D.F., M.A. Godkin, D. Alves, K. Lissemore, K. Leslie, B. McEwen and C. Church, 1998. Sentinel herds to monitor udder health and milk quality in the province of Ontario. National Mastitis Council. Meeting Proceedings of the 37th Annual Meeting of the National Mastitis Council.

[2] Wolfinger, R., and M. O'Connell, 1993. Generalized linear mixed models: a pseudo-likelihood approach. J. Statist. Comput. Simul. 48, 233-243.

A mathematical model for the dynamics of digital dermatitis in groups of cattle to study the efficacy of group-based therapy and prevention strategies

D. Dopfer[1], M.R. van Boven[2], and M.C.M. de Jong[2]
[1]Quantitative Veterinary Epidemiology Group, Animal Sciences Group of Wageningen, The Netherlands; [2]Quantitative Veterinary Epidemiology Group, Animals Sciences group of Wageningen AB Lelystad, The Netherlands

(Papillomatous) Digital dermatitis, (P)DD, is an infectious disease in cattle that leads to painful ulcerative lesions along the coronary band of the claws. Herd health managers have failed to eradicate this classical production disease of cattle and this may be due to the 'paradox of modern animal health manager'. Complex hierarchical dynamic modelling can help in developing new preventive and therapeutic strategies with the aim to characterize the 'manageable endemic state' of disease. The evaluation of therapy and prevention is commonly accomplished by generic data analysis of incidence or prevalence data of lesions alone. The mathematical model presented in this study aims at providing a hierarchical transition model for the population transmission dynamics in closed groups of cattle incorporating the complex nature of host pathogen interactions.

During the pathogenesis of (P)DD, four different types of lesions (M1 to M4) have been described and recorded before[1], while three types of individual dynamics (negative, single lesion and the problem animal with repeated lesions) and three types of group dynamics (no outbreaks, single outbreak and repeated outbreaks on problem farms) are being used to construct a hierarchical mathematical model. Threshold values for the transitions between different individual and group dynamics will be discussed and an effort will be made to characterize the steady states of the model that will predict manageable, endemic states for groups of cattle. Interventions can be simulated and the model can be fitted to real-world data from outbreaks and intervention trials.

Many if not all intervention studies are aimed at eliminating (P)DD together with its risk factors and pathogens (the knock-out strategy) and not at modulating the complex population dynamics of disease. Animal health managers are expected to confine complex animal husbandry systems to states defined by 'animal health and welfare' under economic constraints. The human brain on the other hand has no intuitive insight into consequences of complex interactions because human cognition was conditioned by evolution in a partially predictable ecosystem at steady state[2]. This results in the 'paradox of modern animal husbandry'. Intensive animal husbandry systems are not balanced ecosystems and may not be predicted by intuition alone. This paradox should lead to a change of consciousness in production medicine still aspiring intuitive insight or immediate insight after "measuring" health or disease. Applied mathematics may provide structures, which when interpreted in terms of real entities may result in a cognitive aid, that is a structurising algorithm, able to disentangle paradoxical situations and help to develop management strategies for animal husbandry.

References
[1] Döpfer, D., A. Koopmans, F.A. Meijer, I. Szakall, Y.H. Schukken, W. Klee, R.B. Bosma, J.L. Cornelisse, A.J.A.M. van Asten and A.A. ter Huurne, 1997. Histological and bacteriological evaluation of digital dermatitis in cattle, with special reference to spirochaetes and *Campylobacter faecalis*. Vet. Rec. 140, 620-3.
[2] Vollmer, G. 1986. Was können wir wissen? Die Erkenntnis der Natur, vol. 2, Hirzel Verlag Stuttgart, Stuttgart/Germany.

Diagnostic test evaluation without gold standard using two different bayesian approaches for detecting verotoxinogenic *Escherichia coli* (VTEC) in cattle

D. Dopfer[1], L. Geue[2], W. Buist[1] and B. Engel[1]
[1]Quantitative Veterinary Epidemiology Group, Animals Sciences group of Wageningen UR, P.O.Box 65, 8200 AB Lelystad, The Netherlands, [2]Federal Research Centre for Virus Diseases of Animals, Institute of Epidemiology, Seestrasse 55, D-16868 Wusterhausen, Germany

Pre-screening PCR and multiple hybridisations were used to detect seven virulence factors of verotoxin producing *Escherichia coli* (VTEC) isolated from bovine faeces sampled between birth and slaughter[1]. The test results of three farms with 29 to 30 kalves each were used for this study. The seven virulence factors were verotoxin 1 and/or 2, intimin, haemolysin, katalase P, serine esterase P and colicin combined into 5 tests. A diagnostic test evaluation was performed to compare % positive test results, sensitivity, and specificity when calculated in three different ways:

1. After defining a gold standard for VTEC having to be positive for vt1 and/or vt2 and positive for eae, sensitivity and specificity were calculated conventionally as the fraction of test positive or negative out of the other respective positive or negative test results.
2. In addition, prevalence, sensitivity and specificity were estimated in the absence of a gold standard test using a Bayesian approach including a Gibbs Sampler as implemented in WinBUGS[2]. The test parameters were estimated using a linear regression for individual animal test data correcting for interdependence between tests, clustering of data, and missing values.
3. The third method included estimation of the test parameters without gold standard using a Bayesian approach based on herd level prevalence data similar to the method applied by Hui and Walter[3].

The programs for the Gibbs sampler can be adapted to manage interdependence of test results and clustering of data. The choice of priors is crucial to the generation of sensitive test parameters by the Bayesian approaches. All test parameters calculated were in the same order of magnitude. An analogue of receiver operating characteristic (ROC) curves was produced using the test parameters for the different virulence factors that deviated the most from the random line. The areas under the curve are calculated to test for the differences between the ROC's and the random line[4].

Using the three procedures we arrived at similar estimates for the test parameters. The greatest limitation of the conventional procedure is that a gold standard test must be available. The implications of test independence are an important issue of discussion during the process of creating procedures for diagnostic test evaluation. In the future, Bayesian algorithms will estimate test parameters without gold standards while controlling for test dependence, data clustering, and include a larger number of tests simultaneously. The interpretation of a conjunction of test parameters using analogues of ROC curves should be considered.

References

[1] Geue, L., M. Segura-Alvarez, F.J. Conraths, T. Kuczius, J. Bockemühl, H. Karch and P. Gallien, 2002. A long-term study on the prevalence of shiga-toxin-producing Escherichia coli (STEC) on four German cattle farms. Epidmiol. Infect. 129, 173-185.

[2] Spiegelhalter D., A. Thomas and N. Best, 2000. WinBUGS Version 1.3. MRC Biostatistics Unit, Institute of Public Health, Robinson Way, Cambridge CB2 2SR, UK.

[3] Hui, S.L. and S.D. Walter, 1980. Estimating the error rates of diagnostic tests. Biometrics 36, 167-171.

[4] Hanley, J.A. and B.J. McNeil, 1982. The meaning and use of the area under a receiver operating characteristic (ROC) curve. Radiology 143, 29-36.

A relative comparison of diagnostic tests for bovine paratuberculosis (Johne's disease)

S.H. Hendrick[1], T.F. Duffield[1], D.F. Kelton[1], K.E. Leslie[1], K. Lissemore[1] and M. Archambault[2]
[1]Department of Population Medicine, Ontario Veterinary College, Guelph, ON, CAN; [2]Animal Health Laboratory (AHL), University of Guelph, Guelph, Ontario, Canada

Prevention and control of Johne's disease would be much improved if diagnostic tests were able to reliably determine the infection status of subclinically infected cows both rapidly and economically. Serology-based tests offer the convenience of quick diagnosis at a low cost, but their precision and accuracy remain questionable. The objective of this study is to evaluate a commercially available milk ELISA test relative to other diagnostic tests and determine it offers any improved test characteristics.

Herds with a suspected high prevalence of Johne's disease were enrolled in this trial. Fecal and serum samples were collected from all milking and dry cows from 32 dairy herds in south-western Ontario. Serum samples were tested in duplicate for antibodies with an IDEXX enzyme-linked immunosorbant assay (ELISA) (AHL, Guelph, ON, Can.). Cows with a corrected optical density (OD) ratio greater than 0.25 were considered positive for Johne's disease. Milk samples preserved with bronopol were collected at the following Dairy Herd Improvement (DHI) test day. These milk samples were sent to AntelBio (Lansing, MI, USA) for an in-house milk ELISA test. Cows with a corrected OD of greater than 0.1 were considered positive for Johne's disease. Cows identified as positive on either the serum or milk ELISA test had their corresponding fecal sample tested. Feces were tested with all three of the following: traditional fecal culture (AntelBio), an IDEXX fecal PCR probe (AHL), and radiometric fecal culturing using the BACTEC culturing system (AHL). For nine of the herds enrolled in this study, traditional fecal cultures were completed for all of the 874 milking and dry cows. The sensitivity and specificity of the milk and serum ELISA was calculated relative to this one traditional fecal culture. A Receiver-Operator Characteristic (ROC) curve was generated for each of the two ELISAs to evaluate their overall performance.

2148 serum samples were evaluated with 286 of these being ELISA positive (13.4%). Only 1699 cows were milking on DHI test day, with 124 of these samples testing positive on the milk ELISA (7.3%). The kappa statistic between the milk and serum ELISA for the 1699 cows tested was 0.45. 326 cows were identified as positive on one or both of the ELISA tests. Of the ELISA positive cows (either milk and/or serum), 144 cows were positive on traditional fecal culture (44%), 89 cows were identified positive on fecal PCR (27%) and 186 cows positive through BACTEC culturing (57%). These results indicate that radiometric culture is more sensitive than PCR or traditional culture. The sensitivity and specificity of the milk ELISA was 61% and 95%, respectively. Similarly, the sensitivity of the serum ELISA was 75% and the specificity was 88%. The area under the ROC curves of the milk and serum ELISAs were not significantly different (P=0.36). The milk ELISA test appears to be a reasonable approach to predicting fecal shedding status. The performance of the fecal-based tests (*i.e.* culture and PCR) remains superior, but the convenience and economics of the milk ELISA may make it a more favourable screening test.

Etiology, pathophysiology and prevention of fatty liver in dairy cows

R.R. Grummer
University of Wisconsin, Department of Dairy Science, Madison, WI 53706, USA

Abstract

Triacylglycerol (TAG) accumulation in the liver of dairy cows occurs when blood nonesterified fatty acids (NEFA) increase at parturition, during negative energy balance in early lactation, and when energy intake is compromised due to disease. Decreases in feed intake prior to and during parturition partially account for the increase in NEFA; hormonal changes associated with parturition and lactogenesis also contribute. Negative consequences of fatty liver are difficult to identify. *In vitro* studies indicated that TAG accumulation in hepatocytes compromises ureagenesis, hormonal responsiveness, and hormone clearance. Effects on gluconeogenesis are inconsistent. Strategies to prevent or alleviate fatty liver include suppression of NEFA mobilization from adipose tissue or enhancing TAG secretion from the liver as very low density lipoprotein. The latter option is preferred because NEFA mobilization is a response by the animal to provide energy or precursors for milk synthesis. Ruminants have a limited capacity to export TAG as VLDL and little research has been conducted to identify the factor(s) that restrict export. Choline may be a limiting nutrient for VLDL synthesis and secretion; therefore, supplementation in a ruminally protected form may prevent and alleviate fatty liver.

Keywords: fatty liver, nonesterified fatty acids, very low density lipoprotein, ureagenesis, choline

Introduction

Fatty liver occurs during intense fatty acid mobilization from adipose tissue. This can occur at parturition due to decreases in feed intake and hormonal changes associated with parturition and lactogenesis or during periods of negative energy balance. Negative energy balance occurs when energy intake is not sufficient to meet energy requirements for maintenance and production. This naturally occurs during early lactation in modern dairy cows or may occur secondary to metabolic disorders or infectious diseases. In a normal "healthy" dairy cow, the period of greatest hepatic triacylglycerol (TAG) accumulation is during parturition. Plasma nonesterified fatty acids (NEFA) concentrations often exceed 1000 uEq/l. Hepatic uptake of NEFA is dependent primarily on rate of blood flow to the liver and NEFA concentration in blood. The most desirable fate of NEFA taken up by the liver is complete oxidation to CO_2 or reesterification and export as TAG within a very low density lipoprotein (VLDL). Extra-hepatic tissues such as the mammary gland can utilize VLDL TAG. Unfortunately, the liver has a finite amount of energy it requires through oxidation and it has a limited capacity to export VLDL. Hence, excessive uptake of NEFA results in partial oxidation to ketones and storage of reesterified fatty acids as TAG. The consequence of fatty liver in dairy cattle is largely unknown. Induction protocols for fatty liver *in vivo* are confounded and do no permit examination of the independent effects of elevated TAG on hepatic function. Limited studies are available from *in vitro* models to assess the effects of TAG accumulation on hepatic function, and extrapolation of *in vitro* results to the live animal is risky. Consequently, the importance of preventing or alleviating fatty liver in dairy cattle is unknown. The two most logical strategies for prevention and alleviation of fatty liver are to reduce adipose tissue mobilization or increase hepatic VLDL secretion. The latter seems to be the best approach since fatty acid mobilization from adipose tissue provides a useful metabolic fuel and precursor for milk synthesis.

(Please note that the following text is not intended to be a literature review. Rather, it is a synopsis of research that has been conducted in our laboratory.)

Etiology of fatty liver

Analysis of liver biopsies obtained from dairy cattle every other day during the periparturient period indicates that the most rapid rate of hepatic TAG infiltration in the liver is during parturition (Vazquez-Anon *et al.*, 1994). This coincides with the most rapid and extensive rise in plasma NEFA concentration during the same period. This "spontaneous" development of fatty liver occurs to a variable extent in all dairy cattle. Data from our herd indicates that approximately 50% of cattle develop moderate to severe fatty liver during this time (Grummer, 1993). Additional elevation of hepatic TAG commonly occurs after calving and is dependent on the severity of negative energy balance associated with copious milk production and the occurrence of metabolic disorders and diseases. Elevated NEFA at calving is partially caused by a reduction in feed intake. Bertics *et al.* (1992) demonstrated that force feeding cows to maintain feed intake through calving reduced but did not eliminate the elevation in NEFA and liver TAG at calving. Consequently, hormonal changes associated with parturition and lactogenesis most certainly contribute to the surge in plasma NEFA at calving.

Force feeding cows through calving caused a reduction in fatty liver (Bertics *et al.*, 1992), consequently, it was commonly recommended that feed intake should be maximized prior to calving to prevent fatty liver. However, numerous studies have indicated that prepartum feed intake was not always strongly related to liver TAG immediately after calving and the original interpretation of the data from Bertics *et al.* (1992) was too simplistic. For example, feed intake was lower when cows consumed high fiber diets, yet liver TAG was not higher (Rabelo 2002, Rabelo *et al.*, 2003). Likewise, heifers consume less feed than mature cows (even when expressed as a percentage of body weight), yet they have lower hepatic TAG at calving (Rabelo 2002, Rabelo *et al.*, 2003). Cows fed high fiber diets and heifers demonstrate a relatively flat intake curve prior to calving, *i.e.*, less intake depression as calving approaches. Therefore, were the benefits of force feeding observed by Bertics *et al.* (1992) due to maximizing feed intake, or were they due to the avoidance of feed intake depression? We pooled data from three of our studies to try and answer this question. Plasma NEFA or hepatic TAG at calving were more closely associated with the magnitude of intake depression than to the absolute level of feed intake prior to depression (measured between 21 and 14 days prior to expected calving; Figure 1 and 2; Mashek *et al.*, unpublished). Further studies are needed to verify that change in feed intake (or energy balance) is a more important metabolic signal for the development of fatty liver than the absolute level of feed intake (or energy intake). If this is true, certain management factors (diet composition, diet changes, grouping changes, cow comfort) that contribute to a decline in feed intake prior to calving may contribute to the development of fatty liver.

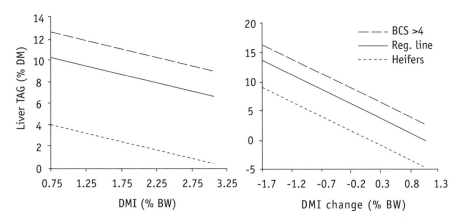

Figure 1. The relationship between prefresh dry matter intake (DMI) during days −21 to −14 prior to calving or DMI change during the final 21 days prior to calving on liver triglyceride (TG) at 1 day after calving. The solid line represents the regression generated using data from all animals (P = 0.32 for DMI, P = 0.0001 for DMI change). The long dashed line represents cows with a body condition score (BCS) > 4 (BCS effect: P = 0.11 for DMI, P = 0.07 for DMI change) and the short dashed line represents heifers (parity effect: P = 0.0001 for DMI, P = 0.0001 for DMI change). There were no interactions with BCS or parity, thus the slopes of all three lines within a graph are similar (slope = -1.53 for DMI, -4.96 for DMI change).

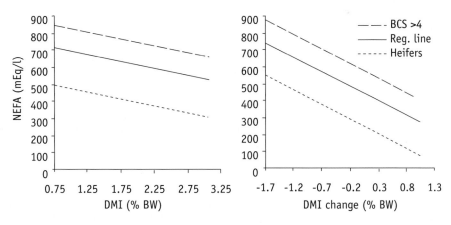

Figure 2. Effects of prefresh dry matter intake (DMI) during days −21 to −14 prior to calving or DMI change during the final 21 days prior to calving on plasma NEFA concentrations at 1 day after calving. The solid line represents the regression generated using data from all animals (P = 0.32 for DMI, P = 0.0001 for DMI change). The long dashed line represents cows with a body condition score (BCS) > 4 (BCS effect: P = 0.09 for DMI, P = 0.09 for DMI change) and the short dashed line represents heifers (parity effect: P = 0.006 for DMI, P = 0.01 for DMI change). There were no interactions with BCS or parity, thus the slopes of all three lines within a graph are similar (slope = -79 for DMI, -172 for DMI change).

Pathophysiology of fatty liver

Assessment of the effects of TAG accumulation on hepatic function is difficult to make. Intuitively, based on the importance of the liver to animal physiology, one could justifiably suggest that alteration of the structural integrity of the liver would have negative consequences. Elevation in blood of hepatic enzymes suggests that structural damage accompanies fatty liver. However, the vast majority of nutrition-related studies that have resulted in treatment differences in liver TAG do not report corresponding decreases in lactation performance. What, if any, are the effects of fatty liver on animal performance? Certainly, there are statistical relationships between fatty liver and reproductive parameters, immune responsiveness, *etc.* However, how does one demonstrate that there is indeed a cause and effect relationship? If fatty liver is induced, *e.g.*, by feed restriction or by over conditioning cows, any subsequent effects on hepatic function may be direct effects of TAG or indirect effects of the multitude of perturbations on whole animal metabolism or endocrine status that may have resulted from the induction protocol.

We set up an *in vitro* model for the development of fatty liver. Briefly, hepatocytes obtained from male calves are cultured in monolayer for an extended period of time, typically 2 to 3 days. Usually, for the first 48 hours after plating, cells are incubated in a medium with various concentrations of fatty acids (0 to 2 mM) to induce varying degrees of TAG loading. High concentration of fatty acid in the media for 48 hours causes a level of TAG accumulation that is consistent with moderate fatty liver (Strang *et al.*, 1988a, b). Media is then discarded, and for a shorter second incubation period (usually 3 hours) cells are exposed to 0 or a high concentration of fatty acid. It is during the second incubation period that estimates of hepatocyte function are assessed. This experimental design allowed us to evaluate the effect of previous (TAG loading) or concurrent exposure to fatty acid on cell function.

Effects of TAG loading on hepatic gluconeogenesis are inconsistent. Our initial experiment indicated that TAG loading and concurrent exposure to 2 mM C18:1 decreased the rate of gluconeogenesis from propionate (Cadorniga-Valino *et al.*, 1997). Results from a second experiment using a physiological mixture of fatty acids instead of C18:1 were contradictory, and even suggested that TAG loading slightly increased the rate of gluconeogenesis (Strang *et al.*, 1998a). Concurrent exposure to NEFA did not affect gluconeogenesis. Finally, Mashek *et al.* (2003) incubated hepatocytes with 1 mM C16:0 plus 1 mM C16:0, C18:1, C18:2, C18:3, C20:5, or C22:6. Addition of C18:1 as the second fatty acid increased the rate of gluconeogenesis from propionate; all other treatments had similar rates of gluconeogenesis.

We also studied the responsiveness of "lean" (cells incubated in media without long chain fatty acids) or TAG loaded cells to insulin and glucagon (Strang *et al.*, 1988a). Insulin suppressed gluconeogenesis more in lean cells than engorged cells suggesting that hormonal responsiveness may be impaired in cows with fatty liver. Glucagon effects on gluconeogenesis were not affected by TAG loading. Albumin and total protein synthesis were not affected by TAG loading (Strang *et al.*, 1998b). However, TAG loaded cells were less responsive to hormonal stimulation of albumin and protein synthesis than lean cells. Additionally, insulin clearance rates were lower in TAG loaded cells (Strang *et al.*, 1998b).

TAG loading reduced the rate of endogenous urea production as well as urea production from increasing concentrations of ammonium chloride in the media (Strang *et al.*, 1998a). Concurrent exposure to NEFA did not affect ureagenesis. Ammonia detoxification in ruminants occurs by two major pathways in the liver. Ureagenesis is a low affinity system (Km of 1-2 mM for ammonia) and works "upstream" in periportal cells. Glutamine (GLN) synthesis (from glutamate [GLU])is a high affinity system (Km of.2 mM) and works downstream in the pericentral cells. If ureagenesis is compromised due to fatty liver, more ammonia should escape downstream and be removed via glutamine synthetase. Circulating ammonia concentrations should only increase if both systems approach saturation. We hypothesized that animals with fatty liver should have elevated circulating ammonia or GLN % [GLN% = GLN*100/(GLN + GLU)] (Zhu *et al.*, 2000). Fourteen Holstein cows were monitored from 27 d prepartum to 35 d postpartum. There was a rise in circulating ammonia and GLN% at calving, suggesting an increase in ammonia passing to and through the liver. This experiment indicated that some of the effects of TAG on hepatic function observed *in vitro* might also be expressed *in vivo*.

Prevention and alleviation of fatty liver

Developing strategies for the prevention and alleviation of fatty liver depends in part on what one views as the primary metabolic problem. Is fatty liver a problem of excess fatty acid mobilization from adipose tissue or limited capacity to completely oxidize fatty acids or export fatty acids as VLDL TAG? Utilization of energy reserves from adipose tissue to support lactation is common in mammals and can be viewed as a desirable process. It should not be considered a problem unless one makes the argument that genetic selection for copious milk production has exacerbated the process to the detriment of cow health. Complete oxidation of fatty acids is probably a saturable process in hepatic tissue of all animals and is related to size and metabolic activity of the liver, *i.e.*, the energy requirements of the tissue. In strict sense, it should not be viewed as "limiting". Some have proposed that increasing complete oxidation of fatty acids be a strategy for preventing fatty liver. However, this is probably not a feasible solution unless increased oxidation can be uncoupled from ATP production, *e.g.*, energy is released as heat. But what is the advantage of increasing heat production to the animal? Increasing rates of hepatic VLDL TAG secretion seems like a logical strategy to reduce fatty liver. From a metabolic viewpoint, it would be advantageous to deliver the fatty acids to extra-hepatic tissues where they could be utilized.

Inhibition of adipose tissue mobilization

Because adipose tissue lipolysis can be viewed as desirable, inhibiting it probably does not represent the most logical approach to preventing fatty liver. However, if employing this strategy represents a short-term treatment that preserves animal health and consequently provides long term production benefits, it can be justified. Propylene glycol (PG) is a glucogenic precursor and can be administered to increase blood insulin, an antilipolytic hormone. Although cows may become "insulin insensitive" during the periparturient period, the data from oral drenching of PG clearly suggests there is some sensitivity to insulin. Oral drenching of one liter of PG one time per day for the final 7 to 10 days prior to calving can significantly reduce TAG in liver sampled within two days following calving (Studer *et al.*, 1993). Further studies using feed restricted, far-off dry cows as a model for transition cows indicated that lower doses (300 ml) could be effective as long as it was administered to the cow in a short period (*e.g.*, oral drench or when offered with a small amount of grain) (Grummer *et al.*, 1994; Christensen *et al.*, 1997). Oral drenching PG is labor intensive.

We examined the feasibility of a single injection of slow-release insulin (SRI) three days after calving to reduce liver TAG in 43 Holstein cows (Hayirli *et al.*, 2002). Concurrent IV infusion of glucose was not considered due to the extra labor requirement. Our hope was to identify a dose of SRI that may reduce blood NEFA and liver TAG without lowering blood glucose concentration. Prepartum injection of SRI would have been most desirable, but inability to accurately predict calving dates made that strategy impractical. Single doses administered at 3 days postcalving were 0, 0.14, 0.29, or 0.43 IU of SRI/kg body weight (BW). One cow injected with 0.29 and two cows injected with 0.43 IU SRI could not complete the trial due to sever hypoglycemia. Increasing dose of SRI caused a linear decrease in blood glucose and a quadratic decrease in blood NEFA and beta-hydroxybutyrate (BHBA) and accumulation of TAG in the liver. The low dose of SRI can be considered for prophylactic use against fatty liver.

Chromium (Cr) is an active component of glucose tolerance factor, and may potentiate the action of insulin (Hayirli *et al.*, 2001). Therefore, we hypothesized that it may have an antilipolytic effect when fed to periparturient cattle. Treatments were 0, 0.03, 0.06 and 0.12 mg of Cr as Cr-methionine/kg of $BW^{0.75}$ from 28 days prior to expected calving until 28 days postpartum. Chromium, particularly at the dose of 0.06 mg/kg $BW^{0.75,}$ caused rather dramatic increases in pre- and postpartum dry matter intake and milk production. However, no beneficial effects of Cr on liver TAG were noted.

Enhancing VLDL export from the liver

VLDL formation and exportation from the liver is a very complex process. This is partially due to the complex nature of the particle. Consisting of a variety of proteins, cholesterol, cholesterol esters, phospholipids, and triglyceride, the formation of these constituents and packaging into a particle encompasses a large number of biochemical pathways and cellular organelles. Basic research has demonstrated that blocking the synthesis of VLDL constituents also blocks the secretion of VLDL particles from the liver. Identification of the factor or combination of factors that contribute to slow rates of hepatic VLDL secretion is a challenging research endeavor.

There are two main schools of thought concerning possible explanations for slow VLVL export. One is that there is a limiting nutrient that prevents the synthesis of one or more of the VLDL constituents. A parallel is that a car will not run because it is out of gas. The other is that there is a "biochemical deficiency" that limits the synthesis of constituents or packaging of constituents into a VLDL particle. For example, perhaps the expression of an important gene is limiting. A parallel would be that a car will not run because there is a mechanical problem, *e.g.*, it needs a spark plug.

Limiting nutrients

Because of microbial degradation and synthesis of nutrients in the rumen, prediction of the amounts of nutrients that pass to the lower gut for absorption is very difficult. It also makes it very challenging to estimate nutrient requirements for ruminants. Various amino acids and vitamins have often been suggested to be limiting for milk production. Is it possible for them or other nutrients to be limiting for VLDL secretion?

Methionine is often cited as a potentially limiting amino acid for ruminants. Some have hypothesized that methionine supply to the liver may limit apolipoprotein synthesis and, therefore, limit VLDL secretion. Likewise, choline is a vitamin-like compound that serves as a precursor for phosphotidylcholine synthesis (a constituent of VLDL). As a methyl donor, choline may also spare methionine for apolipoprotein synthesis. Choline, like methionine, is extensively degraded in the rumen and estimated flow to the intestine (adjusted for BW) is far below that of nonruminants. We conducted a series of experiments to determine if methionine or choline were limiting nutrients, contributing to accumulation of TAG in the liver.

To conduct these experiments, we used far-off dry cows. In the first of two studies, cows are energy restricted to approximately 30% of requirements for maintenance and pregnancy. This is done to mimic feed intake depression prior to calving and it allows for lipid mobilization and development of fatty liver. During the energy restriction, cows are fed an unsupplemented diet or one supplemented with ruminally protected methionine or choline. This protocol allows us to determine if either nutrient has a role in the *prevention* of fatty liver. For the second experiment, cows are energy restricted for 10 days similar to that in the first experiment. During this time, all cows are fed the same diet. Following the 10 day energy restriction, cows are fed ad libitum for 6 days. During that time cows are fed an unsupplemented diet or one supplemented with ruminally protected methionine or choline and depletion of TAG from the liver is monitored. This protocol allows us to determine if either nutrient has a role in the *alleviation* of fatty liver. For these experiments we do not use transition cows because there is considerable variation among animals and large animal numbers are needed to detect significant treatment effects. Although such a model can be criticized because it does not utilize transition cows, it still enables us to test the principle of limiting nutrients contributing to fatty liver disease.

Methionine, provided as methionine hydroxy analog at the rate of 13 g/day did not prevent or alleviate fatty liver (Bertics and Grummer, 1999). In contrast, 15 g choline/day fed in a ruminally protected form prevented fatty liver and possible alleviated fatty liver (Cooke *et al.*, unpublished). Supplementation during energy restriction reduced liver TAG after 10 days (Figure 3). Plasma NEFA were also reduced by choline supplementation (Figure 4), therefore, it can not be distinguished whether the beneficial effects of choline on liver TAG were due to direct effects on the liver or indirect effects on lower plasma NEFA. It is not known how choline affects plasma NEFA. The results of the second experiment are a bit more difficult to interpret. At the end of the 10 day energy restriction and prior to application of treatments, the cows destined to receive choline supplementation during ad libitum feeding had 12.7 ug TAG/ug DNA in the liver. Cows destined to not receive supplementation during ad libitum feeding had 6.8 ug TAG/ug DNA in the liver. Since treatments had not been applied yet, one would have expected the values to be similar. In an attempt to adjust for this discrepancy, the values for liver TAG after the 10 day energy restriction but prior to treatment were used as covariates in statistical analysis. Covariately adjusted liver TAG after 3 and 6 day of treatment and ad libitum feeding are shown in Figure 5. Additionally, results were expressed as a percentage of that after the induction phase. Liver TAG was 60.4 and 52.2% of that after induction on day 3 after ad libitum feeding and treatment for the control and choline supplemented groups and 48.5 and 29.9% on day 6. Plasma NEFA were not affected by treatment in this trial (Figure 4). The liver TAG results supported the findings from the first trial and suggest that choline deficiency may contribute to fatty liver.

Niacin is antilipolytic vitamin and has been examined closely to determine if it has a role in prevention or treatment of fatty liver and ketosis. When feeding diets that are not supplemented

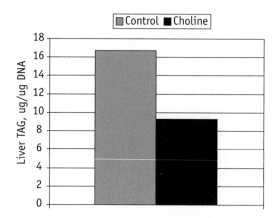

Figure 3. Effect of choline supplementation on liver TAG after 10 days of feed restriction of far-off dry cows (Experiment 1) (Cooke et al.*, unpublished).*

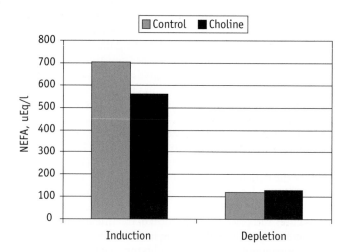

Figure 4. Plasma NEFA during choline supplementation. Induction corresponds to measurements made after 10 days of feed restriction of far-off dry cows (Experiment 1). Depletion corresponds to measurements made during ad libitum feeding that followed 10 days of feed restriction (Experiment 2; Choline was not supplemented during feed restriction) (Cooke et al.*, unpublished).*

with niacin, there is a net synthesis of niacin in the rumen. Nevertheless, some have suggested that supply of niacin to the intestine may be limiting in high producing dairy cows. Niacin is fed in an unprotected form, but estimates are that less than one third of supplemental niacin reaches the intestine. Amounts fed are usually pharmacological doses ($\geq$ 6 g per day). In two trials in which we fed 12 g niacin/day beginning 2 to 3 weeks prior to calving, no reduction in liver TAG was observed at 1 to 2 or 28 to 35 days postpartum. In a review of the literature

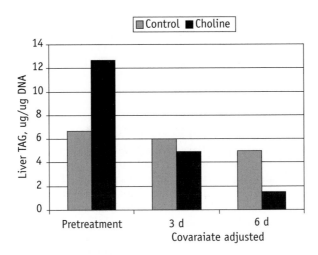

Figure 5. Liver TAG after a 10 day feed restriction (pretreatment with choline) and after three or 6 days of ad libitum feeding during which time cows were or were not supplemented with choline (Experiment 2). Due to the discrepancy in liver TAG pretreatment, the pretreatment values were used for covariate adjustment of data (Cooke et al.*, unpublished).*

published in the Nutrient Requirements of Dairy Cattle (NRC, 2001), a summary of 14 treatment comparisons in which niacin was fed indicated plasma NEFA concentration was significantly reduced once, increased twice, and not changed 11 times. If restricted to studies in which niacin was fed prepartum, plasma NEFA was significantly reduces once, increased twice, and not altered 8 times. Evidence does not support feeding 6 to 12 g niacin/day to prevent or treat fatty liver (Skaar *et al.*, 1989; Minor *et al.*, 1998).

Biochemical deficiency

As mentioned above, synthesis and secretion of VLDL from the liver is a biochemically complex process. Identifying a factor or factors that limit secretion can be challenging, although modern techniques in molecular biology (*e.g.*, microarray assays) may speed up the process. Indeed, if limiting factors can be identified, then innovative approaches (*e.g.* transgenics) eventually may be employed to prevent fatty liver.

We speculated that inadequate microsomal triglyceride transfer protein (MTP) activity might be responsible for low rates of VLDL secretion in ruminants (Bremmer *et al.*, 1999a). MTP is required for transfer of TAG into the lumen of the endoplasmic reticulum and for assembly of apolipoprotein B and lipid constituents into VLDL. MTP is absent in humans that cannot secrete apolipoprotein B containing lipoproteins from the liver and suffer from a genetic disorder called abetalipoproteinemia.

For our first experiment, we compared MTP activity in hepatic tissue from species that have variable rates of VLDL secretion form the liver (Bremmer *et al.*, 1999b). There was no association between MTP activity and rates of VLDL secretion. In a second experiment we monitored MTP activity and mass in liver of Holstein cows as they progressed from the dry period to lactation (Bremmer *et al.*, 1999b). The hypothesis was that hepatic MTP activity and mass would increase at calving during elevated blood NEFA and that hepatic MTP activity and mass would be inversely related to liver TAG. Neither hypothesis was supported by the data. Finally, we examined if high grain feeding or PG administration could alter MTP activity, but it did not (Bremmer *et al.*, 1999b).

If there is a biochemical deficiency in VLDL secretion, one approach would be to "fix" it. Another approach would be to recognize it, and alter hepatic metabolism in a manner that avoids the ramifications of the deficiency. Data from laboratory animals suggests that the fate of fatty acids in the liver might be dependent on the type of fatty acids presented to the liver. More specifically, polyunsaturated fatty acids were more likely to be oxidized than esterified and stored as TAG in the liver. The effectiveness of this approach was not known in ruminants and is dependent on the capacity of the liver for enhanced oxidation, either coupled or uncoupled to ATP production.

Mashek *et al.*, (2003) incubated monolayers of hepatocytes in media containing 1 mM C16:0 and an additional 1 mM fatty acid as C16:0, C18:0, C18:1, C18:2, C18:3, C20:5, or C22:5. Lowest TAG accumulation occurred when C16:0 or C18:3 was the second fatty acid in the media and highest accumulation occurred when C22:6 was the second fatty acid. Beta-hydroxybutyrate secreted into the media was high when C16:0 was the second fatty acid (14 ug BHBA/ug cell DNA), but it was low when C18:3 or C22:6 was the second fatty acid (6 to 7 ug BHBA/ug cell DNA). Based on these results, an *in vivo* trial was conducted to determine if type of fatty acid administered to the cow could influence liver TAG (Mashek *et al.*, unpublished). For this experiment, far-off dry cows were fasted for 4 days to induce fatty acid mobilization and fatty liver. During that time, 380 g lipid/day as tallow, linseed oil, or fish oil were infused intravenously as lipid emulsions. Tallow was chosen as the control because it reflects the fatty acid profile that is mobilized from adipose tissue. Linseed oil was selected because it was high in C18:3. Fish oil was selected because it is a rich source of C22:6. At the end of the fast, cows infused with linseed oil had the lowest liver TAG, fish oil the highest TAG, and tallow intermediate (Figure 6). Cows infused with linseed oil had the lowest plasma NEFA concentrations during the four day fast (Figure 7). Therefore, it is not clear if the beneficial effects of linseed oil infusion were due to direct effects on the liver or indirect effects on plasma NEFA concentration and uptake by the liver. Results indicate there is a potential for feeding certain fatty acids, protected from ruminal hydrogenation, to prevent fatty liver.

Conclusion

In many ways, fatty liver is a lipid related metabolic disorder that is easy to understand. It is likely to occur when blood NEFA are elevated and uptake of NEFA increases to the point in which the capacity of the liver to oxidize fatty acids or export fatty acids as VLDL TAG is exceeded. There are strategies to prevent fatty liver, such as drenching PG or possibly feeding protected choline. Yet there are several questions that need to be answered. The reason they have not been answered is because finding the answers is difficult. What concentration of NEFA is too high? When is the liver overwhelmed by NEFA uptake to the point where TAG occurs? How much variation is there among animals and what accounts for the variation. We know it is different for heifers and mature cows. Why? What is rate limiting for VLDL export? Lastly, what if any negative consequences occur to the animal from fatty liver? If there are negative consequences,

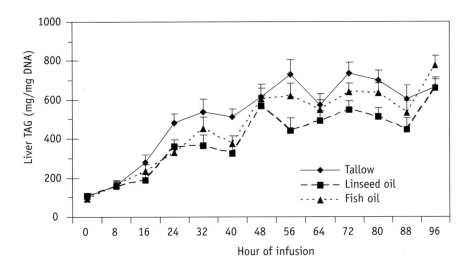

Figure 6. Liver TAG in far-off dry cows after a 4 day fast during which time cows were infused IV with a TAG emulsion (380 g lipid/day) derived from tallow, linseed oil, or fish oil (Mashek et al., unpublished).

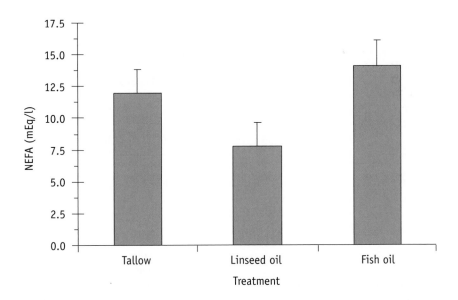

Figure 7. Plasma NEFA in far-off dry cows after a 4 day fast during which time cows were infused IV with a TAG emulsion (380 g lipid/day) derived from tallow, linseed oil, or fish oil (Mashek et al., unpublished).

at what concentration of liver TAG does it occur? The last question couple of questions are very important from a practical, on-farm basis, but also when evaluating how much priority should be placed on fatty liver research.

Acknowledgements

The author appreciates the creativity and the blood, sweat, and tears gleaned the following collaborators: Louis Armentano, Sandy Bertics, Robert Aiello, Barry Kleppe, Todd Skaar, Brian Strange, Lihua Zhu, Vaughn Studer, Doug Minor, Doug Mashek, Darin Bremmer, and Reinaldo Cooke.

References

Bertics, S.J., and R.R. Grummer. 1999. Effects of fat and methionine hydroxy analog on prevention and alleviation of fatty liver induced by feed restriction. J. Dairy Sci. 82, 2731-2736.

Bertics, S.J., R.R. Grummer, C. Cadorniga-Valino, D.W. LaCount, and E.E. Stoddard. 1992. Effect of prepartum dry matter intake and liver triglyceride concentration and early postpartum lactation. J. Dairy Sci. 75, 1914-1922.

Bremmer D.R., R.R. Grummer, and S.J. Beritcs. 1999a. Differences in hepatic microsomal triglyceride transfer protein among species. Comp. Biochem. Phys. Part A 124, 123-131.

Bremmer D.R., R.R. Grummer, and S.J. Beritcs. 1999b. Differences in hepatic microsomal triglyceride transfer protein among species. Comp. Biochem. Phys. Part A 124, 123-131.

Cadorniga-Valino, C., R.R. Grummer, L.E. Armentano, S.S. Donkin, and S.J. Bertics. 1997. Effects of fatty acids and hormones on fatty acid metabolism and gluconeogenesis in bovine hepatocytes. J. Dairy Sci. 80, 646-656.

Christensen, J.O., R.R. Grummer, F.E. Rasmussen, and S.J. Bertics. 1997. Method of propylene glycol delivery influences plasma metabolites of feed-restricted cattle. J. Dairy Sci. 80, 563-568.

Grummer, R.R. 1993. Etiology of lipid related metabolic disorders in periparturient dairy cattle. J. Dairy Sci. 76, 3882-3896.

Grummer, R.R., J.C. Winkler, S.J. Bertics, and V.A. Studer. 1994. Effect of propylene glycol dosage level on plasma energy metabolites of prepartum Holstein heifers experiencing feed restriction. J. Dairy Sci. 77, 3618-3623.

Hayirli, A., D.R. Bremmer, M.T. Socha, and R.R. Grummer. 2001. Effect of chromium supplementation on production and metabolic parameters in periparturient dairy cows. J. Dairy Sci. 84, 1218-1230.

Hayirli, A., S.J. Bertics, and R.R. Grummer. 2002. Effects of slow-release insulin on liver triglycerid and metabolic profiles of Holsteins in early lactation. J. Dairy Sci. 85, 2180-2191.

Mashek, D.G., and R.R. Grummer. 2003. Effects of long chain fatty acids on lipid and glucose metabolism in monolayer cultures of bovine hepatocytes. J. Dairy Sci. 2003, 2390-2396.

Minor, D.J., S.L. Trower, B.D. Strang, R.D. Shaver, and R.R. Grummer. 1998. Effects of nonfiber carbohydrate and niacin on periparturient metabolic status and lactation of dairy cows. J. Dairy Sci. 80, 189-200.

National Research Council. 2001.Nutrient requirements of dairy cattle. 7th rev. ed. Natl Acad Press, Washington, DC.

Rabelo, E.R. 2002. Effects of dietary energy density on metabolic status and lactation performance of periparturient dairy cows. Ph.D. Thesis. University of Wisconsin, Madison.

Rabelo, E., R.L. Rezende, S.J. Bertics, and R.R. Grummer. 2003. Effects of transition diets varying in dietary energy density on lactation performance and ruminal parameters of dairy cows. J. Dairy Sci. 86, 916-925.

Skaar, T.C., R.R. Grummer, MR. Dentine, and R.H. Stauffacher. 1989. Seasonal effects of pre- and postpartum fat and niacin feeding on lactation performance and lipid metabolism. J. Dairy Sci. 72, 2028-2038.

Strang, B.D., S.J. Bertics, R.R. Grummer, and L.E. Armentano. 1998a. Effects of long-chain fatty acids on triglyceride accumulation, gluconeogenesis, and ureagenesis in bovine hepatocytes. J. Dairy Sci. 81, 728-739.

Strang, B.D., S.J. Bertics, R.R. Grummer, and L.E. Armentano. 1998b. Relationship of triglyceride accumulation to insulin clearance and hormonal responsiveness in bovine hepatocytes. J. Dairy Sci. 81, 740-747.

Studer, V.A., R.R. Grummer, S.J. Bertics, and C.K. Reynolds. 1993. Effect of prepartum propylene glycol on periparturient fatty liver in dairy cows. J. Dairy Sci. 76, 2931-2939.

Vazquez-Anon, M., S. Bertics, M. Luck, R.R. Grummer, and J. Pinjeiro. 1994. Peripartum liver triglyceride and plasma metabolites in dairy cows. J. Dairy Sci. 77, 1521-1528.

Zhu, L.H., L.E. Armentano, D.R. Bremmer, R.R. Grummer, and S.J. Bertics. 2000. Plasma concentration of urea, ammonia, glutamine around calving and the relation of hepatic triglyceride to plasma ammonia removal and blood acid-base balance. J. Dairy Sci. 83, 734-740.

Relationships between mild fatty liver and health and reproductive performance in Holstein cows

G. Bobe, B.N. Ametaj, R.A. Nafikov, D.C. Beitz and J.W. Young
Department of Animal Science, Iowa State University, Ames, IA, USA

Fatty liver is a metabolic disease that affects up to 50% of dairy cows in early lactation[1]. Severe fatty liver is associated with an increased incidence and duration and decreased treatment success of infectious and metabolic diseases, such as mastitis, metritis, and ketosis, as well as decreased reproductive performance in affected cows[2]. The objective of the current study was to determine the relationships between mild fatty liver and health status (mastitis, elevated body temperature, and ketosis) and reproductive performance (days to first observed estrus, first service, days open, number of services, conception rate after first service, and overall conception rate).

For the experiment, 32 multiparous Holstein dairy cows were used. For the last 4 weeks prepartum, each cow was offered individually, in addition to hay and the total mixed ration that was formulated to meet NRC requirements, 6 kg of cracked corn daily to potentially increase the incidence of mild fatty liver. At day 8 postpartum, cows were divided into two groups depending on their liver triacylglycerol concentrations (cutoff value of 1% liver wet weight) at day 8 postpartum. Chi-square tests were used to compare treatment group means of incidence rates for health status (mastitis, body temperatures >39.5 °C, and urinary concentrations >40 mg/dl) and reproductive performance parameter (conception rate after the first service, and overall conception rate). A t-test was used to compare average lengths of time with body temperatures >39.5 °C of treatment groups. Days to first heat, days to first service, and days open were assumed Weibull distributed, and a chi-square test was used to compare treatment means.

Mild fatty liver was associated with an increased number of days with elevated body temperatures from days 9–22 postpartum (P=0.05). A higher incidence of subclinical infections could explain partly the increased incidence of mastitis from days 23–150 postpartum (both P=0.05) in cows with mild fatty liver, despite the fact that both groups did not differ at days 23 – 42 postpartum in liver triacylglycerol concentrations. Mild fatty liver was associated with more days to first estrus and service, more days open, and decreased conception (all P=0.05). The decreased reproductive performance was associated with a higher incidence of mastitis and a longer period of time in negative energy balance (both P=0.05).

The current study suggests that mild fatty liver is associated with decreased general health and reproductive performance in dairy cows. Therefore, prevention of fatty liver might improve health and reproductive performance of lactating dairy cows, thereby preventing loss of income of dairy farmers.

Acknowledgements
The research was supported partly by grant number 99-35005-8576 from US Department of Agriculture and was part of regional research project NC-1009.

References
[1] Jorritsma R., H. Jorritsma, Y.H. Schukken, *et al.*, 2001. Prevalence and indicators of post partum fatty infiltration of the liver in nine commercial dairy herds in The Netherlands. Livest. Prod. Sci. 68, 53-60.

[2] Wensing T., T. Kruip, M.J.H. Geelen, *et al.*, 1997. Postpartum fatty liver in high-producing dairy cows in practice and in animal studies. The connection with health, production and reproduction problems. Comp. Haematol. Int. 7, 167-171.

Prevention of fatty liver in transition dairy cows by subcutaneous glucagon injections

R.A. Nafikov[1], B.N. Ametaj, G. Bobe, K.J. Koehler[2], J.W. Young and D.C. Beitz
[1]Department of Animal Science; [2]Department of Statistics, Iowa State University, Ames, IA 50011, USA

The objective of this study was to evaluate the use of glucagon as a preventive for fatty liver development in transition dairy cows. Twenty-four multiparous Holstein cows were used in this study. Cows were assigned randomly to Saline (control), 7.5 mg/d, or 15 mg/d of glucagon treatment groups with 8 cows per group. During the final 4 weeks of gestation, all cows were supplemented with 6 kg of cracked corn in addition to their regular NRC-based diet to stimulate the development of fatty liver disease postpartum. Beginning at day 2 postpartum, cows were injected subcutaneously with saline or 7.5 mg/d or 15 mg/d of glucagon for 14 days. Liver samples were obtained by puncture biopsy at -4, 2, 6, 9, 16, 20, 27, 34, and 41 days postpartum and analyzed for lipid composition and glycogen. Blood samples were collected from the coccygeal vein at days 1 through 17, 20, 27, 34, and 41 postpartum and analyzed for blood metabolites. Feed intake was monitored twice daily from day 1 through day 17 postpartum. Milk samples were collected at the days of liver biopsies and analyzed for different milk components. Milk production was recorded throughout the entire sampling period. Data were analyzed as repeated measures using the mixed models procedure of SAS.

Glucagon administration at 15 mg/d dosage prevented liver triacylglycerol accumulation during treatment period ($P<0.02$). Plasma glucose concentration was increased in 7.5 mg/d and 15 mg/d of glucagon treatment groups ($P<0.0001$) in a dosage-dependent manner ($P<0.02$). Plasma non-esterified fatty acid (NEFA) concentration was decreased during treatment period in 7.5 mg/d and 15 mg/d of glucagon treatment groups ($P<0.05$ and $P<0.007$ respectively). Glucagon injections did not significantly change plasma β-hydroxybutyrate and urea concentrations. Plasma insulin concentration was increased in 7.5 mg/d and 15 mg/d of glucagon treatment groups ($P<0.01$ and $P<0.001$ respectively). Glucagon administration did not cause any significant changes in feed intake and milk production. Milk lactose, fat, and protein concentrations were not altered significantly by glucagon injections. However, there was a decrease in milk urea concentration in 7.5 mg/d and 15 mg/d of glucagon treatment groups ($P<0.02$ and $P<0.002$ respectively).

In conclusion, glucagon can be used to prevent fatty liver development in transition dairy cows because of its NEFA lowering effect and potency to increase blood glucose and insulin concentrations without altering cows' feed intake and milk production.

Clinical relevance and therapy of fatty liver in cows

M. Fürll, H. Bekele, D. Röchert, L. Jäckel, U. Delling and Th. Wittek
Medizinische Tierklinik Leipzig, An den Tierkliniken 11, 04103 Leipzig, Germany

The fatty liver is the most important disturbance of liver function in dairy cows. According to literature it is not easy or impossible to cure a severe fatty liver. Therefore the fatty liver syndrome is one of the main reasons for cow loss especially after parturition. However, there are doubts about the significance of primary and secondary fatty liver.

We measured the liver fat content
1. in 35 healthy cows with different milk yield after calving,
2. in 100 slaughtered cows,
3. and the state of liver cell organelles in 10 cows 1 and 8 weeks after calving,
4. in 38 cows with abomasal displacement,
5. and therapeutic results in 40 cows with „severe liver damages".

Our results are
1. The liver fat content differed significantly between healthy cows with a milk yield of 4500 kg/a and 9500 kg milk/a 1 and 8 weeks after calving.
2. In the livers of 100 forced slaughtered cows the most frequent alterations were fat infiltration (66%), degeneration (16%) and reactive inflammation (29%). The highest liver fat contents were found in cows with nephritis (110 g/kg) and mastitis (92 g/kg).
3. Electron microscopically we did not find a relationship between alterations of cell organelles (mitochondrions and lysosoms) and liver fat content. However, the nucleus alterations correlated significantly with liver fat in cows with high as well as with low liver fat contents.
4. In 38 cows with abomasal displacement the liver fat content was always elevated; 37% had a simple fat infiltration, 13% a degenerative fatty liver and 29% necroses. All cows were cured independent of the liver fat content.
5. 40 cows of Leipzig Veterinary Medical Hospital with „severe liver damages" (AST> 200 U/l, GLDH> 100 U/l, Bilirubin > 50 µmol/l) were selected and studied with regard to therapeutic results and prognostic value of laboratory tests. The therapy was performed with continuous infusions of glucose (750g/day), 0,9% NaCl, propylene glycol, antiphlogistics, furthermore in some cows antioxidants and glucocorticoids. 65% of these cows were cured, 35% of the cows had to be euthanized because of primary diseases with a poor prognosis (botulism, mastitis, muscle rupture, nephritis, peritonitis, ulcera or pulmonary thrombosis).

The fatty liver is the most frequent pathological liver alteration; however, the clinical significance is only moderate. Without a primary disease a fatty liver had almost always a good prognosis. Decisive is a consequent and prognostic therapy.

The inflammation could have a role in the liver lipidosis occurrence in dairy cows

G. Bertoni[1], Trevisi Erminio[2], Calamari Luigi[2] and Bionaz Massimo[2]
[1]Head of Istituto di Zootecnica, Facoltà di Agraria, U.C.S.C., Piacenza, Italy; [2]Istituto di Zootecnica, Facoltà di Agraria, U.C.S.C, Piacenza, Italy

The liver lipidosis is a very dangerous metabolic situation that can occur, in a more or less serious manner, in the so called transition period of dairy cows. Among the possible causes, the more frequently suggested are: liver NEFA overload[1], some nutrient deficiency (*i.e.* choline)[2] and inadequate apolipoprotein synthesis[3]. The last one agrees with our hypothesis, but we suggest that the reduced apolipoprotein synthesis could be a consequence of inflammatory conditions around calving which impair the common liver activity.

To verify this hypothesis, 8 multiparous dairy cows have been monitored 30 days before and after calving. Cows were individually fed almost ad libitum (2 meals of forages and 2 or 8 of concentrate by auto feeder; type and amount of forages and concentrates were different in pregnancy *vs.* lactation). They were routinely checked for: DMI and health status (daily), body weight and BCS (every 14 days), rectal temperature and blood samples (every 3-4 days and daily around calving) and milk yield (at each milking). Liver biopsies were taken approximately at -27, -6, 3, 15, 24 and 27 DIM for triacylglycerols (TAG) content[4]. On blood, an extended metabolic profile was carried out, however some parameters were more carefully considered: haptoglobin, ceruloplasmin, zinc, total cholesterol, phospholipids, albumin, NEFA, AST, GGT and paraoxonase. At the end, cows were divided according to the highest post-calving TG content of the liver, observed between 5 and 15 DIM; namely 4 of them had a higher (HLT) and 4 a lower (LLT) TAG content. The borderline value was 10% on wet basis which is considered to discriminate between mild and serious lipidosis.

Interestingly, cows of both groups were roughly normal: similar milk yield and DMI, despite slightly worse in the HLT group, in which the 1st month body weigh loss was significantly higher (69.2 *vs.* 46,5 kg, P<0.06). Moreover, the liver integrity were not sensibly impaired because the cytolytic enzymes (AST, GGT) were only slightly increased. On the contrary, the hepatic activity was dramatically modified: there was an increase of acute phase protein synthesis with higher blood levels of haptoglobin (0.84 *vs.* 0.46 g/l for HLT *vs* LLT at 6 DIM, n.s.) and ceruloplasmin (3.73 *vs.* 2.80 µmol/l for HLT *vs.* LLT at 9 DIM, P<0.05) and lower levels of zinc (stored as metallothionein into the liver; 8.46 *vs.* 12.64 µmol/l for HLT *vs.* LLT at 9 DIM, P<0.05). On the contrary, a parallel reduction in blood of usually synthesized proteins was observed: less albumins (33.2 *vs.* 37.7 g/l at 21 DIM, P<0.05, for HLT *vs.* LLT respectively), total cholesterol (2.43 *vs.* 3.22 mmol/l for HLT *vs.* LLT in the 1st month of lactation, P<0.07) and phospholipids – both indices of lipoproteins - as well as paraoxonase (39.4 *vs.* 54.0 U/ml at 21 DIM, P<0.07, for HLT *vs.* LLT respectively), a liver enzyme considered as a negative acute phase protein[5].

These results seem to confirm that liver lipidosis is related to the occurrence of inflammatory conditions at the end of pregnancy and early lactation, when lipolysis is also high. In fact, the cytokines responsible of the acute phase response act on the liver with the increase of the positive acute phase proteins, but with a contemporary decrease of some usual liver proteins: *i.e.* paraoxonase, albumins and likely apolipoproteins. Therefore the paradox should be of a very active liver with a lower usual functionality namely of TAG excretion; this would suggest any effort to avoid inflammatory phenomena in this stage.

Acknowledgements

This research was supported by P.F. R.A.I.Z.

References

[1] Bertics, S.J., et al., 1999. Effects of fat and methionine hydroxy analog on prevention or alleviation of fatty liver induced by feed restriction. J. Dairy Sci. 82, 2731-2762.

[2] Grummer, R.R., 1993. Etiology of lipid-related metabolic disorders in periparturient dairy cows. J. Dairy Sci. 76, 3882-3896.

[3] Gruffat, D., et al., 1997. Hepatic gene expression of apolipoprotein B100 during early lactation in underfed, high producing dairy cows. J. Dairy Sci. 80, 657-666.

[4] Rukkwamsuk, T., et al., 1999. Relationship between overfeeding and overconditioning in the dry period and the problems of high producing dairy cows during the postparturient period. Vet. Q. 21, 71-77

[5] Feingold, K.R., et al., 1998. Paraoxonase activity in the serum and hepatic mRNA levels decrease during the acute phase response. Atherosclerosis 139, 307

Effect of propylene glycol on fatty liver development and hepatic gluconeogenesis in periparturient dairy cows

T. Rukkwamsuk[1], Apassara Choothesa[2], Sunthorn Rungruagn[3] and Theo Wensing[4]
[1]Department of Medicine, Faculty of Veterinary Medicine, Kasetsart University, Kampangsaen, Nakhon-Pathom, Thailand 73140; [2]Department of Physiology, Faculty of Veterinary Medicine, Kasetsart University, Bangkok, Thailand 10900; [3]Pakthongchai Dairy Farm, Nakhon-Rachasima, Thialand 30150; [4]Faculty of Veterinary Medicine, Utrecht University, 3508 TD Utrecht, The Netherlands

Negative energy balance or fatty liver has adverse effects on health, production, and reproduction in dairy cows, partly due to some consequences of intensive lipolysis causing fatty liver or hepatic lipidosis[1]. Several researches have tried to alleviate this problem using different methods[2]. The present study attempted to determine the effect of propylene glycol on fatty liver development and hepatic gluconeogenesis in periparturient dairy cows.

Twenty-three Holstein Friesian cows were randomly allocated into 2 groups: a control group of 9 cows and a treated group of 14 cows that were drenched with 400 ml of propylene glycol once daily from 7 (6 ± 4) days before anticipated calving date until 7 days after calving. At -2, 2, and 4 wk from parturition, blood samples were collected from all cows for determination of serum glucose, non-esterified fatty acid, and β-hydroxybutyrate concentrations; biopsied liver samples were collected from all cows for determination of triacylglycerol and glycogen concentrations. At all sampling times, hepatic fructose 1,6 bisphosphatase activity was also measured. Milk yields were recorded daily.

Compared with the concentrations at -2 wk, serum glucose concentrations decreased sharply at 2 and 4 wk, and the concentrations did not differ between the two groups at all intervals. Serum non-esterified fatty acid concentrations did not differ between the two groups at -2 wk, the concentrations at 2 wk increased 269% and 118% for control and treated groups, respectively. The concentrations were higher at 2 and 4 wk for control group than for treated group. Serum β-hydroxybutyrate concentrations did not differ between the two groups at -2 wk, the concentrations increased after calving and were higher for control than for treat group at 2 wk. Triacylglycerol concentrations in the liver increased 245% at 2 wk in the control group, but only 125% in the treated group. At -2 wk, hepatic triacylglycerol concentrations did not differ between the two groups; however, the concentrations were higher at 2 and 4 wk for the control group than for the treated group. Hepatic glycogen concentrations decreased at 2 wk when compared with the concentrations at -2 wk; however, the concentrations did not differ between the two groups at all intervals. Hepatic fructose 1,6 bisphosphatase activity did not change throughout the experimental period in both groups, and the activities were similar in both groups at all intervals. Average 30-d milk yields were 29.1 ± 5.3 and 30.2 ± 4.1 kg/d for control and treated groups, respectively, and the milk yields did not differ between the two groups of cows.

Our results indicated that dairy cows drenched with propylene glycol at a dosage of 400 ml/cow daily could improve negative energy balance and could alleviate an excessive mobilization of fat, consequently lower triacylglycerol accumulation in the liver. In conclusion, propylene glycol giving between 7 days before expected calving date and 7 days postpartum could be used in practice to alleviate fatty liver problems and their consequences in postparturient dairy cows.

Acknowledgements
The authors thank Thailand Research Fund (TRF) for the financial support of this research.

References

[1] Gerloff, B.J., T.H. Herdt and R.S. Emery, 1986. Relationship of hepatic lipidosis to health and performance in dairy cattle. J. Am. Vet. Med. Assoc. 188, 845-850.

[2] Hippen, A.R., P. She, J.W. Young, *et al.*, 1999. Alleviation of fatty liver in Dairy Cows with 14-day intravenous infusions of glucagons. J. Dairy Sci. 82, 1139-1152.

The report of classical swine fever in Romania between april 2001 and april 2003

C. Pascu, V. Herman and N. Catana
Faculty of Veterinary Medicine Timisoara, Romania

The classical swine fever immunoprophilaxy with live attenuated vaccines was compulsory in Romania until December 31, 2001[1,2]. However, an outbreak of classical swine fever was declared in Harghita county (in the middle of country) in April 2001, though the last declared outbreak in Romania was in 1974. Beginning with January 1st 2002 the vaccination in classical swine fever was prohibited only in 12 counties from the west side of Romania. The present paper has the objective to emphasize the effects of application of the Veterinary Agency and for Food Safety's decisions and propose measures to establish the present situation concerning classical swine fever evolution in Romania.

To achieve the purpose of this paper the official statistics data about the evolution of classical swine fever in Romania between April 2001 and April 2003 were taken, as sent by the Romanian Veterinary Agency at the International Office of Epizootics (O.I.E.).

The report of classical swine fever in Romania between April 2001 and April 2003 is presented in Table 1. In 2001 only one outbreak occurred (in Harghita County). In 2002 all the 38 outbreaks appeared in 9 counties, of which 5 were in regions where the vaccination was prohibited and in 4 counties where the vaccination was compulsory. In the first four months of the year 2003, 92 outbreaks were declared, both in counties where the vaccination was prohibited and in counties where the vaccination was compulsory. Because the actual situation must be brought to normal, we propose and sustain a return to mass compulsory vaccination of all pigs in the whole Romania, which is the basic measure in the present context of growing pigs in Romania.

Section F.

Table 1. The geographic allocation and the outbreaks of classical swine fever evolution in Romania.

Month	# of outbreaks	Counties affected[a]
April	1	Harghita
Total 2001	1	one county
April	1	Arad
July	1	Bihor
August	5	Salaj
September	5	*Giurgiu, Olt,* Salaj, Timis
October	12	Salaj, Satu-Mare, *Teleorman*
November	6	Bihor, *Giurgiu, Olt*
December	8	Arad, Bihor, *Giurgiu, Ialomita,* Satu-Mare
Total 2002	38	9 counties
January	12	Arad, Bihor, *Brasov,* Caras-Severin, Satu-Mare, Timis
February	36	Bihor, Caras-Severin, Cluj, *Giurgiu,* Gorj, Satu-Mare, Salaj, Timis
March	28	Alba, Arad, *Arges,* Bihor, Caras-Severin, Cluj, *Dolj,* Gorj, *Harghita,* Satu-Mare, Salaj, *Vaslui*
April	16	Alba, Arad, Bihor, Satu-Mare, Salaj, *Suceava*
Total 2003	92	16 counties

[a] Counties were vaccination was compulsory are shown in italic.

References

[1] Herman V., 2003. Evolutia pestei porcine clasice în România în anul 2002. Lucr. St. Med. Vet. Timisoara 36, 342-345.
[2] Leman A.D., B.E. Straw, W.L. Mengeling, S. D`Allaire and D.J. Taylor, 1992. Diseases of Swine. 7-ed, Iowa State University Press, Ames, Iowa, USA.

Epidemiological survey on the downers in Korea

H.R. Han, C.W. Lee, H.S. Yoo, B.K. Park, Bo Han and D. Kim
Department of Clinical Sciences, Seoul National University, Seoul, 151-742, Korea

The downers explosively occurred in the 538 Holstein dairy cows and 58 Korean native cattle in 307 herds in South Korea from July to September 2000, regardless of the age of animals. The Epidemiological survey was carried out to find out causes of the downers in cattle raised in South Korea.

During 24 months from July 2001 to July 2003, 232 cases were confirmed as downers. We have done epidemiological and clinical investigations, isolation and identification of infective agents, nutritional and metabolic disorders and analysis of toxin including fungal and heavy metal toxin. Meanwhile, differential diagnosis was required to differentiate downers usually showing similar clinical signs. Those methods included RT-PCR and PCR to detect viral genome, direct EM examination, FA to conform specific antigen, sequencing of chromosomal DNA to identify pathogens.

Total 232 cows were confirmed as downers from July 2001 to July 2003 in South Korea. Those were included parturient paresis 96(41.38%), hypocalcemia with hypophosphatemia 24(10.35%), Ketosis 15(6.47%), unknown 10(4.31%). Traumatic injury 9(3.88%), hypocalcemia (non-associated calving) 9(3.88%), nutritional deficiency 8(3.45%), fatty liver syndrome 8(3.45%), hypomagnesemia 6(2.6%), gangrenous mastitis 5(2.16%), hepatic and renal failure 5(2.16%), hip joint luxation 4(1.72%), severe foot rot 4(1.72%), chronic pneumonia 4(1.72%), hypokalemia 3(1.29%), sporadic bovine encephalomyelitis 3(1.29%), renal failure 3(1.29%), Akabane disease 2(0.86%), chronic coronavirus enteritis 2(0.86%), navel ill 2(0.86%), acute diffuse peritonitis 2(0.86%), RDA 2(0.86%), systemic weakness with nutritional deficiency 2(0.86%), hepatic toxicosis 2(0.86%), hypochloremia 1(0.43%), calcium overdose injection 1(0.43%), internal hemorrhage bleeding 1(0.43%), intestinal torsion 1(0.43%), intestinal obstruction 1(0.43%), acute tympany 1(0.43%), vertebral fracture 1(0.43%), bacillary hemoglobinuria 1(0.43%), protein losing enteropathy 1(0.43%), and the overall outcome of downer cases was that 52.6% of them recovered, 34.5% of them were slaughtered, and 12.9% of them died.

This study was carried out to investigate the causes of downers in cattle that occurred whole country under high humidity and temperature environmental conditions from July to September 2000 year. However, the etiology of downers was not clear yet, because the available evidences and clinical experiences suggested that the disease was different from the downers outbreak previously except being just down. Most of cases, which occurred during the period of 2000, only showed knuckling of the fetlock and prolonged recumbency. But they were usually alert and bright and, although the appetite was reduced slightly, the cow did eat and drink moderately well, also there was no relationship on age predilection and parturition. Only the antibody titers of Akabane virus (>128), IBR (<2), BVD (>128), BEF (>128), Aino (<2), Ibaraki (<2), and Chuzan virus (<2) were detected very variable levels in each case. Especially, antibody titers against Akabane virus were higher in the downers outbreak in 2000 year and the virus was showed highly homology with S gene of Akabane OBE-1 stain and the antibody titers against the isolation were detected in the experimentally infected cows. However, clinical signs could not be observed in the cows. The results obtained from this study will be suggested that it was vector-born disease in downers 2000 in Korea. Therefore, it is possible to outbreak the downers in cattle in 2004 or 2005 year epizootically because of absence of the antibodies against the agents. Based on the results and suggestions, epidemiological investigation on the syndrome should be continued.

Section G.
Alternatives to growth promoting antibiotics

Alternatives to antibiotic feed additives

Chad R. Risley and John Lopez
Chr. Hansen Inc., 9015 West Maple Street, Milwaukee, Wisconsin 53214, USA

Abstract

Livestock producers continue to reevaluate their usage of antibiotics to determine their cost and benefit opportunities. The reasons behind this evaluation include improved production practices that rely less on antibiotic feed additives; increased concern over antibiotic resistant pathogens; more regulatory oversight in the feed industry; and the increased availability of economical alternatives to antibiotics. Today, producers have many options to improve animal performance without using antibiotic feed additives, including the modifying of management strategies like improving biosecurity and sanitation, controlling animal and people movement, optimizing vaccine management and implementing continuing education efforts. In addition to changing management practices, altering existing nutritional programs can also be effective. This paper will address various alternatives to antibiotic feed additives that are currently available to the livestock industry. Specially, the proposed mode of actions and performance and economic benefits will be discussed for acidifiers, prebiotics, novel ingredients and direct fed microbials (probiotics).

Keywords: antibiotics, probiotics, acids, prebiotics

Antibiotics

Antibiotics are substances naturally produced by bacteria that, in adequate concentrations, inhibit or kill bacteria. In addition to the naturally produced antibiotics, chemically synthesized chemobiotics have been developed; these with antibiotics are referred to as antimicrobial agents. After their discovery in the 1920s, antibiotics were used at therapeutic dosages to combat disease caused by bacteria. During the 1940s, scientists discovered that chickens grew better when they were fed sub therapeutic levels of antibiotics compared to chickens that were not fed antibiotics, even in the absence of overt disease (Moore, *et al.*, 1946). Subsequent research on a variety of species, confirmed that healthy animals grew faster and more efficiently when fed a relatively low level of antibiotic (Hays, 1981). Continuing research efforts showed that the magnitude of the antibiotic effect was influenced by a wide variety of factors including, age, environment, stress, diet quality and nutrient adequacy (Cromwell, 2001) (Table 1).

Due to their ability to improve performance, antimicrobial agents can lower the cost of production and due to the FDA-CVM approving these compounds as feed additives, their use in livestock diets has become commonplace. NAHMS (2000) has reported that antimicrobials, outside of disease treatment, are currently used in 83% of swine starter feeds and 73% of the grower feeds. Additionally, approximately 61% of US sow herds use feed antibiotics to treat or prevent disease

Several theories have been put forth on why antibiotics improve animal performance, but three theories prevail. These are: 1) Nutrient sparing effect, 2) Subclinical disease control and 3)

Table1. Efficacy of antibiotics as growth promoters for pigs (Cromwell, 2001).

Stage	Control	Antibiotic	Improvement (%)
Starting phase (7-25 kg)			
Daily gain (kg)	0.39	0.45	16.4
Feed/Gain	2.28	2.13	6.9
Growing phase (17-49 kg)			
Daily gain (kg)	0.59	0.66	10.6
Feed/Gain	2.91	2.78	4.5
Grow/Finish phase (24-89 kg)			
Daily gain (kg)	0.69	0.72	4.2
Feed/Gain	3.30	3.23	2.2

Metabolic effect theory (Hays, 1991). Nutrient sparing effect is the result of a shift in intestinal bacteria populations that result in increased nutrient availability and utilization by the animal (Hays, 1969). This theory also includes the decrease in total intestinal bacteria that leads to less competition for limited nutrients resulting in more nutrients available for the animal as well as a thinning of intestinal lining improving nutrient absorption (Visek, 1978). Subclinical disease control is based on the fact that antimicrobials decrease or kill bacteria and that unhealthy animals do not grow as well as healthy animals (Stahly, 1996). Metabolic effect theory is that antimicrobials cause changes in the microbial population of the intestinal tract leading to an alteration of growth factors/hormones produced by the host animal, for example, increased protein accretion (Moser *et al.*, 1980).

Unfortunately, with the good, comes the bad. Antibiotic use is under review due to concerns that food animal antibiotic resistance might be transferred to humans (Swann, 1969; NRC, 1999; CAST, 1981). Therefore, the animal agriculture industry is and has adopted policies outlining prudent antibiotic use (Sunberg, 2003; Szkotnicki, 2003) and is actively investigating alternatives to the use of antibiotics for growth promotion.

Management

The first option to producers for achieving the beneficial effects of growth promoting antibiotics without using them is to improve management and the health status of the animal. As mentioned previously, antibiotics, as do most feed additives, generally work better in environments that are less than ideal. Work conducted by Iowa State University demonstrated that healthy animals can outperform antigen challenged (unhealthy) pigs (Stahly, 1996). Anything the producer can do to improve the environment and health status of the herd will greatly increase performance and decrease the need for antibiotics or other feed additives.

Management factors to consider are, but not limited to, implementing all in/all out techniques, new production systems like wean to finish barns, optimal ventilation rates, increased focus on weaning weights, and better genetics. Increasing sanitation and biosecurity (controlling animal, people and feed flow) will also limit possible pathogen transmission through a facility and thereby increase animal health. Additionally, proper vaccine management can improve the ability of the pig to ward off disease and when applied to the breeding herd, can increase weanling pig performance via optimal passive immunity via colostrum (Doyle, 2002; Lawrence and Hahn, 2001).

Outside of feed additives, diet manipulation can also be an effective way to decrease the negative performance experienced when antibiotics are removed from the diet. Europe has led this area of research and has shown that restrictive feeding, low protein diets and fermented feeds can be appropriate ways to lessen the problems associated with the removal of antibiotics, especially in weanling pigs (Kjeldsen, 2002).

Organic Acids

Organic acids have been shown to improve animal performance, especially during stressful periods, such as weaning (Partanen, 2001; Ravindran and Kornegay, 1993; Roth *et al.*, 1998). The acids and salt forms of the acids, formic, lactic, sorbic, fumaric, citric and have been the primary acids of choice. Acids are theorized to improved animal performance by lowering intestinal pH, which increases protein digestion and mineral absorption as well as inhibiting pathogenic bacteria.

Some of the first trials on feeding organic acids occurred in the 1960's when lactic acid was fed to weanling pigs (Burnett and Hanna, 1963). Lactic acid was chosen due to the fact that lactic acid is formed by intestinal bacterial fermentation of lactose in the stomach and is consider important in protein enzyme activation and for lowering the population of *E. coli* in suckling piglets. Since these early trials, numerous trials have been conducted to study the effect of dietary organic acids in weaned and grow-finish pigs (Partanen, 2001; Ravindran and Kornegay, 1993; Walsh *et al.*, 2004d) (Table 2). More recent research has focused on the effect of various acidifiers on intestinal microflora (Canibe *et al.*, 2001; Palacios *et al.*, 2004; Walsh *et al.*, 2004a, b) as well as their effect on general growth performance (Giesting *et al.*, 2004; Mroz, 2003; Walsh *et al.*, 2004c).

The antibacterial activity of organic acids is based on pH depression and the ability of the acid to dissociate which is determined by the acid's pKa-value. Undissociated organic acids are lipophilic and therefore can diffuse across the cell membrane and enter the cell. The lower the extracellular pH, the more undissociated acid will enter the cell. Once inside the cell, the more alkaline (neutral) environment will cause the acid to dissociate which disrupts the cell function by interfering with the proton pump (Cherrington *et al.*, 1991). In general, organic acids pKa are between 3 – 5 and hence work more efficiently in the stomach, and become more dissociated in the lower intestinal tract (Thompson and Hinton, 1997) (Figure 1).

The killing ability of an organic acid depends, like a disinfectant, on the target organism(s), time of exposure, concentration and type of acid. Gram negative bacteria are more sensitive to acids that are less than 8 carbons in length compared to longer chained bacteria; gram positive bacteria are generally more resistant to acidic pH.

Table 2. Efficacy of acidifiers as growth promoters for pigs based on meta-analysis of published data (Partanen, 2001).

	Formic acid	Fumaric acid	Citric Acid	K-diformate
Experiments, N	6	18	9	3
Observations	10	27	19	13
Acid levels, %	0.3-1.8	0.5-2.5	0.5-2.5	0.4-2.4
Feed Intake, (lb/d)				
Control	1.47 ± 0.2	1.35 ± 0.33	1.18 ± 0.61	1.68 ± 0.02
Treatment	1.56 ± 0.16	1.35 ± 0.33	1.16 ± 0.66	1.81 ± 0.08
Gain, (lb/d)				
Control	0.85 ± 0.14	0.79 ± 0.22	0.84 ± 0.27	1.05 ± 0.01
Treatment	0.94 ± 0.14	0.82 ± 0.22	0.87 ± 0.28	1.18 ± 0.06
Feed/Gain				
Control	1.64 ± 0.13	1.59 ± 0.16	1.67 ± 0.25	1.60 ± 0.02
Treatment	1.60 ± 0.14	1.55 ± 0.14	1.60 ± 0.24	1.54 ± 0.04

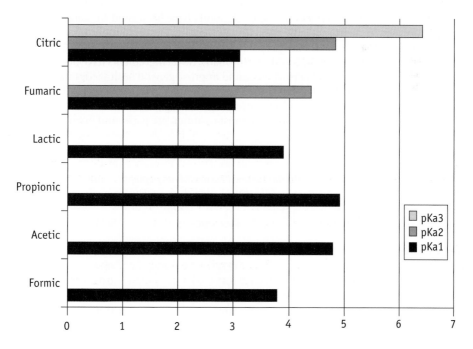

Figure1. pKa of various organic acids (adapted from Mroz, 2003).

Prebiotics

Prebiotics are indigestible carbohydrates that can enhance the performance of animals by being a readily available nutrient source for intestinal bacteria which may supply nutrients to the host animal and more importantly, by stimulating growth of beneficial intestinal bacteria, like *Lactobacillus* and *Bifidobacterium* (Table 3). There are many prebiotics available, but the following are the most common: oligofructose, fructooligosaccharide (FOS), inulin, chicory, and lactulose (Doyle, 2002;Patterson and Burkholder, 2003). Additionally, due their ability to be preferentially used by beneficial bacteria, prebiotics have also been used along with probiotics (direct fed microbials), a practice referred to as synbiotics (Patterson and Burkholder, 2003).

Many research trials have been conducted investigating the effect of prebiotics on the performance and their effect on health and microbial population of livestock. Most studies have indicated that prebiotics can alter the intestinal population by increasing the number of beneficial bacteria (Kerley and Allee, 2001;Smiricky-Tjardes *et al.*, 2003) or decreasing the population of potentially pathogenic bacteria (Letellier *et al.*, 1999; Naughton *et al.*, 2001); their effect on animal performance has been inconsistent (Turner *et al.*, 2001).

Novel Ingredients (nutraceuticals)

Novel ingredients, which is currently the preferred term of the US regulatory agencies, can be herbs, spices, plant extracts, botanicals or essential oils, and have traditionally been referred to as nutraceuticals (Wang *et al.*, 1998). These compounds have been used for centuries for their medicinal properties to cure practically everything (Table 4). The main mode of action for these novel ingredients is their antioxidant properties and their antiviral and antibacterial effects (Table 5) (Kamel, 2000; Lis-Balchin, 2003). Additionally, novel ingredients have been reported to stimulate appetite, increase animal performance and improve general health (Kis and Bilkei, 2003; Sims *et al*, 2004; Utiyama *et al.*, 2004) but these improvements can be variable (Lis-Balchin, 2003; Turner, *et al*, 2001). There are a tremendous variety of novel ingredients available to marketplace, most being labeled as flavoring agents, but the most common are garlic and oregano.

Table 3. Effect of a prebiotic on growth performance and ileal microflora populations in nursery pigs (adapted from Mathew et al., 1997).

	Control	Prebiotic	SEM
Feed intake, (kd/d)	0.79	0.90	0.06
Gain, (kg/d)	0.33	0.37	0.14
Gain/Feed	0.41	0.41	0.03
Lactobacilli (Log_{10}/g)	8.57	8.81	0.52
Streptococci (Log_{10}/g)	9.73	8.69	0.40
E. coli (Log_{10}/g)	7.70	6.15	0.62

Table 4. Suggested functions of Chinese herbal medicines (Wang et al., 1998).

Nutrients
Appetizer and flavor
Pigmentation
Antimold growth
Antioxidants
Enhancement of immune function
Hormone-like effects
Vitamin-like effects
Antistress and adaptation
Antimicrobial effects
Anthelmintic effects
Metabolic regulation

Table 5. Minimum inhibitory concentration of various novel ingredients and an antibiotic against common food pathogens (adapted from Kamel, 2000).

Oil	E. coli	Salmonella typhimurium	Campylobacter	Clostridia perfingens
Oregano	400	400	400	800
Cinnamon	400	400	200	400
Garlic	500	500	NT	100
Olaquindox	10	20	20	NT

All values are ppm.

Direct fed microbials (probiotics)

In the US, the term probiotic has been replaced by the terminology – direct fed microbial (DFM). DFM are labeled as "a source of live (viable), naturally occurring microorganisms." This is an important statement as true DFM cannot make any performance claim, must be alive, and cannot be genetically modified or patented due to their requirement of being naturally occurring. Additionally, DFM cannot be bacterial strains that have been selected to produce antibiotics (AAFCO, 2004) (Table 6).

The majority of the DFM studied have been *Lactobacillus*, *Bacillus* and *Enterococcus*, yeast (*Saccharomyces cerevisiae*) and combinations of these bacteria. *Bacillus* are gram-positive bacteria that form spores that are very stable to environmental extremes of heat, moisture and pH. The spores germinate into active vegetative cells when ingested by the animal. Lactic acid bacteria (LAB) are gram-positive rods or cocci. They are effective due to the ability to produce lactic acid, which, in turn, can suppress pathogenic bacteria. There are three types of LAB bacteria: *Lactobacillus*, *Bifidobacterium* and *Enterococcus* (previously know as *Streptococcus*). Yeast are fungi and have various beneficial effects on the gastrointestinal ecosystem (Risley, 1992).

Table 6. List of microorganisms approved for use in United States animal feeds (AAFCO, 2004).

Bifidobacterium adolescentis	Lactobacillus fermentum	Pediococcus acidilactici
Bifidobacterium animalis	Lactobacillus helveticus	Pediococcus cerevisiae
Bifidobacterium bifidum	Lactobacillus lactis	(damnosus)
Bifidobacterium longum	Lactobacillus plantarum	Pediococcus pentosaceus
Bifidobacterium infantis	Lactobacillus reuterii	Bacteriodes amylophilus
Bifidobacterium thermophilum	Enterococcus cremoris	Bacteriodes capillosus
Lactobacillus brevis	Enterococcus diacetylactis	Bacteriodes ruminocola
Lactobacillus buchneri	Enterococcus faecium	Bacteriodes suis
Lactobacillus bulgaricus	Enterococcus intermedius	Propionibacterium freudenreichii
Lactobacillus casei	Enterococcus lactis	Propionibacterium shermanii
Lactobacillus cellobiosus	Enterococcus thermophilus	Saccharomyces cerevisiae
Lactobacillus curvatus	Leuconostoc mesenteroides	
Lactobacillus delbruekii		

The premise behind why DFM work is that there is a critical balance between beneficial bacteria and potentially pathogenic bacteria. When this balance is disrupted via a stress (weaning, environmental changes, dietary changes, diseases), the health and performance of the animal suffers (Kung, 1992). During these times, a DFM can restore this balance and animal performance (Anderson et al., 2000).

Besides maintaining a balance between beneficial bacteria and potentially pathogenic bacteria, DFM have been reported to improve performance and health of animals by (Fuller, 1989; NFIA, 1991):

1. competing against pathogenic bacteria for nutrients in the gut;
2. competing with pathogens for binding sites on the intestinal wall;
3. producing compounds that are toxic to pathogens;
4. stimulating the immune system so the host is ready to fight an invading pathogen.

Considerable research has been devoted to determining the benefits of using a DFM and the results have been somewhat variable (Simon et al. 2003; Turner et al., 2001). However, as we learn more about how and why DFM function, more consistent performance benefits are being reported (Bontempo et al., 2004; Garcia et al., 2004a, b; Kritas and Morrison, 2004; Min et al., 2004; Murry et al., 2004; Parrott and Rehberger, 2004; Risley, 2003) (Figure 2, 3).

Competitive exclusion, an emerging technology, utilizes a complex mixture of bacteria obtained from a specific pathogen free animal (Nurmi and Rantala, 1973). This culture is typically obtained from the cecal contents or cecal mucosal lining. Research in the area of competitive exclusion indicates that this technology can improve food safety and animal performance (Blankenship et al., 1993; Harvey et al., 2003; Stern et al., 2001) (Table 7). However, true competitive exclusion products are considered a drug by the FDA-CVM, and hence must be approved through a New Animal Drug Application (NADA), which has greatly hampered efforts of bringing this technology to market.

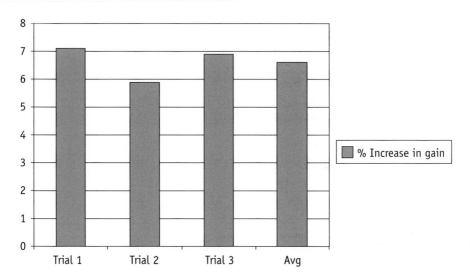

Figure 2. Summary of 3 trials investigating the effect of a Bacillus-based direct fed microbial on weanling pig performance in diets not containing antibiotics (adapted from Risley, 2003).

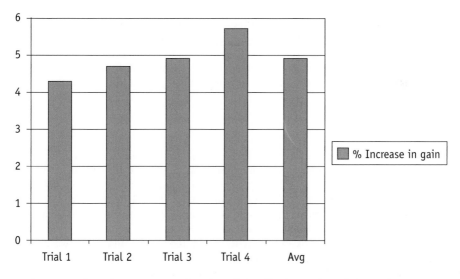

Figure 3. Summary of 4 trials investigating the effect of a bacillus-based direct fed microbial on weanling pig performance in diets containing antibiotics (adapted from Risley, 2003).

Table 7. Results from 4 farms using a competitive exclusion product to decrease production costs associated with E. coli *(adapted from Harvey et al., 2003)*

Farm	Treatment	Number of Pigs	Mort + Cull, %	Medication Cost/pig
Farm A	Control	3,242	9.06	0.695
	CE	10,402	2.80	0
	Difference		6.16	0.695
Farm B	Control	6,318	3.33	0.11
	CE	4,900	2.54	0
	Difference		0.79	0.11
Farm C	Control	3,068	3.30	0.28
	CE	3,127	2.45	0
	Difference		0.85	0.28
Farm D	Control	1,331	9.00	0.40
	CE	1,288	4.20	0
	Difference		4.80	0.40

Conclusion

Many antibiotic alternatives are available on the marketplace, such as organic acids, prebiotics, novel ingredients and direct fed microbials. These alternatives can have variable results depending on the environment in which they are used and cannot be considered a complete replacement for growth promoting antibiotics but are only tools to be used with good management techniques. As producers consider various alternatives to antibiotics, they will find it increasingly more important to grasp a broader knowledge of how to evaluate, select and utilize the many feed additives now available.

References

AAFCO (Association of American Feed Control Officials), 2004. AAFCO Official Publication. P.O. Box 478, Oxford, IN.

Anderson, D.B., V.J. McCracken, R.I. Aminov, J.M. Simpson, R.I. Mackie, M.W.A. Verstegen and H.R. Gaskins, 2000. Gut microbiology and growth-promoting antibiotics in swine. Nutrition and Abstracts and Reviews. Series B, Livestock Feeds and Feeding 70, 101-108.

Blankenship, L.C., J.S. Bailey, N.A. Cox, N.J. Stern, R. Brewer and O. Williams, 1993. Two-step mucosal competitive exclusion flora treatment to diminish salmonellae in commercial broiler chickens. Poult Sci. 72, 1667-72.

Bontempo, V., A. Di Giancamillo, C. Domeneghini, M. Fava, C. Bersani, R. Paratte, E. chevaux, V. Dell'Orto and G. Savoini, 2004. Effects of probiotic supplementation on gut histometry and fecal microflora in weaned pigs. J. Anim. Sci. 82(Suppl 1), M102.

Burnett, G.S., and J. Hanna, 1963. Effect of dietary calcium lactate and lactic acid on faecal *Escherichia coli* counts in pigs. Nature 79, 815.

Canibe, N., S.H. Stein, M. Overland and B.B. Jensen, 2001. Effect of K-diformate in starter diets on acidity, microbiota and the amount of organic acids in the digestive tract of piglets and on gastric alterations. J. Anim Sci. 79, 2123-2133.

CAST, 1981. Antibiotics in animal feeds. Report 88. Council for Agricultural Science and Technology, Ames, IA.

Cherrington, C.A., M. Hinton, G.C. Mead and I. Chopra, 1991. Organic acids: chemistry, antibacterial activity and practical applications. Adv Microb. Physiol. 32, 87-108.

Cromwell, G.L., 2001. Antimicrobial and promicrobial agents. In: Swine Nutrition Second Edition. A.J. Lewis and L.L. Southern (editors). CRC Press, Boca Raton, FL.

Doyle, E., 2002. Alternatives to antibiotic use for growth promotion in animal husbandry. A review of the scientific literature. Food Research Institute, Madison, WI.

Fuller, R., 1989. Probiotics in man and animals. J. Appl. Bacteriol. 66, 365-378.

Giesting D. W., M. J. Pettitt, and E. Beltranena. 2004. Evaluation of organic acid blends and antibiotics for promoting growth of young pigs. J. Anim Sci. 82(Suppl 1), M100.

Gracia, M.I., R.M. Engberg, A.E. Espinet, M. Cores and F. Baucells, 2004a. Bioefficacy of probiotics in broiler diets. J. Anim. Sci. 82(Suppl 1), W37.

Gracia, M.I., S. Hansen, J. Sanchez and P. Medel, 2004b. Efficacy of addition of *B. licheniformis* and *B. subtilis* in pig diets from weaning to slaughter. J. Anim. Sci. 82(Suppl 1), M105.

Harvey, R.B., R.C. Ebert, C.S. Schmitt, K. Andrews, K.J. Genovese, R.C. Anderson, H.M. Scott, T.R. Callaway and D.J. Nisbet, 2003. Use of a porcine-derived, defined culture of commensal bacteria as an alternative to antibiotics to control *E. coli* disease in weaned pigs: field trial results. 9th International Symposium on Digestive Physiology in Pigs 1, 72-74.

Hays, V.W., 1969. Biological basis for the use of antibiotics in livestock production. In: The use of drugs in animal feeds. National Academy of Science- National Research Council Pub. No. 1679, Washington D.C., pp. 11.

Hays, V.W., 1981. The Hays Report. Rachelle Laboratories, Inc. Long Beach, CA.

Hays, V.W., 1991. Effects of antibiotics. In: Growth Regulation in Farms Animals. Advances in Meat Research. Volume 7. A.M. Pearson and T.R. Dutson (editors). Elsevier Applied Science, New York,

Kamel, C., 2000. A novel look at a classic approach of plant extracts. Feed Mix Special, pp. 19.

Kerley J.S. and G. Allee, 2001. Fermenting dietary fiber: When and how to maximize fructooligosaccharides benefits. Feed Management.

Kirchgessner, M. and F.X. Roth, 1982. Fumaric acid as a feed additive in pig nutrition, Pig News Info 3, 259.

Kis, R.K. and G. Bilkei, 2003. Effect of a phytogenic feed additive on weaning-to-estrus interval and farrowing rate in sows. J. Swine Health Prod. 11, 296-299.

Kjeldsen, N. 2002. Producing pork without antibiotic growth promoters: the Danish experience. Advances in Pork Production. 13, 107-115.

Kritas, S.K. and R.B. Morrison, 2004. Can probiotics substitute fro sub therapeutic antibiotics? A field evaluation in a large pig nursery. 18th IPVS Congress, Hamburg, Germany 2, 739.

Kung, L. Jr., 1992. Direct-fed microbial and enzyme feeds additives. In: 1993 Direct-fed Microbial, Enzyme, and Forage Additive Compendium, Miller Publishing Co., Minnetonka, MN, pp. 17-21.

Lawrence, B. and J. Hahn, 2001. Swine feeding programs without antibiotics. 62nd Minn Nutr. Conf.

Letellier A., S. Messier, L. Lessard and S. Quessy, 1999. Assessment of different treatments to reduce *Salmonella* in swine. 3rd International symposium on the epidemiology and control of *Salmonella* in pork. Washington, D.C.

Lis-Balchin, M., 2003. Feed additives as alternatives to antibiotic growth promoters: botanicals. 9th International Symposium on Digestive Physiology in Pigs 1, 333-352.

Mathew, A.G., C.M. Robbins, S.E. Chattin and J.D. Quigley III, 1997. Influence of galactosyl lactose on energy and protein digestibility, enteric microflora, and performance of weanling pigs. J. Anim. Sci. 75, 1009-1016.

Min, B.J., O.S. Kwon, K.S. Son, J.H. Cho, W.B. Lee, J.H. Kim, B.C. Park and I.H. Kim, 2004. The effect of *bacillus* and active yeast complex supplementation on the performance, fecal bacillus counts and ammonia nitrogen concentrations in weaned pigs. J. Anim. Sci. 82(Suppl 1), M104.

Moore, P.R., A. Evenson, T.D. Luckey, E. McCoy, C.A. Elvehjen and E.B. Hart, 1946. Use of sulfasuxidine, streptothricin and streptomycin in nutritional studies with the chick. J. Bio. Chem.165, 437.

Moser, B.D., E.R. Peo and A.J. Lewis, 1980. Effect of carbadox on protein utilization in the baby pig. Nutr. Rep. Int. 22, 949.

Mroz, Z., 2003. Organic acids of various origin and physio-chemical forms as potential alternatives to antibiotic growth promoters for pigs. 9th International Symposium on Digestive Physiology in Pigs 1, 267-293

Murry, A.C., A. Hinton and R.J Buhr, 2004. Effect of a probiotic containing two *Lactobacillus* strains on growth performance and population of bacteria in the ceca and carcass rinse of broiler chickens. J. Anim. Sci 82(Suppl 1), W38.

NAHMS (National Animal Health Monitoring System), 2000. Part II: Reference of swine health and health management in the United States.

Naughton, P.J., L.L. Mikkelsen and B.B. Jensen, 2001. Effects of nondigestible oligosaccharides on *Salmonella enterica* sevoar *typhimurium* and nonpathogenic *Escherichia coli* in the pig small intestine *in vitro*. Appl. Environ. Micro. 67, 3391-3395.

NFIA (National Feed Ingredient Association), 1991. Direct fed microbials in animal production. West Des Moines, Iowa.

NRC (National Research Council), 1999. The use of drugs in food animals: Benefits and risks. National Academy Press, Washington D.C.

Nurmi E. and M. Rantala, 1973. New aspects of *Salmonella* infection in broiler production. Nature 241, 210-211.

Palacios, M.F., E.A. Flickinger, C.M. Grieshop, C.T. Collier and J.E. Pettigrew, 2004. Effects of lactic acid and lactose on the digestive tract of nursery pigs. J. Anim. Sci. 82 (Suppl 1), 250.

Parrott, D.S. and T.G. Rehberger, 2004. Isolation of *Bacillus* strains to inhibit pathogenic *E. coli* and enhance weanling pig performance. J. Anim Sci. 82(Suppl 1), M106.

Partanen, K., 2001. Organic acids – their efficacy and modes of action in pigs. In: Gut Environment of Pigs. A. Piva, K.E. Bach Knudsen and J.E. Lindberg (editors). Nottingham University Press, Nottingham, United Kingdom. pp. 201-218.

Patterson, J.A. and K.M. Burkholder, 2003. Prebiotic feed additives: Rationale and use in pigs. 9th International Symposium on Digestive Physiology in Pigs 1, 319-331.

Ravindran, R and E.T. Kornegay, 1993. Acidification of weaner pig diets: A review. J. Sci Food Agric. 62, 313-322.

Risley, C.R., 1992. An overview of basic microbiology. In: Direct-fed Microbial, Enzyme, and Forage Additive Compendium, Miller Publishing Co., Minnetonka, MN, pp. 11-13.

Risley, C.R. 2003. Bacillus-based DFM and weanling pig performance. Feed Management.

Roth, F.X., W. Windisch and M. Kirchgessner, 1998. Effect of potassium diformate (Formi™ LHS) on nitrogen metabolism and nutrient digestibility in piglets at graded dietary lysine supply. Agribiol Res. 51,167-175.

Simon, O., W. Vahjen and L. Scharek, 2003. Micro-organisms as feed additives - probiotics. 9th International Symposium on Digestive Physiology in Pigs 1, 95-318.

Sims, M., P. Williams, M. Frehner and R. Losa, 2004. CRINA® poultry essential oils and BMD in he diet of broilers exposed to *Clostridium perfringens*. J. Anim Sci. 82(Suppl 1), 674.

Smiricky-Tjardes, M.R., C.M. Grieshop, E.A. Flickinger, L.L. Bauer, and G.C. Fahey, Jr., 2003. Dietary galactooligosaccharides affect ileal and total-tract nutrient digestibility, ileal and fecal bacterial concentrations, and ileal fermentative characteristics of growing pigs. J. Anim. Sci. 81, 2535-2545.

Stahly, T., 1996. Impact of immune system activation on growth and optimal regimens of pigs. In: Recent advances in animal nutrition. P.C. Garnsworthy, J. Wiseman and W. Haresign (editors), Nottingham University Press, Nottingham, pp. 197-206.

Stern N.J., N.A. Cox, J.S. Bailey, M.E. Berrang and M.T. Musgrove, 2001. Comparison of mucosal competitive exclusion and competitive exclusion treatment to reduce *Salmonella* and *Campylobacter* spp. colonization in broiler chickens. Poult. Sci. 80, 156-60.

Sunberg, P., 2003. Antimicrobials: Can we continue using them? U.S. Checkoff programs and perspectives. Advances in Pork Production 14, 73-81

Swann, 1969. Joint committee on the use of antibiotics in animal husbandry and veterinary medicine. Her Majesty's Stationery Office, London.

Szkotnicki, J., 2003. Bugs, drugs and you. Advances in Pork Production 14, 83-89.

Thompson, J.L.and M. Hinton, 1997. Antibacterial activity of formic and propionic acids in the diet of hens on salmonella in the crop. Brit. Poul. Sci. 38, 59-65.

Turner, J.L., S.S. Dritz and J.E. Minton, 2001. Review: Alternatives to conventional antimicrobials in swine diets. Prof. Anim. Scientist 17:217-226.

Utiyama, C.E., L.L. Oetting, P.A. Giani, U.S. Ruiz and V.S. Miyada, 2004. Antimicrobials, probiotics, prebiotics and herbal extracts as growth promoters on performance of weanling pigs. J. Anim Sci. 82(Suppl 1), M103.

Visek, W.J., 1978. The mode of growth promotion by antibiotics. J. Anim. Sci. 46, 1447.

Wang, R., D. Li and S. Bourne, 1998. Can 2000 years of herbal medicine history help us solve problems in the year 2000? In: Biotechnology in the Feed Industry. T.P. Lyons and K. Jacques (editors). Nottingham University Press, Nottingham, United Kingdom. pp. 273.

Walsh, M., D. Sholly, K. Saddoris, R. Hinson, A. Sutton, S. Radcliffe, B. Harmon, R. Odgaard, J. Murphy and B. Richert, 2004a. Effects of diet and water acidification on weanling pig growth and microbial shedding. J. Anim. Sci. 82 (Suppl 1), 245.

Walsh, M., D. Sholly, K. Saddoris, R. Hinson, A. Yager, A. Sutton, S. Radcliffe, B. Harmon and. B. Richert. 2004b. Effects of diet acidification and antibiotics on weanling pig growth and microbial shedding. J. Anim. Sci. 82 (Suppl), 246.

Walsh, M., D. Sholly, K. Saddoris, R. Hinson, A. Yager, A. Sutton, S. Radcliffe, B. Harmon and. B. Richert. 2004c. Effects of diet acidification and buffering capacity on weanling pig performance. J. Anim. Sci. 82 (Suppl 1), 244.

Walsh, M.C., B.T. Richert, A.L. Sutton, J.S. Radcliffe and R. Odgaard. 2004d. Past, present, and future uses of organic and inorganic acids in nursery pig diets. Amer. Assoc. of Swine Veterinarians. pp. 155-158.

Continuous use of a dry disinfectant/antiseptic to enhance health and well being in food-animal production facilities

Bud G. Harmon
Purdue University, West Lafayette IN, USA

A product in dry form manufactured from processed mineral salts was sprayed as a disinfectant/antiseptic (D/A) in poultry and swine facilities. Processing the D/A greatly increases surface area and drying capacity as it absorbs 8 times its weight in water. In extensive studies with individual microorganisms, D/A has been shown to effectively kill viruses, bacteria, fungi, protozoa, and insect larvae. The D/A is designed to treat the environment and not the animals. D/A is sprayed throughout the environment usually on a weekly basis and is safe for animals and humans. Studies reported here were conducted to evaluate the effectiveness of the D/A to (1) reduce environmental bacterial load in an egg laying facility and (2) influence productivity in farrowing units.

In the first study, dry D/A was sprayed throughout the egg transport and elevator equipment area and a small section of the housed layers of a single production room within a large egg laying complex. Hens had been in production for about 8 months prior to initiating the study. Dry D/A was sprayed at 50 gm/sq m of space and repeated for a total of 4 weekly sprayings. Bacterial sampling was made prior to the first spraying and weekly for 7 successive weeks. Eight surface samples were collected on each sampling day and tested for aerobic plate counts, coliforms, and *Salmonella*. Aerobic counts in $\log_{10}$ cfu/6.5 cm^2 were: 0 weeks, 6.13; 1 week, 6.21; 2 weeks, 6.01; 3 weeks, 5.78; post spraying at 4 weeks, 5.62; 5 weeks, 5.60; 6 weeks, 5.83; and 7 weeks, 5.65. Actual aerobic counts dropped by more than 70% during the 4 week spraying period and then stabilized over the next 4 weeks post spraying. Coliforms followed a similar pattern, but at a much lower count, measuring 2.50 $\log_{10}$ cfu/100 cm^2 prior to spraying and dropping to 1.9 $\log_{10}$ cfu/6.5 cm^2 at the termination of spraying a drop of 81% in actual count. *Salmonella* counts were extremely low throughout the study. Regular use of dry D/A reduced aerobic and coliform counts.

In a swine study, dry D/A was used as a hygiene product at the Purdue swine farrowing complex. Farrowing rooms were cleaned and sprayed with a liquid disinfectant between groups of sows. Experimental treatment rooms were sprayed with 50 gm/m^2 of dry D/A on the day prior to placing sows in the rooms and weekly thereafter until weaning. Rooms not sprayed with dry D/A constituted the control treatment. In total, 238 sows and litters were utilized in this study involving 7 groups of sows. Pigs were weaned at an average of 17.8 days of age. Standard management practices were used including antibiotic treatment at the discretion of the unit manager. Pigs from litters within the environment utilizing dry D/A were 0.5 lbs heavier at weaning. Number of pigs/litter treated was 3 times greater (0.88 *vs.* 0.28) in the control rooms and death loss as pigs/litter was 1.8 times greater (0.47 *vs.* 0.26) in the control rooms. Regular spraying of dry D/A in the environment until weaning, improved performance and reduced individual treatment of pigs and sows. The D/A originated in Denmark 35 years ago and is currently used in 45 countries to improve performance, health, well being, and reduce the need for medications.

Bovine colostrum as an alternative to feed additive in weaning diet improves gut health of piglets

I. Le Huërou-Luron, A. Huguet and J. Le Dividich
Institut National de la Recherche Agronomique - Unité Mixte de Recherches sur le Veau et le Porc, France

Adding colostrum in diets for weaner pigs improves feed intake and growth rate[1,2]. It also tends to improve the sanitary status of pigs reared in an "unclean" environment[1]. We hypothesize that bovine colostrum supplementation of the weaning diet could favour the maintenance of the small intestinal integrity.

At 21d of age, twelve piglets were weaned and fed either a standard diet (C group) or a standard diet supplemented with bovine colostrum (5% of dry matter, Col group) for 7 days. A third group of six piglets were sow-reared up to 28 d of age (S group). Piglets were reared in a high-sanitary environment. They were euthanized at 28 d of age. The proximal small intestine was collected for histological measurements and brush-border enzyme activity analysis. Data were analysed by variance analysis using the GLM procedure of SAS.

During the experimental period, food intake and growth rate were similar in both groups of weaned pigs. In weaned piglets, pH of gastric contents was lower (P<0.01) in Col group (1.71) than in C group (2.55). The coliform counts in Col group were 17% lower than in C group (P<0.05), while no difference was found in lactobacilli counts, indicating that colostrum supplementation prevented proliferation of pathogenic bacteria. Compared to C group, supplementation with bovine colostrum enhanced the intestinal weight (33.6 *vs.* 31.4 g/kg BW; P<0.05). In Col group, villous height was 19% higher (P=0.08) than in the weaned C group, but was 22% lower (P<0.05) than in (S group). Lactase, maltase and aminopeptidase N activities, markers of digestive maturation of the small intestine, were not significantly different between S, C and Col groups, suggesting that the functional properties of the small intestinal mucosa did not seem to be affected by colostrum supplementation. In conclusion, supplementation of a weaning diet with a bovine colostrum improves gut health around weaning in pigs.

Acknowledgements
This study was supported by a grant from the french project «Porcherie Verte»

References
[1] Pluske J.R., G. Pearson, P.C.H. Morel, *et al.*, 1999. A bovine colostrums product in a weaner diet increases growth and reduces days to slaughter. In: P.D. Cranwell (ed.), Manipulating Pig Production VII, Australasian Pig Science Association, Werribee, Australia, pp. 256.

[2] Le Huërou-Luron I., A. Huguet, J. Callarec, T. Leroux and J. Le Dividich, 2004. Supplementation of a weaning diet with bovine colostrum increases feed intake and growth of weaned piglets. Journées Rech. Porcine 36, 103-108.

Production responses from antimicrobial and rendered animal protein inclusions in swine starter diets

B.V. Lawrence, S.A. Hansen and D. Overend
Hubbard Feeds Inc, Mankato, MN, USA

A 2 X 2 factorial experiment (288 Genetiporc pigs, 21-d of age, 4.18 ± 0.08 kg) evaluated the effect of diets with rendered animal proteins (RAP) *vs.* without (NORAP), where RAP consists of plasma, fishmeal, and poultry meal and antimicrobials (Tiamulin®, included at 35 g/ton and chlortetracycline, at 400 g/ton (MED) *vs.* unmedicated (NOMED)) on pig performance through 41-d post-weaning. Pigs were sorted visually by size to 32 pens (9 pigs/pen, 8 pens/treatment). Treatments were fed for 21-d (7-d of Phase 1 starter, 14-d of Phase 2 starter), followed by the feeding of a common corn-soy Phase 3 diet. Pen weights were recorded weekly, as was daily feed delivery. Pigs were individually weighed on d-0, 21, and 41 to evaluate dietary treatment on uniformity of growth and body weight. The coefficient of variation (CV) was the response criteria. Data were analyzed using the GLM procedures of SAS for main effects of MED and RAP inclusion and any interactions.

Gain was improved (P<0.05) for MED *vs.* NOMED pigs from d-0 to 35 (13.8%), and from d-0 to 41 (11.1%). The improvement in gain was associated with an improvement in intake (P<0.05), with a trend for the MED pigs to have higher intake from d-35 to 41. Gain:feed was greater (P<0.05) for the MED pigs from d-14 to 21 but was not different (P>0.10) during the remainder of the trial. There was no difference (P>0.10) in performance of the RAP *vs.* NORAP fed pigs. There were however, significant MED X RAP interactions. From d-0 to 7, gain and intake of the NOMED/NORAP and MED/NORAP fed pigs was not different (P>0.10) while gain and intake of the MED/RAP fed pigs was improved by 59.5 and 56.6% respectively compared with the NOMED/RAP fed pigs. The improvement in gain observed the first 7-d, continued through d-28 of the trial when gain increased by 18.6, 32.0, and 26.0% from d-7 to 14, 14 to 21, and 21 to 28 respectively for the MED/RAP *vs.* NOMED/RAP fed pigs. These increases in gain were associated with 16.0, 19.8, and 17.3% improvements (P<0.05) in feed intake. In contrast, through d-28, there was no MED influence on gain or intake of the NORAP pigs. There was no MED X RAP interaction for gain:feed during any period of the trial (P>0.10). However, small numerical differences in each period resulted in a significant positive effect of MED, on cumulative gain:feed for the RAP fed pigs while there was no effect of MED when RAP were excluded from the diets from d-0 to 21 post-weaning (P>0.01). Inclusion of RAP had no effect on uniformity of either growth or pig weight during the trial. MED inclusion however, significantly reduced (P<0.01) growth rate variance from d-0 to 21 and d-0 to 41 (CV = 19.9 NOMED *vs.* 15.3 MED, 17.2 NOMED *vs.* 12.9 MED), as well as pen uniformity measured at day 21 (CV = 13.5 NOMED *vs.* 10.9 MED) and day 41 (CV = 14.2 NOMED *vs.* 10.9 MED). The improved CV of the MED *vs.* NOMED pigs at d-41 suggests carryover benefits from the early medication program on performance of starter pigs. When diets contained RAP a greater benefit of MED against uniformity of growth and morphology was observed. These results indicate that RAP in starter diets can constrain digestive performance, enhancing the necessity for antimicrobials in post-weaning diets.

Section H.
Macromineral metabolism and production diseases

Strategies for controlling hypocalcemia in dairy cows in confinement and pasture settings

Jorge M. Sánchez[1] and Jesse P. Goff[2]
[1]Animal Nutrition Center-CINA, Universidad de Costa Rica, San Jose, Costa Rica; [2]USDA-ARS, National Animal Disease Center, Ames, IA 50010, USA

Introduction

Milk fever is a disorder affecting about 6% of dairy cows each year in the United States. Sub-clinical milk fever, defined as blood calcium concentration falling below 8 mg Ca/dl, occurs in up to 50% of older cows during the days immediately following calving (see Horst *et al.*, 2004 in these proceedings). The decline in blood calcium concentration around parturition represents a breakdown in the calcium homeostatic mechanisms of the body. Blood Ca in the adult cow is maintained around 8.5-10 mg/dl. There are 3 g Ca in the plasma pool and only 8-9 g Ca in all the extracellular fluids (outside of bone) of a 600 kg cow. The fluid within the canaliculi of bone may contain another 6-15 g Ca; the size of this Ca pool being dependent on the acid-base status of the animal; larger during acidosis and smaller during alkalosis. Dairy cows producing colostrum (containing 1.7-2.3 g Ca/kg) or milk (containing 1.1 g Ca/kg) withdraw 20 -30 g Ca from these pools each day in early lactation. In order to prevent blood Ca from decreasing, which has a variety of severe consequences to life processes beyond parturient paresis, the cow must replace Ca lost to milk by withdrawing Ca from bone or by increasing the absorption of dietary Ca. While this is potentially damaging to bones (lactational osteoporosis typically results in loss of 9-13% of skeletal Ca in dairy cows, which is reversible in later lactation), the main objective - to maintain normocalcemia - can be achieved. Bone Ca mobilization is regulated by parathyroid hormone (PTH) which is produced whenever there is a decline in blood Ca. Renal tubular reabsorption of Ca also is enhanced by PTH. However the total amount of Ca that can be recovered is usually relatively small. A second hormone, 1,25-dihydroxyvitamin D, is required to stimulate the intestine to efficiently absorb dietary Ca. This hormone is made within the kidney from vitamin D in response to an increase in blood PTH. Put simply, hypocalcemia and milk fever occur when cattle do not extract enough Ca from their bones and diet to replace the Ca lost to milk. Several nutritional factors are involved in the breakdown of Ca homeostasis that results in milk fever.

Metabolic alkalosis causes disruption of parathyroid hormone function

Metabolic alkalosis predisposes cows to milk fever and subclinical hypocalcemia (Craige, 1947). Metabolic alkalosis blunts the response of the cow to PTH (Gaynor *et al.* 1989; Phillippo *et al*, 1994; Goff *et al.*, 1991). *In vitro* studies suggest the conformation of the PTH receptor is altered during metabolic alkalosis rendering the tissues less sensitive to PTH (Bushinsky, 1996; Martin *et al.*, 1980). Lack of PTH responsiveness by bone tissue prevents effective utilization of bone canaliculi fluid Ca, sometimes referred to as osteocytic osteolysis, and prevents activation of osteoclastic bone resorption. Failure of the kidneys to respond to PTH reduces renal reabsorption of Ca from the glomerular filtrate. More importantly, the kidneys fail to convert 25-hydroxyvitamin D to 1,25-dihydroxyvitamin D. Therefore enhanced intestinal absorption of dietary Ca that normally would help restore blood Ca to normal, fails to be instituted. Metabolic alkalosis is largely the result of a diet that supplies more cations (K, sodium (Na), Ca, and Mg) than anions (chloride (Cl), sulfate (SO_4), and phosphate (PO_4)) to the blood. In simplest terms, a disparity in electrical charge occurs in animals fed these diets because a greater number of positively charged cations enter the blood than negatively charged anions. To restore electroneutrality to this positively charged blood, a positive charge in the form of a hydrogen ion (H^+) is lost from the blood compartment and the pH of the blood is increased (Stewart, 1983).

It must also be remembered that magnesium plays a role in calcium homeostasis. Hypomagnesemia affects Ca metabolism in two ways: 1. By reducing PTH secretion in response to hypocalcemia, and 2. by reducing tissue sensitivity to PTH. PTH secretion is normally increased greatly in response to even slight decreases in blood Ca concentration. However hypomagnesemia can blunt this response (NRC, 2000). Hypomagnesemia is also capable of interfering with the ability of PTH to act on its target tissues. When PTH binds its receptor on bone or kidney tissues, it normally initiates activation of adenylate cyclase, resulting in production of the second messenger, cyclic AMP. In some tissues, PTH-receptor interactions should also cause activation of phospholipase C, resulting in production of the second messengers diacylglycerol and inositol 1,4,5-triphosphate. Both adenylate cyclase and phospholipase C have a Mg^{++} binding site which must be occupied by a Mg ion for full activity (Rude, 1998). Field evidence suggests that blood Mg concentrations below 1.6 mg/dl in the periparturient cow will increase the susceptibility of cows to hypocalcemia and milk fever (van de Braak et al, 1987).

Strategy to prevent milk fever in confined cows: reducing diet cation-anion difference

In theory all the cations and anions in a diet are capable of exerting an influence on the electrical charge of the blood. The major cations present in feeds and the charge they carry are Na (+1), K (+1), Ca (+2), and Mg (+2). The major anions and their charges found in feeds are Cl (-1), SO_4 (-2), and phosphate (assumed to be -3). Cations or anions present in the diet will only alter the electrical charge of the blood if they are absorbed into the blood.

The difference between the number of cation and anion particles absorbed from the diet determines the pH of the blood. The cation-anion difference of a diet is commonly described in terms of mEq/kg (some authors prefer to use "mEq/100 g" diet) of just Na, K, Cl, and SO_4 (traditionally calculated on S% reported when diet is analyzed by wet chemistry) as follows:

Dietary Cation-Anion Difference (DCAD) = (mEq Na^+ + mEq K^+) - (mEq Cl^- + mEq S^{--}).

While DCAD equations provide a theoretical basis for dietary manipulation of acid-base status they are not necessary for formulation of mineral content of prepartum dairy cow rations because, with the exception of K and Cl, the rate of inclusion of the other macrominerals can be set at fixed rates.

For example, the NRC (2000) requirement for Na in the diet of a late gestation cow is about 0.12%. That is the amount that should be fed. At least two studies have clearly demonstrated that inclusion of Ca in the diet at NRC required levels or several fold above NRC required levels does not influence the degree of hypocalcemia experienced by the cow at calving (Goff and Horst,

1997; Beede *et al*, 2001). It appears from these studies that close-up diet Ca concentration should be maintained between 0.85 and 1.0% Ca.

To ensure adequate concentrations of Mg in the blood of the periparturient cow the dietary Mg concentration should be 0.35-0.4% to take advantage of passive absorption of Mg across the rumen wall.

Dietary P concentration should be fed at a level to meet the NRC requirement for P in the late gestation cow. This is generally about 0.4% P for most cows, though recent studies suggest this may overestimate the true requirement of the cow for dietary P (Peterson and Beede, 2002). A diet supplying more than 80 g P/day, or utilizing dietary P in the form of phosphoric acid as an anion to acidify the blood of the cow (Crill *et al.*, 1996), will block production of 1, 25-dihydroxyvitamin D and cause milk fever. Dietary S must be kept above 0.22% (to ensure adequate substrate for rumen microbial amino acid synthesis) but below 0.4% (to avoid possible neurological problems associated with S toxicity).

Now, with the exception of K and Cl, the "variables" in the various proposed DCAD equations have become "fixed". The key to milk fever prevention (at least with Holstein cows) is to keep K as close to the NRC requirement of the dry cow as possible (about 1.0% diet K). The key to reduction of subclinical hypocalcemia, not just milk fever, is to add Cl to the ration to counteract the effects of even low diet K on blood alkalinity. For formulation purposes the concentration of Cl required in the diet to acidify the cow is approximately 0.5% less than the concentration of K in the diet. In other words, if diet K can be reduced to 1.3% the Cl concentration of the diet should be increased to 0.8%. If dietary K can only be reduced to 2.0% the diet Cl would need to be roughly 1.5% to acidify the cow. This level of Cl in the diet is likely to cause a decrease in dry matter intake. Chloride sources differ in their palatability and since achieving low dietary K can be difficult it is prudent to use a palatable source of Cl when formulating the diet. Ammonium chloride (or ammonium sulfate) can be particularly unpalatable when included in rations with a high pH. At the higher pH of some rations the ammonium cation is converted to ammonia, which is highly irritating when smelled by the cow. Prilling the Cl (and SO_4) salts can reduce the unpleasant taste of the salts. In our experience hydrochloric acid has proved the most palatable source of anions. As with sulfuric acid, hydrochloric acid can be extremely dangerous to handle when it is procured as a liquid concentrate. Several companies now manufacture hydrochloric acid based anion supplements, which are safe to handle.

Urine pH of the cows provides a cheap and fairly accurate assessment of blood pH and can be a good gauge of the appropriate level of anion supplementation (Jardon, 1995) Urine pH on high cation diets is generally above 8.2. Limiting dietary cations will reduce urine pH only a small amount (down to 7.5-7.8). For optimal control of subclinical hypocalcemia the average pH of the urine of Holstein cows should be between 6.2 and 6.8, which essentially requires addition of anions to the ration. In Jersey cows the average urine pH of the close-up cows has to be reduced to between 5.8 and 6.3 for effective control of hypocalcemia. If the average urine pH is between 5.0 and 5.5, excessive anions have induced an uncompensated metabolic acidosis and the cows will suffer a decline in dry matter intake.

Strategy to prevent milk fever in pastured cows: Ca deficient diet to stimulate PTH secretion

Those areas of the world where grazing is the most efficient means of producing milk also tend to be areas where soil potassium is high (which promotes grass growth) which means that metabolic alkalosis is going to be a factor that will not be easily overcome. There is no practical way to add enough anions to the diet of these cows to overcome the metabolic alkalosis created by the high potassium forages. A different strategy can be employed successfully in some areas.

When cows are fed a diet that supplies less Ca than they require, the cows are in negative Ca balance. This causes a minor decline in blood Ca concentration stimulating PTH secretion, which in turn stimulates osteoclastic bone resorption and renal production of 1,25-dihydroxyvitamin D. The key is that prolonged (several weeks) exposure to elevated PTH will overcome the PTH resistance in the tissues associated with metabolic alkalosis. This increases bone Ca efflux and the intestine is ready to absorb Ca efficiently once it becomes available in the lactating cow ration. At parturition the cow's osteoclasts are already active and in high numbers and the lactational drain of Ca is more easily replaced from bone Ca. If provided with Ca in the lactation ration, the previous stimulation of enterocytes by 1,25-dihydroxyvitamin D will allow efficient utilization of dietary Ca and the cow avoids hypocalcemia (Green *et al.*, 1981; Goings *et al.*, 1974).

The absorbable Ca requirement (NRC, 2000) of the late gestation cow is from 14 g/day in Jerseys to about 22 g in large Holsteins. A truly low Ca diet capable of stimulating PTH secretion supplies considerably less absorbable Ca than required by the cow. A 600 kg cow consuming 13 kg DM, typical of confinement cow dry matter intakes, must be fed a diet that is less than.15% absorbable Ca if it is to provide less than 20 g available Ca/day. Low Ca diets have proved more practical under grazing situations. The availability of Ca from forages is just 30% according to the 2001 NRC. In these cases the total dry matter intake of pasture was 6-7 kg DM/day and the grasses being grazed were less than 0.4% Ca, which would provide < 28 g total Ca and somewhere around 9-10 g absorbable Ca/day. It is important to note that after calving the animal is switched to a high Ca diet. Grasses grown under tropical conditions may have lower calcium content than the same grasses grown in temperate conditions. For example, common grasses grown in Costa Rica under tropical conditions such as Kikuyu, African Star, and Brachiaria grasses are 0.37, 0.32, and 0.35% Ca respectively. Kikuyu grass grown under temperate conditions, such as New Zealand experiences, is typically 0.60% Ca. This makes institiution of a low calcium diet for prevention of milk fever more practical under tropical conditions.

Recently two methods have been developed to reduce the availability of dietary Ca for absorption. The first method involves incorporation of zeolite (a silicate particle) into the ration, which binds Ca and causes it to be passed out in the feces. At present the method is unwieldy because very large amounts of zeolite must be ingested each day (1 kg) and the effects of zeolite on P and trace mineral absorption are not clear (Thilsing-Hansen *et al.*, 2002) By chemically modifying the zeolite it is theoretically possible to increase the affinity and the specificity of the zeolite for Ca, which may allow it's practical use. The second method involves administration of vegetable oils which bind Ca to form an insoluble soap preventing absorption of diet Ca (Wilson, 2003). These have been successfully used in cattle fed reduced Ca diets containing 30-50 g/day. However they irreversibly bind enough dietary Ca to cause the reaction typically seen when diet Ca is <15 g absorbable Ca/day.

Conclusion

In animals fed large amounts of dry matter pre-calving in a confinement setting the best option for controlling hypocalcemia appears to be to reduce the offending cause of metabolic alkalosis. This is done thru better agronomy to control potassium intake, and by adding anions to the diet to acidify the cow's blood. In animals at pasture prevention of metabolic alkalosis is extremely difficult. In these settings the best approach may be to trick the parathyroid gland into early secretion of PTH to stimulate bone and kidneys prior to calving. The key is to reduce the available Ca in the diet to a level below the requirement of the cow for several weeks before calving. Tropical growing conditions aid this process and new technologies that bind dietary Ca may also be available in the future to make this approach more successful in temperate climates.

References

Beede, DK, Pilbean TE, Puffenbarger SM and R.J. Tempelman, 2001. Peripartum responses of Holstein cows and heifers fed graded concentrations of calcium (calcium carbonate) and anion (chloride) three weeks before calving. J Dairy Sci. 84 (Suppl 1), 83

Bushinsky, DA., 1996. Metabolic alkalosis decreases bone calcium efflux by suppressing osteoclasts and stimulating osteoblasts. Am. J. Physiol. 271 (1 Pt 2), F216-222.

Craige, AH and IV Stoll, 1947. Milk fever (parturient paresis) as a manifestation of alkalosis. Am. J. Vet. Res. 8, 168.

Crill, RL, Sanchez WK, Guy MA and LA Griffel, 1996. Should phosphorus be included in the dietary cation-anion difference expression? J. Dairy Sci. 85 (Suppl 1), 187.

Gaynor, PJ, Mueller FJ, Miller JK, Ramsey N, Goff JP and RL Horst, 1989. Parturient hypocalcemia in jersey cows fed alfalfa haylage-based diets with different cation to anion ratios. J. Dairy Sci. 72, 2525-2531.

Goff, JP and RL Horst, 1997. Effects of the addition of potassium or sodium, but not calcium, to prepartum ratios on milk fever in dairy cows. J. Dairy Sci. 80, 176-186.

Goff, JP, Horst RL, Mueller FJ, Miller JK, Kiess GA and HH Dowlen, 1991. Addition of chloride to a prepartal diet high in cations increases 1,25-dihydroxyvitamin D response to hypocalcemia preventing milk fever. J. Dairy Sci. 74, 3863-3871.

Goings, RL, Jacobson NL, Beitz DC, Littledike ET and KD Wiggers, 1974. Prevention of parturient paresis by a prepartum, calcium-deficient diet. J. Dairy Sci. 57, 1184-8.

Green, HB, Horst RL, Beitz DC and ET Littledike,1981. Vitamin D metabolites in plasma of cows fed a prepartum low-calcium diet for prevention of parturient hypocalcemia. J. Dairy Sci. 64, 217-26.

Jardon, P. Using urine pH to monitor anionic salt programs. Compend Contin Educ Pract Vet. 1995; 17, 860.

Martin, KJ, Freitag JJ, Bellorin-Font E, Conrades MB, Klahr S and E. Slatopolsky, 1980. The effect of acute acidosis on the uptake of parathyroid hormone and the production of adenosine 3', 5'-monophosphate by isolated perfused bone. Endocrinology 106, 1607-1611.

NRC, 2000. Nutrient Requirements of Dairy Cattle. Washington, D.C., National Academy Press.

NRC, 2001. Nutrient Requirements of Dairy Cattle. Washington, D.C., National Academy Press.

Peterson, AB and DK Beede, 2002. Periparturient responses of multiparous Holstein cows to varying prepartum dietary phosphorus. J. Dairy Sci. 85 (Suppl 1), 187.

Phillippo, M, Reid GW and IM Nevison, 1994. Parturient hypocalcaemia in dairy cows: effects of dietary acidity on plasma minerals and calciotrophic hormones. Res. Vet. Sci. 56, 303-309.

Rude, RK, 1998. Magnesium deficiency: a cause of heterogeneous disease in humans. J. Bone Miner. Res. 13, 749-758.

Stewart, PA, 1983. Modern quantitative acid-base chemistry. Can. J. Physiol. Pharmacol. 61, 1444-1461.

Thilsing-Hansen, T, Jorgensen RJ, Enemark JM and T. Larsen, 2002. The effect of zeolite A supplementation in the dry period on periparturient calcium, phosphorus, and magnesium homeostasis. J. Dairy Sci. 85, 1855-1862.

van de Braak, AE, van't Klooster AT and A. Malestein, 1987. Influence of a deficient supply of magnesium during the dry period on the rate of calcium mobilization by dairy cows at parturition. Res. Vet. Sci. 42, 101-108.

Wilson, G.F., 2003. Development of a novel concept (Calcigard) for activation of calcium absorption capacity and prevention of milk fever. Acta Vet. Scand. Suppl. 97, 77-82.

Phosphorus digestion and metabolism in ruminants: applications to production disease and environmental considerations

Ernst Pfeffer
Department of Animal Nutrition, University of Bonn, Endenicher Allee 15, D53115 Bonn, Germany

Introduction

Of the more than one hundred names found in the periodic system of elements today, only a dozen were known 350 years ago, among them carbon, sulphur, iron, copper, silver and gold. The term 'element' was not used in today's meaning and alchemists were convinced that they could, by experimentation, find the 'philosopher's stone' which should enable them to turn worthless materials into gold. One of these alchemists was Henning Brand in Hamburg who in 1669 heated concentrated urine without admitting air and found a snow-white substance which immediately burned out, thereby illuminating the dark room (van der Krogt, 2003; Childs, 2003). This property of giving light was the base for naming of the substance discovered by Brand, from the Greek words phos = light; and phero = to carry, to bring. Phosphorus (P) thereby was the first element to be identified in modern times. About a hundred years after Brand's discovery, the Swedish chemists Gahn and Scheele found calcium phosphate to be a major constituent of bone.

Today it is common knowledge that P, linked into phosphate (PO_4) or substances containing phosphate plays an essential role in metabolism of plants, animals and micro-organisms. P containing nucleic acids form the basis of genetics, nucleotides enable organisms to utilize chemically bound energy, phospholipids are essential compounds of cellular membranes and, in vertebrates, calcium phosphates as apatites give strength to the skeleton. Inorganic phosphate participates in buffering the pH of biological fluids.

The essentiality of P for animals is beyond question and this element may be limiting animal performance in some areas of the world. Until recently, scientifically based recommendations concerning the supply of P to farm animals appear to have been guided solely by the goal to avoid any risk arising from potential shortages in P intake.

Insufficient replacement of nutrients extracted by plants from the soil of the fields was a major reason for low crop yields with the consequence of increasing poverty and famines at regular intervals in Europe over long periods. In the 19th century, acidulating bones with the aim of increasing the solubility of phosphate was empirically experienced in several places and finally the industrial production of superphosphate, predominantly from bones, was developed. Considerable quantities of plant nutrients were transported from South America to Europe in the form of Guano, excreta of birds on the Peruvian islands, rich in salts of nitric acid and phosphoric acid.

With increasing demands of sustainability in all sections of agriculture, however, the conscience has grown that enrichment of P in the soil may be just as dangerous to the environment as its emaciation. Sustainable farm units have to be managed among other with the objective of zero P balance, *i.e.*, during some defined period of time the amount of P imported into the unit should equal the amount exported out of the unit.

In cattle farms, import of P may occur in the form of fertilizer and feeds whereas P is exported in the form of milk and animal bodies. If import exceeds export of P reducing or even ceasing the use of P containing fertilizer may be seen as a first step towards avoiding the imbalance, and this step has the advantage of lowering the cost of production. But in many areas this may not be sufficient for achieving a zero P balance. Exporting manure may be an adequate means under certain conditions, but as a general strategy it is unlikely to be economically justified. A more promising strategy will be to restrict P concentrations in purchased feeds (Beede and Davidson, 1999; Beede, 2003; Sutton and Beede, 2003).

If the amount of P imported into the farm with purchased feeds does not exceed the amount of P exported in milk and carcasses this does not mean that P intake of the animals has to equal the amount of P transferred into milk and body substance. As long as the amount of P excreted by animals does not exceed that amount consumed in the form of self-produced feeds, zero P balance in the farm unit as a whole is possible.

In animal nutrition research it is now accepted that recommendations of P supply to farm animals must aim at preventing not only insufficiencies but also unnecessary surpluses for the animals. It is the aim of this lecture to review the relevant scientific literature on P metabolism in ruminants as a base for deriving P requirements of cattle and to discuss the importance of phosphate concentrations in blood plasma for well-being of dairy cows.

P balances and differentiation between endogenous and exogenous faecal P

The role of P in the metabolism of animals has attracted the interest of scientists practically from the start of Animal Nutrition as a scientific discipline. Henneberg (1870) found that in adult wethers faecal P excretion practically equalled the quantity of P ingested whereas the urine contained only negligible quantities of this element. This is a remarkable difference in comparison to monogastric animals. Wildt (1874) showed that substantial secretion of P into the forestomach of sheep must take place, amounting to almost the quantity ingested and that absorption of a corresponding amount must take place in the small intestine. Thus, the essential characteristics of P metabolism in ruminants were published before the end of the 19th century; namely that the gut is the dominant pathway for excretion of P and that the difference between P intake and faecal P excretion reflects the difference between quantities of P secreted into and quantities absorbed from the digestive tract.

If the quantity of P excreted in faeces is less than P intake, the difference indicates the extent of net absorption, if P excreted exceeds P intake, the difference is identical to the extent of net secretion. Net secretion by definition is equivalent to negative net absorption.

The use of ^{32}P for studying kinetics of P in ruminants was initiated at the University of California-Davis, and the first of a series of publications concerning this subject (Kleiber *et al.,* 1951) begins by defining the terms endogenous and exogenous. Endogenous faecal P is described as P originating from blood or body tissues and exogenous faecal P is of dietary origin and has passed unabsorbed through the digestive tract.

This original definition of endogenous and exogenous P will be used here throughout, irrespective of various changes in semantics that have occurred repeatedly in the course of the years. The term "endogenous", therefore, only indicates that part of faecal P originating from the animal's body – it does not indicate a necessity of excretion of this P.

A key principle of the isotope dilution method is that body P is labelled by parenteral administration of the isotope. Because all endogenous P is derived from blood plasma, the specific activity of P must be identical in secreted P and in phosphate (P_i) of blood plasma, because plasma P_i is the sole source of secreted P. When endogenous and exogenous P are mixed in the digestive tract, the degree of "dilution" of specific activity of secreted P is a function of the ratio between exogenous and endogenous P. If the lag between time of secretion into the digestive tract and time of excretion in faeces is taken into account properly, then that proportion of faecal P which is of endogenous origin can be calculated from measured specific activities in plasma phosphate and in faecal P.

Subsequent to the work of Kleiber *et al.* (1951), a large number of studies were carried out using radio-phosphorus. Table 1 shows the endogenous faecal P as a percentage of total faecal P of cattle and sheep found in 29 papers published between 1951 and 1996. One of the studies was conducted with weaned calves weighing less than 100 kg (Gueguen, 1963) and seven were with growing cattle weighing between 100 and 250 kg (Tillman and Brethour, 1958a; Brüggemann *et al.*, 1959; Tillman *et al.*, 1959; Challa and Braithwaite, 1988a, b; Coates and Ternouth, 1992; Bortolussi *et al.*, 1996). One study was performed with Holstein cows weighing 650 kg (Martz *et al.*, 1990). In two of the studies using lambs (Valdivia *et al.*, 1982; Garcia-Bojalil *et al.*, 1988), the endogenous proportions of faecal P ranged below the 43% found in one of the two cows of Kleiber *et al.* (1951). An exceedingly low proportion of only 36% also was found for one of five treatments studied by Lofgreen (1960). No reason can be given to explain why these results do not fit into the general range of the majority of the published work. On the other hand, the minimum endogenous proportion of faecal P was more than 50% in 21 of the 29 publications, and the respective maximum was equal to or exceeded the 70% of the second cow of Kleiber *et al.* (1951) in seventeen of the studies. About one-third of all results ranged between 70% and 95%. Ignoring the apparent outlier results, it can be assumed generally that about two-thirds to three-quarters of faecal P in ruminants are of endogenous origin. In milk-fed calves, endogenous faecal P amounted to 26% of total faecal P and this proportion was further reduced by supplementation of phosphate in milk (Challa and Braithewaite, 1989).

If the endogenous proportion of faecal P is known, then the exogenous proportion must be known as well according to the logic of the definition. The difference between P intake and exogenous faecal P indicates the degree of absorption of dietary P and, consequently, the proportion of dietary P absorbed is referred to as "efficiency of P absorption. Table 1 also shows efficiencies of P absorption calculated by the respective authors of the cited papers. In 23 of the publications these values range above 0.50, the majority of values is found between 0.60 and 0.80.

P in secretions into the digestive tract

McDougall (1948) showed that saliva of sheep contained large quantities of phosphate and thus proved that at least a major part, if not all, of the P secretion into the reticulo-rumen mentioned by Wildt (1874) took place via salivary secretion. Kay (1960b) showed that secretion of the

Table 1. Ranges of endogenous P as percentage of total P in ruminant faeces and absorption efficiencies (given in the original papers) as well as recalculated relative quantities of P secretion and P absorption.

Animals	% of faecal P endogenous	P absorption efficiency	as % of P intake		Reference
			Secretion	Absorption	
Cows	43 – 70	0.50 – 0.64	75 – 130	88 – 111	Kleiber *et al.*, 1951
Lambs	75 – 95	0.81 – 0.96	97 – 104	115 – 131	Lofgreen and Kleiber, 1953
Wethers	90 – 93	0.93 – 0.96	920 – 1462	948 – 1500	Lofgreen and Kleiber, 1954
Steers	58 – 63	0.75 – 0.76	139 – 166	180 – 202	Tillman and Brethour, 1958a
Wethers	67 – 73	0.65 – 0.80	203 – 274	191 – 261	Tillman and Brethour, 1958b
Wethers	82 - 85	0.87 – 0.90	357 – 462	379 – 498	Tillman and Brethour, 1958c
Steers	55 – 56	0.68 – 0.78	120 – 126	150 – 176	Tillman *et al.*, 1959
Bulls	48 – 89	0.71 – 0.88	94 – 369	122 – 423	Brüggemann *et al.*, 1959
Wethers	46 – 82	0.48 – 0.82	87 – 390	89 – 395	
Wethers	36 – 66	0.43 – 0.69	56 – 193	67 – 200	Lofgreen, 1960
Lambs	54 – 63	0.20 – 0.55	117 – 396	100 – 408	Lueker and Lofgreen, 1961
Wethers	57 – 62	0.51 – 0.58	134 – 161	122 – 146	Guéguen, 1962
Calves	45	0.80	83	146	Guéguen, 1963
Lambs	70 – 80	0.81 – 0.86	244 – 392	254 – 423	Preston and Pfander, 1964
Wethers	60 – 71	0.53 – 0.79	169 – 266	167 – 290	Young *et al.*, 1966c
Wethers	52 – 69	0.42 – 0.68	118 – 192	97 – 181	Potthast *et al.* 1976
Lambs	48 – 63	0.57 – 0.82	92 – 171	110 – 201	Field *et al.*, 1982
Lambs	15 – 39	neg. – 0.33	neg. – 57	neg. – 54	Valdivia *et al.*, 1982
Lambs	68 – 78	0.77 – 0.84	219 – 345	245 – 374	Boxebeld *et al.*, 1983
Sheep	45 – 56	0.44 – 0.55	100 – 122	100 – 122	Braithewaite. 1984
Lambs	58 – 67	0.17 – 0.77	139 – 206	42 – 235	Braithewaite, 1985
Lambs	60 – 74	0.80 – 0.87	149 – 289	192 – 335	Field *et al.* 1985
Ewes	53 – 64	0.55 – 0.69	114 – 155	129 – 171	Braithewaite, 1986
Calves	81 – 83	0.56 – 0.84	107 – 210	150 – 260	Challa and Braithwaite, 1988a-c
Lambs	24 – 43	0.30 – 0.54	38 – 76	41 – 95	Garcia-Bojalil *et al.*, 1988
Sheep	75 – 78	0.55 – 0.69	114 – 155	129 – 173	Ternouth, 1989
Cows	75 – 78	0.68 – 0.72	292 – 360	282 – 313	Martz *et al.*, 1990
Heifers	64 – 87	0.66 – 0.92	204 – 933	185 – 950	Coates and Ternouth, 1992
Sheep	81 – 83	0.83 – 0.85	425 – 503	448 – 503	Rajaratne *et al.*, 1994
Steers	66 – 84	0.50 – 0.82	200 – 505	205 – 393	Bortolussi *et al.*, 1996

parotid glands in kids develops in the first three months of life. Rumination is the most effective stimulus of secretion of parotid saliva in cattle (Bailey and Balch, 1961; Kaufmann and Orth, 1966). Inorganic P concentrations in ruminal fluid are determined principally by the extent of salivary P secretion (Tomas, 1973; Tomas *et al.*, 1967).

Table 2 shows ranges of phosphate concentrations in different types of saliva. Differences among secretions from different glands may be one source of variation in P concentration of mixed saliva. Yet, there are substantial ranges in P concentration of specific types of saliva.

Table 2. Ranges of P concentration in saliva of ruminants.

Animals	Type of saliva	Salivary Pm mmol/l	Plasma P_i mmol/l	S/P[a]	Reference
Sheep	mixed	17 – 28			McDougall, 1948
	parotid	6 – 42			
Sheep	parotid	12 – 40			Kay, 1960a
	submaxillary	1 – 88			
	sublingual	0.3 – 2.0			
	labial	1 – 5			
	inferior molar	22 – 26			
	palatine	12			
	residual	22 – 35			
Calves	parotid	8 – 24			
	submaxillary	0.2 – 2.0			
	inferior molar	9 – 27			
	residual	10 – 17			
Goats	parotid	15 – 40			Kay, 1960b
Steer	parotid	16 –32	1.4	5	Bailey and Balch, 1961
Cows	mixed	15 – 66			Bailey, 1961
Lambs	parotid	16 – 20			Tribe and Peel, 1963
	residual	9 – 16			
	mixed	11 – 23			
Wethers	parotid	14 – 20			
	residual	12 – 16			
	mixed	21 – 35			
Sheep	parotid	13 – 23	1.4 – 2.7	8.4 – 10.2	Tomas et al., 1967
Sheep	mixed	16 – 32	2.1 – 3.4	9 – 12	Perge et al., 1982
Sheep	parotid	18 – 75	1.3 – 8.0	8 – 23	Manas-Almendros et al.,1982
Heifers	mixed	7 – 14	1.0 – 1.8	6 – 9	
Sheep	mixed	11.7 ± 2.6	1.7 ± 0.2	7.5 ± 1.7	Gartner et al., 1982
		4.7 ± 1.0	0.7 ± 0.1	6.8 ± 1.2	Breves et al., 1987
Cows	mixed	0.9 – 2.6	4.3 – 12.1	3.5 – 6.8	Valk et al., 2002

[a] S/P = Salivary P/Plasma P_i

Perge et al. (1982) found the phosphate concentration of mixed saliva to be influenced not only by intake of P and of Ca but also by the time of sampling in sheep fed once daily. Varying plasma P_i concentrations within wide ranges by dietary depletion or phosphate loading via infusions directly influenced salivary P concentration but not salivary flow in cattle (Riad et al., 1986) or sheep (Scott, 1978; Wright et al., 1984; Breves et al., 1987). Bailey (1961) estimated that salivary flow in cows consuming about 10 kg DM daily ranged between 100 and 190 litres per day.

It was noted for sheep that chloride is the predominant anion contained in bile, pancreatic juice and secretions into the upper jejunum, and that about equal concentrations of chloride and bicarbonate are found in secretions into lower jejunum, ileum, caecum and spiral colon, whereas only very small concentrations of phosphate are found in all of these secretions (Kay and Pfeffer, 1970). It is, therefore, concluded that the quantity of P secreted into the intestine of ruminants is almost irrelevant in comparison to salivary P secretion.

Revised interpretation of isotope studies

The original illustration given by Lofgreen and Kleiber (1953) for interpreting results of their ^{32}P studies gives the impression that digesta first pass through a section of the digestive tract in which absorption of P takes place and then through a section in which endogenous P is added. This, however, does not correctly illustrate the movements of P in the digestive tract, because most of the P secreted into the digestive tract is contained in saliva.

Endogenous faecal P excretion is not a direct measure of total P secretion into the digestive tract. It only represents that part of the secreted quantity that is not reabsorbed during the passage along the tract. Further derivations are possible if it is legitimate to assume that secretion of P takes place mostly proximal to the site of absorption and that complete mixing between endogenous and exogenous P can be taken as certain. Young *et al.* (1966c) reported identical specific activities of P in the solid and the liquid phases of faeces after sheep were given parenteral doses of radio-phosphorus. They concluded that mixing of endogenous and exogenous P in the digestive tract must be complete. Practically identical patterns were found of the respective specific activities of P in ruminal contents and in faeces after correction for the time lag due to passage through the digestive tract when sheep were given an intravenous dose of [32]P (Potthast *et al.*, 1976). This not only confirmed complete mixing of P in the forestomach, it also indicated that quantities of P secreted into the post-ruminal part of the digestive tract must have been insignificant, relative to the P in salivary secretions. Results of Thévis and Francois (1985) indicate that secretion of P into the small intestine after the entrance of pancreas and bile must be comparatively small. Therefore, the dominant role of salivary P secretion into the reticulo-rumen is emphasized. Because complete mixing of P of the different origins is assumed, it is concluded that the absorption efficiency for dietary P must be applicable also for endogenous P secreted into the digestive tract as well.

When an animal is fed the same diet over time and is in steady state, then the sum of P entering the digestive tract from feed and from secretions must be equal to the sum leaving, either by absorption or by excretion in faeces. The efficiency of P absorption is defined as the absorbed fraction of the sum of P entering the digestive tract.

Secretion and absorption of P were recalculated for the 29 cited publications. The ranges for both, relative to the respective P intake, are shown in Table 1. It is obvious that in most of the experiments with cattle and sheep secretion as well as absorption of P exceeded P intake, in some cases by several fold. Thus, it can be stated generally that P absorption largely exceeds P net absorption from the digestive tract of ruminants. Consequently, identification of sites of P secretion into and absorption from the digestive tract was regarded as a consideration of major scientific relevance.

P flow at specific sites of the digestive tract

The method of surgically fitting cannulas into well defined sites of the intestine led to a better understanding of net movements of mineral elements through the major segments of the digestive tract of ruminants. Table 3 shows flow rates of total P through the proximal duodenum as well as faecal excretion reported in 15 publications involving sheep and six with cattle.

Flow of P through the proximal duodenum exceeded P intake in all but one of the experiments reported. The flow of P at the proximal duodenum substantially exceeded P intake in almost all experiments, which is consistent with the previously mentioned findings about secretion of P in saliva. Faecal excretion of P was lower than P intake in most of the experiments reported, but P net absorption efficiency from the total tract varied widely among different experiments.

It can be concluded that P net absorption from the intestine of ruminants is much larger than P net absorption from the whole digestive tract, and that intestinal P net absorption efficiency may exceed 0.80 under given circumstances. The coincidence of high rates of salivary P secretion and

Table 3. Relative flows of P into the small intestine and net absorption efficiencies of P from the whole tract and from intestinal sections of sheep and cattle.

	Duodenal flow, % of P intake	P net absorption efficiency			Reference
		From the total tract	From the intestines		
			Small	Large	
Sheep	500	0.17	0.82	0.10	Bruce et al., 1966
	192 – 352	-0.06 – 0.20	0.34 – 0.64	0.02 – 0.30	Pfeffer et al., 1970
	118 – 248	0.07 – 0.26	0.22 – 0.58	0.00 – 0.20	Grace et al., 1974
	85 – 119	0.25 – 0.61	0.29 – 0.57		Leibholz, 1974
	315	0.16	0.71	0.08	Ben-Ghedalia et al., 1975
	180	0.68	0.84	0.00	Dillon and Scott, 1979
	238 – 314	0.30 – 0.40	0.71 – 0.75	0.00 – 0.12	Greene et al., 1983a
	337	<0.01	0.60	0.25	Théwis and Francois, 1985
	218 – 425	0.09 – 0.26	0.68 – 0.80	-0.04 – 0.18	Wylie et al., 1985
	160 – 296	-0.24 – 0.62	0.42 – 0.55	-0.11– 0.06	Breves et al., 1985
	169 – 227	0.07 – 0.47	0.59 – 0.69		Scott and Buchan, 1985
	323 – 428	-0.06 – 0.00	0.65 – 0.75		Grings and Males, 1987
	215 – 243	0.23 – 0.36	0.67 – 0.70		Scott and Buchan, 1988
	159 – 174	0.48 – 0.56	0.68 – 0.75	-0.12 – 0.00	Khorasani and Armstrong 1990
	306 – 316	0.32 – 0.34	0.75 – 0.77	0.04 – 0.15	Kirk et al., 1994
Cattle	131 – 195	0.04 – 0.40	0.48 – 0.68		Pfeffer and Kaufmann, 1972
	137 – 216	0.24 – 0.48	0.35 – 0.72	0.00 – 0.24	Bertoni et al., 1976
	167 – 201	0.36 – 0.52	0.68 – 0.74	-0.07 – 0.05	Greene et al., 1983b
	157 – 268	0.31 – 0.53	0.58 – 0.78	-0.18 – 0.18	Khorasani and Armstrong, 1992
	127 – 148	0.32 – 0.42	0.50 – 0.54		Rahnema et al., 1994
	132 – 184	0.20 – 0.39	0.51 – 0.59		Khorasani et al., 1997

of extensive absorption of P in the intestine results in the endogenous cycle of P between blood and the lumen of the digestive tract typical for all ruminant species.

Flow of P through the terminal ileum was measured in most experiments on sheep and in three of the six publications with cattle. From these results it must be concluded that intestinal net absorption of P takes place mainly between proximal duodenum and terminal ileum. In most, though not all experiments, P flow through the terminal ileum was slightly higher than faecal P excretion. Therefore, it can be concluded that the large intestine contributes a small proportion to total net absorption of P. In a series of perfusion studies into the colon and rectum of sheep some P net secretion into the gut was observed as long as phosphate free buffers were used. However, when buffer solutions contained phosphate, there always was net absorption of P_i which increased with increasing phosphate concentrations (Höller *et al.* 1988).

Further specific definition of the site of P absorption in ruminants is based on studies using sheep with cannulas in different sites of the small intestine (Kay, 1969; Ben-Ghedalia *et al.*, 1975). These authors found that elevation of pH of the digesta is comparably modest during passage through the first third of the small intestine. Correspondingly low pH was found in duodenal contents of cows in which re-entrant cannulas were positioned well distal to the addition of bile and pancreatic secretion (Kaufmann *et al.*, 1972). These values can be compared with pH measured in defined sites of the small intestine of piglets (Eidelsburger *et al.*, 1992; Risley *et al.*, 1992), humans (Fallingborg *et al.*, 1994), horses (Meyer *et al.*, 1997), or mice (Delcenserie *et al.*, 2001). It can be summarized that pH of digesta of the proximal small intestine is lower in ruminants than in non-ruminants. This means that for a considerable part of the intestinal passage, inorganic P of digesta is present mainly as the monovalent dihydrogen phosphate, $H_2PO_4^{1-}$ and no precipitation of calcium phosphates is to be expected. As in blood plasma most of the P_i is present as the bivalent mono-hydrogen phosphate, HPO_4^{2-}, diffusion of the primary phosphate through the gut wall is favoured by the very high concentration gradient of the primary phosphate from the mucosal to the serosal side.

About 92% of the P in duodenal digesta was found in the liquid phase when sheep were fed hay and this was reduced to 87% when concentrates were fed (Scott and Buchan, 1985). In sheep fed diets adequate in P, inorganic P made up 92% of total P flowing through the proximal duodenum and 32% of that flowing through the terminal ileum; during P depletion this proportion was reduced to 46% at the duodenum and not changed at the terminal ileum (Breves *et al.*, 1985).

Théwis *et al.* (1978) analysed total P and phospholipid P in digesta in the major sections of the digestive tract of sheep. The respective portion of total P present in the form of phospholipids was 2.9% in the diet, more than 6% in contents of the reticulorumen, more than 15% in contents of the upper small intestine and about 8% in contents of the large intestine. Whereas the difference between diet and ruminal contents indicates microbial synthesis of phospholipids, the peak proportion in the upper small intestine probably is a resultant of bile secretion and of preferential absorption of P_i in the upper part and increased absorption of phospholipids in the lower sections of the small intestine. These results, however, do not allow conclusions regarding the quantity of bile phospholipids secreted per unit of time.

When the flow of P at the proximal duodenum was measured, it was found in almost all of the experiments to exceed P intake by an order of magnitude that may be attributed to salivary flow. This occasionally has led to the interpretation of P net secretion prior to the duodenum as a direct measurement of salivary P secretion (Scott and Buchan 1985, 1988; Challa and Braithwaite, 1988a,b; Challa *et al.*, 1989). Yet, the ruminal mucosa is not to be regarded as impermeable to phosphate. Scarisbrick and Ewer (1951) injected ^{32}P into the rumen and after 10 minutes found more isotope in the ruminal vein than in the carotid artery. Also, they found arterio-venous differences after short term changes in phosphate concentration in rumen contents.

Net secretion into the rumen was found in sheep fed diets either adequate or deficient in P when P_i concentrations on the mucosal side did not exceed 2.2 mmol/l, whereas net absorption was found when concentrations exceeded 4.1 mmol/l. These findings were recorded irrespective of the P status of the sheep (Breves *et al.*, 1988; Beardsworth *et al.*, 1989).

There is one paper which indicates that substantial net absorption of P may take place in the omasum of ruminating calves (Edrise and Smith, 1986). This finding, however, has not been confirmed by other researchers, presumably due to the specific complications for research work concerned with that organ.

Regulation of P absorption

Depletion of Ca increases plasma concentrations of calcitriol (also known as 1,25-(OH)$_2$ vitamin D$_3$ or vitamin D hormone) and increases Ca absorption efficiency in non-ruminant and ruminant animals (Braithwaite, 1974; Fox and Care, 1978; Fox and Ross, 1985; Abdel-Hafeez *et al.*, 1982).

P depletion increased plasma calcitriol concentrations and absorption efficiencies of both Ca and P in rats (Hughes *et al.*, 1975; Ribovich and DeLuca, 1978) and pigs (Fox and Care, 1978; Fox *et al.*, 1978; Fox and Ross, 1978). Contrary to this finding, however, P depletion increased absorption efficiency of P in ruminants without raising that of Ca (Young *et al.*, 1966b, Abdel-Hafeez *et al.*, 1982).

Differences in vitamin D metabolism of different species are of relevance in this context. During P depletion no increase in plasma concentration of calcitriol was found in sheep (Abdel-Hafeez *et al.*, 1982; Breves *et al.*, 1985; Maunder *et al.*, 1986) or in lactating goats (Müschen *et al.*, 1988). In addition, Maunder *et al.* (1986) found that neither metabolic clearance rate nor production rate of calcitriol were altered by dietary P depletion. However, the finding of Riad *et al.* (1987) is not in accord with these results, namely that intramuscular injections of 1⊠-OH-vitamin D$_3$ in heifers increased plasma P_i concentrations and concurrently decreased salivary P_i concentrations and thus secretions.

Neither the substantial net secretion of P into the digestive tract prior to the duodenum nor the net absorption of P from the intestines were ever accompanied by comparable respective net movements of Ca. This explains why in tracer studies using isotopes of both elements, absorption of P in most cases exceeded that of Ca. Uncoupling of absorptive processes for P and Ca, respectively, allows maximum absorption of P during periods of P deficiency and at the same time decreased absorption of Ca (Young *et al.*, 1966a; Breves *et al.*, 1985). This fact marks a substantial biological difference between ruminants and non-ruminants.

The independence between absorption of P and Ca explains why P-depleted dairy goats increased faecal excretion of Ca (Müschen *et al.*, 1988) and why additional dietary supplementation of Ca did not increase the negative consequences of P deficiency in lactating goats (Deitert and Pfeffer, 1993) and in weaned kids (Pfeffer *et al.*, 1996). Obviously, excessive dietary Ca concentrations do not negatively affect P absorption and dietary Ca/P-ratio is not regarded as having impact on the amount of P required by ruminants.

When *in vivo* and *in vitro* measurements were combined for growing kids fed diets varying in dietary P and/or Ca (Pfeffer *et al.*, 1995; Schröder *et al.*, 1995), reducing only P intake drastically lowered retention of both elements and caused hypophosphataemia in combination with hypercalcaemia without affecting plasma concentrations of Parathyroid Hormone (PTH) or calcitriol and drastically increased unidirectional as well as net P_i fluxes from the mucosal to the serosal side of duodenal and jejunal preparations. Reducing only Ca intake caused increased plasma P_i concentrations and gave rise to greater plasma concentrations of PTH and calcitriol with no significant changes in P_i fluxes through the mucosa of duodenum or jejunum, measured *in vitro*. Kids fed diets low in both elements produced results very similar to those fed the diet deficient only in P with the exception of plasma calcitriol which was elevated as in kids fed the diet low in Ca and adequate in P. These results fully confirm that P is absorbed independent of plasma calcitriol concentration.

Faecal excretion of P

Animals excrete P in the faeces for one of three potential reasons:

1. A fraction of the P contained in feeds may be present in a chemical binding that cannot be absorbed. As a consequence due to the nature of the feed, this fraction has to be excreted. This fraction may be substantial in non-ruminants, but is practically negligible in ruminants due to microbial breakdown of phytates in the rumen.

2. Some P inevitably is lost in faeces independent of P intake. These inevitable losses result from metabolism either of the host animal or of the micro-organisms in the digestive tract and are obligatory to normal basal functions of the animal.

3. If intake of absorbable P exceeds that needed for inevitable losses and requirements for growth, reproduction, or lactation, then this surplus P is excreted to maintain homoeostasis. Normally in ruminants most of this surplus is excreted with faeces, whereas in non-ruminants renal excretion predominates.

There is no possibility of separately quantifying these three types of faecal P excretion. This is a challenge for nutritionists in developing more complete understanding of P metabolism and excretion of ruminants.

Absorbability of dietary P

If animals would, under all circumstances, absorb the maximum possible fraction of dietary P, this would mean that exogenous faecal P excretion always would be minimal and a function of the respective sources of dietary P. As a consequence, any homoeostatic regulation of total faecal P excretion could be achieved only by changes of endogenous faecal P excretion. Based on information discussed above it appears unlikely that changes in P secretion alone could be an efficient means of P homoeostasis. Therefore, it would be consequent to assume different efficiencies of absorption for dietary and secreted P, respectively. This differentiation between dietary P and secreted P is, however, hardly justifiable based on the cited evidence for complete mixing of exogenous and endogenous P in the forestomach (Young *et al.*, 1966c, Potthast *et al.*, 1976).

Therefore, as a clear alternative of interpreting isotope studies it is assumed that mixing of P in the rumen is complete. If this is the case, then there is no logic in differentiating between absorption efficiencies of endogenous and exogenous P, respectively, they are identical. As a consequence it has to be accepted that endogenous P as a fraction of total faecal P indicates only the ratio at which dietary and secreted P are being mixed in the digestive tract, *i.e.*, practically the ratio between dietary and salivary P.

Isotope dilution studies in animals provided with sufficient or even excessive amounts of P, therefore, is not an adequate method for differentiating between availabilities of varying sources of dietary P, if P supply is adequate or excessive. Such a differentiation would only be correct under conditions that force animals to maximize P absorption.

Koddebusch and Pfeffer (1988) fed rations to dairy goats containing either less than 1 g P or about 2 g P per kg dry matter. This was achieved by feeding a P-deficient basal diet and supplementing with one of four sources of P (dried grass, wheat bran, mono calcium phosphate or dicalcium phosphate. No isotope of phosphorus was used in the study, but the important point was that even the supplemented diets supplied P only at a level where goats were forced to maximize absorption. As shown in Table 4 net absorption efficiency of P was 0.23 for the basal diet and 0.46 for the supplemented diets. Calculating the differences between the respective diets proves that net absorption efficiency of supplemented P under these conditions exceeded 0.9. No differences due to the source of supplemental P were detected.

As such extremely high net absorption efficiencies of P from supplements are verified, it must be concluded that faecal P excretion during P deficiency is primarily caused not by a low "availability" depending on the specific dietary source of P, but rather because of other reasons. As these losses are not determined by the nature of the feed components fed, they must be caused by factors associated with the animal.

During the pre-ruminant phase in early life net absorption efficiency of P contained in milk exceeded 0.95 in calves (Gueguen, 1963; Challa and Braithewaite, 1989), in lambs (Dillon and Scott, 1979), and in kids (Boeser, unpublished).

Table 4. Intake, faecal excretion and net absorption efficiency of P in dairy goats fed diets very low or mildly low in P (Koddebusch and Pfeffer, 1988)

Level of dietary P supply	Very low	Mildly low	Difference
P intake, mg/kg body weight daily	33.6 ± 8.6	53.7 ± 12.4	20.1
Faecal P, mg/kg body weight daily	27.2 ± 6.0	28.7 ± 8.8	1.5
Net absorption efficiency of P	0.23 ± 0.11	0.46 ± 0.05	0.93

The lack of influences of the source of dietary P on P absorption efficiency must be seen as a consequence of microbial metabolism in the forestomach, whereas in non-ruminants such differences in availability of dietary P from varying sources are of great relevance.

Inevitable faecal losses of P

When animals are fed diets very low in P they may not be able to reduce faecal P excretion to such an extent as would be necessary to achieve an equilibrium P balance (zero balance). These losses are defined as inevitable or obligatory losses and it is assumed that they are caused neither by the "quality" or absorbability of dietary P sources nor by the level of P intake, but rather by the physiology of the host animal and/or by microbial metabolism.

As inevitable faecal P losses may be caused by or a result of either the host animal or the microbes in the digestive tract, it is worth comparing P excretion in the pre-ruminant and ruminant phases.

Walker and Al-Ali (1987) fed milk replacers low in P to lambs in their first three weeks after birth and observed daily faecal P excretions in the order of 4 mg per kg of body weight. Kids raised on goat´s milk up to about 17 kg body weight retained 88% of the P ingested with milk, but separate determinations in faeces and urine were not carried out in that work (Pfeffer and Rodehutscord, 1998). In recent balance studies with kids in their first three weeks of life it was found that daily P excretions per kg body weight were 1.4 mg in faeces and 25 mg in urine, respectively. This comparably high urinary P excretion obviously results from Ca being the limiting factor in milk for accretion of both elements in the bodies of kids. When calcium citrate was added to goat`s milk, faecal P excretion increased to 2.8 mg and renal P excretion decreased to only 0.5 mg per kg of body weight per day, respectively (Boeser, unpublished). It is concluded that only a very small fraction of the urinary P of milk-fed pre-ruminants is inevitable loss of P from the body, as most of the loss could be avoided by increasing Ca supply.

The technique of intra-gastric infusion excludes any influences of microbial metabolism on balance results, because nutrients are supplied solely in the form of solutions into the rumen and the abomasum. Using this technique, Rajaratne *et al.* (1996) lowered the daily P supply of three adult sheep weighing about 40 kg from 1.29 to 0.13 g. At this marginal P supply sheep reduced their daily faecal P excretion to rates ranging between 6 and 30 mg per kg of live weight. The rates of these losses agree fairly closely with the 7 mg per kg daily calculated for pigs fed diets insufficient in P (Rodehutscord *et al.*, 1998). It must be concluded that the isolated metabolism of the ruminant host animal can conserve P under conditions of low P supply as efficiently as non-ruminant mammals.

Whereas regressing inevitable faecal P losses on body weight generally is accepted for non-ruminants, the validity of such a regression must be doubted for ruminants. In a series of investigations in Bonn, negative P balances were found in 138 trials carried out with lactating goats, 24 trials with pregnant and non-lactating goats, and 8 trials with non-pregnant and non-lactating goats. Renal P excretion was negligible in each of these goats. No significant correlation was found between faecal P excretion and live weight, whereas a highly significant correlation existed between faecal P excretion and dry matter intake indicating that per kg DMI 0.88 g P were inevitably lost in faeces of these goats (Pfeffer, 1989). From a reanalysis of 158 data points of growing cattle Ternouth *et al.* (1996) concluded that excretions of endogenous faecal P in growing cattle consuming forage diets were correlated most closely with dry matter intake and on average amounted to 0.505 g per kg DMI. Because on average 77% of the faecal P of cattle on the respective diets is endogenous (Coates and Ternouth, 1992; Bortolussi *et al.,* 1996, see Table 1) it may be speculated that regressing total faecal P excretion to dry matter intake on these low P diets might result in roughly 0.66 g P per kg DMI (computed as 0.505 g per kg DMI divided by 0.77).

Spiekers *et al.* (1993) fed a low P diet of constant composition at rates of either 16.9 or 10.0 kg DM daily to dairy cows. Irrespective of the treatment, cows excreted about 1.2 g P per kg dry matter ingested which agrees remarkably well with the results Recalculated from calf studies carried out by Challa and Braithewaite (1988 a,b,c, 1989).

The Technical Committee on Responses to Nutrients (AFRC, 1991) expressed theoretical considerations starting its reappraisal of maintenance requirements for Ca and P. Deviating from the view of AFRC (1991), a strict differentiation is preferred between the terms "endogenous faecal P" on one side and "inevitable" or "obligatory" faecal loss on the other side. In ruminant animals, the latter is seen primarily as a consequence of microbial metabolism and may have been captured from either endogenous P, or from potentially available P from exogenous sources.

AFRC (1991) assumed that sloughing of intestinal epithelial cells is the major reason for inevitable faecal P losses in ruminants and that this phenomenon is caused mainly by fibrous plant material of low digestibility. It is hypothesized that, Therefore, a negative correlation between metabolizability of energy (= q) and inevitable faecal P losses was hypothesized.

Rodehutscord *et al.* (2000) challenged this hypothesis by adding either sawdust as indigestible organic matter or starch as completely digestible organic matter to a diet for dairy goats with low basal P. Supplementing sawdust did not affect the amount of faecal P significantly, whereas supplementing starch caused a significant increase in faecal P excretion. Therefore, it is concluded that the digestible and not the indigestible part of organic matter intake has the most influence on inevitable faecal P losses in ruminants. On the basis of this finding the difference between the regression of Ternouth *et al.* (1996) for growing cattle fed forage diets and the results of Spiekers *et al.* (1993) for dairy cows fed mixed rations containing 58% concentrates appears less dramatic. Dry matter digestibility in most of the diets fed to the growing cattle was well below 0.6 (Coates and Ternouth, 1992; Bortolussi *et al.*, 1996), whereas that of the dairy cows was 0.75 (Spiekers *et. al.*, 1993).

More research is needed for fully understanding factors influencing inevitable faecal P losses in ruminants.

Surplus P excretion in faeces

When P supply to sheep was increased by infusion of phosphate solutions either into the blood or into the rumen, daily quantities of P net secretion prior to the duodenum and P net absorption from the intestine were not influenced significantly, but faecal P excretion was increased by lowering intestinal P net absorption efficiency (Scott *et al.*, 1984a,b). This shows that adaptation of intestinal absorption efficiency to the level of P supply contributes substantially to homoeostatic regulation P in ruminants. The increase in faecal P excretion did not affect digestibility of dry matter. However, it did significantly decrease faecal dry matter concentration and thus increase faecal water excretion.

Adding solutions of monosodium dihydrogen orthophosphate to milk fed to pre-ruminant calves caused faecal P and urinary P excretion to increase at a ratio of about 1:2 and P retention to decrease concurrently (Challa and Braithwaite, 1989). This result is not in agreement with current studies carried out in Bonn in which faecal P excretion of kids was not affected as a consequence of supplementation of phosphate to milk and all extra P was excreted in the urine (Boeser and Loof, unpublished).

Renal excretion of P

Phosphorus balances were reported in several of the papers cited in Tables 1 and 2. Well above 90% of the total P excretion was via the faeces in most of these studies, leaving only marginal importance to excretion in urine (Grace *et al.*, 1974; Bertoni *et al.*, 1976; Boxebeld *et al.*, 1983; Braithwaite, 1984, 1985, 1986; Wylie *et al.*, 1985; Martz *et al.*, 1990; Khorasani and Armstrong, 1992; Bortolussi *et al.*, 1996). Comparable balances were reported in dairy cow studies (Morse *et al.*, 1992; Delaquis and Block, 1995; Knowlton *et al.*, 2001; Valk *et al.*, 2002; Knowlton and Herbein, 2002). Ignoring renal P excretion completely would, therefore, not appreciably bias the calculated P balances in these experiments. This situation must, however, not be generalized because there are situations in which P excretion may be elevated, even in ruminants.

In the 1960s beef production turned to feeding very high concentrate diets. Reed *et al.* (1965) reported that in steers fed such diets P excreted in urine could exceed the quantity of P excreted in faeces. Partitioning of P excretion by ruminants was found to deviate from the predominant faecal route with a substantial contribution by the kidneys in several investigations. It appears that renal P excretion is not changed immediately at weaning from urine to faeces (Dillon and Scott, 1979) and this phenomenon is being further investigated at present in Bonn.

Factorial derivation of P requirements in cattle

The ARC (1965) employed a factorial approach for the assessment of requirements of ruminant animals for major mineral elements which continues to be used in establishment of the most recent requirement estimates (AFRC, 1991; GfE, 1993; NRC, 2001; Valk and Beynen, 2002).

Daily retentions during particular phases of growth and at specified stages of pregnancy as well as amounts contained in the milk yielded as parts of "net requirements" of an element are accepted unanimously. There is, however, some controversy with regard to the inevitable or obligatory losses (maintenance). ARC (1965) also used the term "net endogenous requirement" for the latter. Judged on the basis of knowledge of today the two-fold use of the term "endogenous" is regrettable because it contributes to misunderstanding.

Deposition of P in the body during growth

P concentrations in empty bodies of Friesian bulls fed to gain on average 1 kg live weight daily and slaughtered at one of five weights between about 150 kg and 575 kg, respectively, were close to 7.0 g/kg empty body weight which corresponded to about 5.8 g/kg live weight (Schulz et al., 1974).

Hoey et al. (1982) analysed body composition of Hereford heifers fed a basal diet containing only 0.9 g P/kg DM. This diet was fed either without P supplement or an additional 12 g P per day as mono-ammonium phosphate. A third group of heifers received the supplement, but the quantity of diet was restricted to intake of the unsupplemented group. Empty body P concentrations ranged from 6.0 g/kg in the group without supplemental P to 6.8 g/kg in the group receiving the supplement but at restricted feed intake. These results show that in growing cattle empty body P concentrations may vary to some degree depending upon P intake. The assumption of an accretion of 6.5 g P per kg gain of empty body weight in growing cattle appears appropriate.

Deposition of P in products of conception

It generally is accepted that conceptus growth follows an exponential curve in mammalian species. By far the greatest portion of tissue accretion occurs in the last third of pregnancy.

Mineral element accretion in foetuses and conceptus of single- and twin-bearing ewes was studied by Grace et al. (1986). P accretion has been studied in foetuses of beef heifers (Ferrell et al., 1982) and of Holstein cows (House and Bell, 1993). Although the difference in P deposition between these two cattle populations appears remarkable, it is emphasized that quantities deposited are relatively small compared with inevitable losses, with P accretion during growth, or with transfer of P into milk during lactation. It is concluded that daily P deposition in the conceptus even at the very end of gestation will not exceed 4 g in the beef breeds and 7 g in Holstein cows, respectively.

P concentration in milk

Table 5 shows P concentrations in milk found by some authors within the last fifteen years. Values for colostrum generally are higher than for milk, but they fall within the first week of lactation. No influence of the level of P intake was found on P concentration in milk. Influences of the stage of lactation were not evident in any of the cited studies. A single subcutaneous injection of slow-release bovine somatotropin seven days before the expected calving date significantly reduced P concentration in colostrum, presumably due to increased colostrum volume, but this difference was no longer evident in milk of days 6-9 (Law et al., 1993). It is reasonable to assume generally a P concentration in milk of 0.9 g/kg over the course of a lactation. Especially in high yielding dairy cows the quantity of P secreted into milk must be regarded as the dominating factor for P requirements.

Inevitable losses of P

Urinary P excretion obviously can be reduced to negligible amounts in ruminants. For this reason it is reasonable to ignore inevitable urinary losses as a factor determining P requirements of

Table 5. P concentration in bovine milk.

Breed	n	Stage of lactation	P in milk (g/kg)	Reference
Brahman	36	Parturition	1.27	Salih *et al.*, 1987
		After three months	0.90	
British Friesian	134	"Indoor period"	0.81 – 0.95	Brodison *et al.*, 1989
	217	"Pasture period"	0.85 – 1.10	
Friesian and Friesian-Jersey cross bred	34	Day of parturition	1.73 – 1.46	Law *et al.*, 1993
		Day 1 of lactation	1.32	
		Days 6-9 of lactation	1.01	
Holstein-Friesian	52	Complete lactation	0.90	Brintrup *et al.*, 1993
Holstein-Friesian	26	Complete lactation	0.85 – 0.88	Wu *et al.*, 2000
Holstein-Friesian	36	Early and mid lactation	0.68 – 0.78	Knowlton *et al.*, 2001
Holstein-Friesian	24	Complete lactation	0.87 – 0.94	Valk *et al.*, 2002
Holstein-Friesian	13	11 weeks of lactation	0.89	Knowlton and Herbein, 2002

cattle. Inevitable losses of P from the body of cattle are difficult to predict, because effects of low P intake on the host animal´s metabolism must be differentiated from those taking place in microbial metabolism.

Inevitable P losses are not considered as part of "net requirements" in the sense of ARC (1965), as they are measured as faecal excretion and, therefore, need not be corrected by "availability" to "give the requirement in terms of dietary amount". For dairy herds it is generalized that mixed rations based on silages of maize or grass and comparatively high proportions of concentrates will cause inevitable faecal P losses of 1.2 g per kg DMI.

Availability of dietary P

From what was discussed above, it has to be concluded that quite differing values for "availability of dietary P" may be derived from experimental data, depending on the definition of "availability" chosen. In a factorial approach of deriving recommendations for mineral supply, however, this factor very much dominates the result because all data concerning net requirements have to be transformed into gross requirement by dividing by the "availability" or absorption efficiency.

Findings of Lofgreen (1960) are often regarded as evidence that differences in absorption efficiency of P from different inorganic sources exist, as well as between dicalcium phosphate and calcium phytate. On the other hand, no differences in overall utilization of P could be established between mono-calcium phosphate and calcium phytate (Tillman and Brethour 1958b) or between monosodium phosphate and acid sodium pyrophosphate (Tillman and Brethour 1958c). From the work of Koddebusch and Pfeffer (1988) it is concluded that ruminants will utilize P from the different relevant sources with very high efficiency (greater than 0.9), if P intake is limiting performance.

Currently, for example the availability values or absorption efficiencies for dairy cattle diets used by the various working parties are 0.58, 0.70, and 0.70 for AFRC (1991), GfE (1993), and NRC (2001), respectively. Valk and Beynen (2002) use 0.72 for dry cows and 0.77 for lactating cows.

Uncertainty concerning the proper method of deriving availabilities as criteria typical for feed components may have caused working parties to decide for comparably low values of availability in order to include safety margins in the recommendations. Critical interpretation of the literature justifies the general assumption of an availability of not less than 0.9 for P contained in the feeds commonly fed to cattle.

Evaluation of derived recommendations on the basis of feeding trials

According to the approach of ARC (1965), factorial estimates of requirements have to be compared with results of feeding trials in which the element has been given in two or more amounts and the resultant performance of the animals measured.

Table 6 summarizes the results of nine investigations published in the last two decades. Although based on diverse basal rations, dietary P supply was the variable within each study. These studies were carried out with dairy cows yielding between about 5,000 and more than 12,000 kg of milk per cow and year and were sufficiently long to allow general conclusions. Neither milk yield nor health status of the cows were affected as long as the P concentration was not reduced to less than 2.7 g/kg DM. Dietary P concentrations of 2.4 g/kg (Call et al., 1989) or 2.3 g/kg (Valk and Sebek, 1999) on the other hand produced symptoms of deficiency, mainly reduced feed intake and, consequently, reduced milk yield. The greatest number of cows were included in the study of Lopez et al. (2004a) and from their work it is obvious that rations for dairy cows need not contain more than 0.37% P in their dry matter if feed intake is adequate to meet the needs of animals for energy and other nutrients. No influence on health or fertility of cows was found in that work (Lopez et al., 2004b).

Based on the findings discussed above, recommendations for P supply of dairy cows have been revised in Germany (GfE, 1993), the USA (NRC, 2001) and the Netherlands (Valk and Beynen, 2002). Table 6 shows the amounts of P recommended by each of the working parties and P concentration in rations resulting from these recommendations. Official recommendations in Germany and the USA do not differ greatly, whereas the Dutch standards are notably lower. In each of the three systems, dietary concentration of P has to increase with increasing milk yield. At very high yields, the 3 g/kg which, according to Table 7 are sufficient, are remarkably exceeded in the German and the American system. On the other hand, dietary P concentration recommended for the lower yielding cows may be insufficient for the ruminal microbes in the Dutch system.

In addition, Table 7 shows total P excretions to be expected in dairy cows fed according to each of the recommendations. This excretion value has been calculated on the assumption of a "zero-balance" of P which may not be correct for each of the specific phases of lactation, because P mobilization from the body probably occurs during peak lactation and deposition into the body may occur during advanced phases. It is, however, assumed that mobilization and deposition of P occurring in the course of lactation rather compensate each other.

Table 6. Performance of dairy cows fed diets varying in P concentration over extended periods

No. of cows	Duration	Dietary P % of DM	DMI kg/d	Milk yield kg/d	Reference
13	1 lactation	0.42		21.2	Call et al., 1987
8		0.32		22.2	
13		0.24		17.3	
39	1 year	0.42 – 0.46		17.1	Brodison et al., 1989
39		0.34 – 0.36		16.7	
26	1 lactation	0.39	17.4	24.5	Brintrup et al., 1993
26		0.33	18.1	25.4	
8	3 lactations	0.33	20.2 – 22.6	24.5 – 33.0	Valk and Sebek, 1999
8	4 lactations	0.27	19.9 – 22.5	24.1 – 34.2	
8	1 lactation	0.23	20.4	23.2	
9	44 weeks of lactation	0.49	23.4	36.2	Wu et al., 2000
9		0.40	22.4	36.5	
8		0.31	23.0	35.0	
21	1st year	0.48	20.4	28.8	Wu and Satter, 2000
21		0.38	20.7	29.6	
27	2nd year	0.48	23.4	32.1	
26		0.38	23.2	32.0	
13	2 years	0.47	24.6	39.3	Wu et al., 2001
14		0.39	25.0	38.7	
10		0.31	25.0	42.3	
5	11 weeks	0.67	24.1	45.8	Knowlton and
5		0.51	26.6	48.4	Herbein, 2002
4		0.34	25.3	49.5	
124	165 days of lactation	0.57		34.9	Lopez et al., 2004a
123		0.37		35.1	

Table 7. Three recommendations for P supply to dairy cows and relative total P excretions to be expected from cows fed according to each of the recommendations.

Milk kg/d	Recommended P supply[a] (g/d)			Total P excretion[b] (g/kg milk)		
	D	US	NL	D	US	NL
15	46	51	40	2.2	2.5	1.8
25	65	65	55	1.7	1.7	1.3
35	84	83	69	1.5	1.5	1.1
45	103	96	83	1.4	1.2	0.9
55	121	114	97	1.3	1.2	0.9

[a] Recommendations calculated according to: D – GfE, 1993; US – NRC, 2001; NL – Valk and Beynen, 2002

[b] $\dfrac{\text{recommended P supply (g/d)} - 0.9 \text{ (g/kg)} * \text{milk yield (kg/d)}}{\text{milk yield (kg/d)}}$

(assumes animals are not pregnant and not growing)

If fed according to the respective recommendations, P excretion per kg milk produced will decrease with increasing milk yield in each of the three systems. Feeding according to the Dutch standards will cause least P loading to the land. The fact that the German system causes the highest P excretion results mainly from the assumption of a P concentration of 1.0 g/kg which is not justified by the results shown in Table 5. In addition, there appears to be too much "safety margin" in the "assumed efficiency of utilization" of 0.70.

Indication of phosphate concentrations in blood plasma

The importance of plasma phosphate concentration for health and well-being of ruminants is discussed controversially. Only about 0.1% of total P in the body are present as inorganic phosphate and this small pool is precursor for several movements. As salivary flow is by no means continuous in the course of the day it has to be assumed that the size of this pool must increase or decrease rapidly at the onset or ceasing, respectively, of rumination as predominant factor determining salivary flow. This is believed to cause substantial diurnal fluctuations of plasma P_i concentrations. So far unpublished studies of our Department showed that in dairy cows fed P according to the official recommendations plasma P_i concentration may change from "hypo-phosphataemic" to "hyper-phosphataemic" within 30 minutes in individual cows. As a consequence, the indicative value of single plasma Pi concentrations is to be questioned.

Conclusion

If high yielding dairy cows are fed rations supplying adequate quantities of energy and protein, sufficient concentrations of P may be assumed in most cases due to the nature of ingredients rich in protein. Supplementing phosphates to such rations is hardly necessary for maintaining performance and health of the animals, but would increase P excretion and thus have an impact on the environment.

References

Abdel-Hafez, H.M., Manas-Almendros, M., Ross, R. and A.D. Care, 1982. Effects of dietary phosphorus and calcium on the intestinal absorption of Ca in sheep. British Journal of Nutrition 47, 69-77.

AFRC [Agricultural and Food Research Council], 1991. Technical committee on response to nutrients, report 6. A reappraisal of the calcium and phosphorus requirements of sheep and cattle. Nutrition Abstracts and Reviews (Series B) 61, 573-608.

ARC [Agricultural Research Council], 1965. The Nutrient Requirements of Farm Livestock, No 2, Ruminants. London: H.M.S.O., 264 pp.

Bailey, C.B.,1961. Saliva secretion and its relation to feeding cattle. 3. The rate of secretion of mixed saliva in the cow during eating, with an estimate of the magnitude of total daily secretion of mixed saliva. British Journal of Nutrition 15, 443-451.

Bailey, C.B. and C.C. Balch, 1961. Saliva secretion and its relation to feeding cattle. 1. The composition and rate of secretion of parotid saliva in a small steer. British Journal of Nutrition 15, 371-382.

Beardsworth, L.J., Beardsworth, P.M. and A.D. Care, 1989. The effect of ruminal phosphate concentration on the absorption of calcium, phosphorus and magnesium from the reticulo-rumen of the sheep. British Journal of Nutrition 61, 715-723.

Beede, D.K., 2003. Ration phosphorus management: Requirements and excretion. Proc. Four-State Applied Nutrition and Management Conference, Midwest Plan Service,Iowa State University,Ames, Iowa, pp. 145-151.

Beede, D.K., and J.A. Davidson, 1999. Phosphorus: Nutritional management for Y2K and beyond. Proc. Tri-State Dairy Nutrition Conference, The Ohio State University, Columbus, Ohio, pp. 51-99.

Ben-Ghedalia, D., Tagari, H., Zamwel, S. and A. Bondi, 1975. Solubility and net exchange of calcium, magnesium and phosphorus in digesta flowing along the gut of sheep. British Journal of Nutrition 33, 87-94.

Bertoni, G., Watson, M.J., Savage, G.P. and D.G. Armstrong, 1976. The movements of minerals in the digestive tract of dry and lactating Jersey cows. 1. Net movements of Ca, P, Mg, Na, K and Cl. Zootecnica e Nutrizione Animale 2, 107-118.

Bortolussi, G., Ternouth, J.H. and N.P. McMenniman, 1996. Dietary nitrogen and phosphorus depletion in cattle and their effects on live weight gain, blood metabolite concentrations and phosphorus kinetics. Journal of Agricultural Science

Boxebeld, A., Gueguen, L., Hannequart, G. and M. Durand, 1983. Utilization of phosphorus and calcium and minimal maintenance requirement for phosphorus in growing sheep fed a low-phosphorus diet. Reproduction, Nutrition, Développement 23, 1043-1053.

Braithwaite, G.D., 1974. The effect of changes of dietary calcium concentration on calcium metabolism in sheep. British Journal of Nutrition 31, 319-331.

Braithwaite, G.D., 1984. Changes in phosphorus metabolism of sheep in response to the increased demands for P associated with an intravenous infusion of calcium. Journal of Agricultural Science, Cambridge, 102, 135-139.

Braithwaite, G.D., 1985. Endogenous faecal loss of phosphorus in growing lambs and the calculation of phosphorus requirements. Journal of Agricultural Science, Cambridge, 105, 67-72.

Braithwaite, G.D., 1986. Phosphorus requirements of ewes in pregnancy and lactation. Journal of Agricultural Science, Cambridge, 106, 271-278.

Breves, G., Ross, R. and H. Höller, 1985. Dietary phosphorus depletion in sheep: effects on plasma inorganic phosphorus, calcium, 1,25-(OH)2-Vit.D3 and alkaline phosphatase and on gastrointestinal P and Ca balances. Journal of Agricultural Science, Cambridge, 105, 623-629.

Breves, G., Rosenhagen, C. and H. Höller, 1987. Die Sekretion von anorganischem Phosphor mit dem Speichel bei P-depletierten Schafen. Journal of Veterinary Medicine A 34, 442-47.

Breves, G., Höller, H., Packheiser, P., Gäbel, G. and H. Martens, 1988. Flux of inorganic phosphate across the sheep rumen wall *in vivo* and *in vitro*. Quarterly Journal of Experimental Physiology 73, 343-351.

Brintrup, R., Mooren, T., Meyer, U., Spiekers, H. and E. Pfeffer, 1993. Effects of two levels of phosphorus intake on performance and faecal phosphorus excretion of dairy cows. Journal of Animal Physiology and Animal Nutrition 69, 29-36.

Brodison, J.A., Goodall, E.A., Armstrong, J.D., Givens, D.I., Gordon, F.J., McCaughey, W.J. and J.R. Todd, 1989. Influence of dietary phosphorus on performance of lactating dairy cattle. Journal of Agricultural Science, Cambridge, 112, 303-311.

Bruce, J. Goodall, E.D., Kay, R.N.B., Phillipson, A.T. and L.E. Vowles, 1966. The flow of organic and inorganic materials through the alimentary tract of the sheep. Proceedings of the Royal Society B 166, 46-62.

Brüggemann, J., Bronsch, K., Lörcher, K. and H. Seuss, 1959. Resorptionsstudien an Nutztieren. III. Die Bestimmung der Phosphorresorption aus Mineralfuttern. Zeitschrift für Tierphysiologie, Tierernährung und Futtermittelkunde 14, 224-241.

Call, J.W., Butcher, J.E., Shupe, J.L., Lamb, R.C., Boman, R.L. and A.E. Olson, 1987. Clinical effects of low dietary phosphorus concentrations in feed given to lactating dairy cows. American Journal of Veterinary Science 48, 133-136.

Challa, J. and G.D. Braithewaite, 1988a. Phosphorus and calcium metabolism in growing calves with special emphasis on phosphorus homoeostasis. 1. Studies of the effect of changes in the dietary phosphorus intake on phosphorus and calcium metabolism. Journal of Agricultural Science, Cambridge, 110, 573-581.

Challa, J. and G.D. Braithewaite, 1988b. Phosphorus and calcium metabolism in growing calves with special emphasis on phosphorus homoeostasis. 2. Studies of the effect of different levels of phosphorus, infused abomasally, on phosphorus metabolism. Journal of Agricultural Science, Cambridge, 110, 583-589.

Challa, J. and G.D. Braithewaite, 1988c. Phosphorus and calcium metabolism in growing calves with special emphasis on phosphorus homoeostasis. 3. Studies of the effect of continuous intravenous infusion of different levels of phosphorus in ruminating calves receiving adequate dietary phosphorus. Journal of Agricultural Science, Cambridge, 110, 591-595.

Challa, J. and G.D. Braithewaite, 1989. Phosphorus and calcium metabolism in growing calves with special emphasis on phosphorus homoeostasis. 4. Studies on milk-fed calves given different amounts of dietary phosphorus but a constant intake of calcium. Journal of Agricultural Science, Cambridge, 113, 285-289.

Challa, J., Braithewaite, G.D and M.S. Dhanoa, 1989. Phosphorus homoeostasis in growing calves. Journal of Agricultural Science, Cambridge, 112, 217-226.

Childs, P.E., 2003. Phosphorus: fire from urine. 1. The discovery of phosphorus. http://www.ul.ie/~childsp/CinA/Issue60/TOC55_Urine.htm [Accessed 14 October 2003]

Coates, D.B. and J.H. Ternouth, 1992. Phosphorus kinetics of cattle grazing tropical pastures and implications for the estimation of their phosphorus requirements. Journal of Agricultural Science, Cambridge, 119, 401-409.

Deitert C. and E. Pfeffer, 1993. Effects of a reduced P supply with adequate or high Ca intake on performance and mineral balances in dairy goats during pregnancy and lactation. Journal of Animal Physiology and Animal Nutrition 69, 12-21.

Delaquis, A.M. and E. Block, 1995. Acid-Base status, renal function, water, and macromineral metabolism of dry cows fed diets differing in cation-anion difference. Journal of Dairy Science 78, 604-619.

Delcenserie, R., Menudier, A., Yzet, T., Sevestre, H. and J.L. Dupas, 2001. Intestinal pH and *Staphylococcus* contamination of pancreas, liver and pleen during experimental acute pancreatitis in mice. Pancreatology 1, 129-199.

Dillon, J. and D. Scott, 1979. Digesta flow and mineral absorption in lambs before and after weaning. Journal of Agricultural Science, Cambridge, 92, 289-297.

Edrise, B.M. and R.H. Smith, 1986. Exchanges of magnesium and phosphorus at different sites in the ruminant stomach. Archives of Animal Nutrition 36, 1019-1027.

Eidelsburger, U., Kirchgessner, M. and F.X. Roth, 1992. Zum Einfluss von Ameisensäure, Calciumformiat und Natriumhydrogencarbonat auf pH-Wert, Trockenmassegehalt, Konzentration an Carbonsäuren und Ammoniak in verschiedenen Segmenten des Gastrointestinaltraktes. 8. Untersuchung zur nutritiven Wirksamkeit von organischen Säuren in der Ferkelaufzucht. Journal of Animal Physiology and Animal Nutrition 68, 20-32.

Fallingborg, F., Christensen, L.A., Jacobsen, B.A., Ingeman-Nielsen, M., Rasmussen, H.H., Abildgaard, K. and S.N. Rasmussen, 1994. Effect of olsazine and mesalazine on intraluminal pH of the duodenum and proximal jejunum in healthy humans. Scandinavian Journal of Gastroenterologgy 29, 498-500.

Ferrell, C.L., Laster, D.B. and R.L. Prior, 1982. Mineral accretion during prenatal growth of cattle. Journal of Animal Science 54,618-624.

Field, A.C., Coop, R.L., Dingwall, R.A. and C.S. Munro, 1982. The phosphorus requirements for growth and maintenance of sheep. Journal of Agricultural Science, Cambridge, 99, 311-317.

Field, A.C., Woolliams, J.A. and R.A. Dingwall, 1985. The effect of dietary intake of calcium and dry matter on the absorption and excretion of calcium and phosphorus by growing lambs. Journal of Agricultural Science, Cambridge, 115, 237 – 243.

Fox, J. and A.D. Care, 1978. Effect of low calcium and low phosphorus diets on the intestinal absorption of phosphate in intact and parathyroidectomized pigs. Journal of Endocrinology 82, 417-424.

Fox, J. and R. Ross, 1985. Effects of low phosphorus and low calcium diets on the production and metabolic clearance rates of 1,25-dihydroxycholecalciferol in pigs. Journal of Endocrinology 105, 169-173.

Fox, J., Pickard, D.W., Care, A.D. and T.M. Murray, 1978. Effect of low phosphorus diets on intestinal calcium absorption and the concentration of calcium-binding protein in intact and parathyoidectomized pigs. Journal of Endocrinology 78, 379-387.

Garcia-Bojalil, C.M., Ammerman, C.B., Henry, P.R., Littell, R.C. and W.G. Blue, 1988. Effects of dietary phosphorus, soil ingestion and dietary intake level on performance, phosphorus utilization and serum and alimentary tract mineral concentrations in lambs. Journal of Animal Science 66, 1508-1519.

Gartner, J.W., Murphy, G. M. and W.A. Hoey, 1982. Effects of induced, subclinical phosphorus deficiency on feed intake and growth of beef heifers. Journal of Agricultural Science, Cambridge, 98, 23-29.

GfE [Ausschuss für Bedarfsnormen der Gesellschaft für Ernährungsphysiologie], 1993.Überarbeitete Empfehlungen zur Versorgung von Milchkühen mit Calcium und Phosphor. Proceedings of the Society of Nutrition Physiology 1, 108-113

Grace, N.D., Ulyatt, M.J. and J.C. Macrae, 1974. Quantitative digestion of fresh herbage by sheep. III. The movement of Mg, Ca, P, K and Na in the digestive tract. Journal of Agricultural Science, Cambridge, 82, 321-330.

Grace, N.D., Watkinson, J.H. and P.L. Martinson, 1986. Accumulation of minerals by the fœtus(es). and conceptus of single- and twin-bearing ewes. New Zealand Journal of Agricultural Research 29, 207-222.

Greene, L.W., Fontenot, J.P. and K.E. Webb, 1983a. Site of magnesium and other macromineral absorption in steers fed high levels of potassium. Journal of Animal Science 57, 503-510.

Greene, L.W., Webb, K.E. and J.P. Fontenot, 1983b. Effect of potassium level on site of absorption of magnesium and other macroelements in sheep. Journal of Animal Science 56, 1214-1221.

Grings, E.E. and R.Males, 1987. Effects of potassium on macromineral absorption in sheep fed wheat straw-based diets. Journal of Animal Science 64, 872-879.

Guéguen, L., 1962. L´utilisation digestive réelle du phosphore du foin de luzerne par le mouton, mesurée a l´aide de 32P. Annales de Biologie animale, Biochimie, Biophysique 2, 143-149.

Guéguen, L., 1963. Influence de la nature du régime alimentaire sur l´excretion fécale de phosphore endogène chez le veau. Annales de Biologie animale, Biochimie, Biophysique 3, 243-253.

Henneberg, W., 1870. Neue Beiträge zur rationellen Fütterung der Wiederkäuer. Deuerlich´sche Buchhandlung, Göttingen, 457 pp.

Hoey, W.A., Murphy, G.M. and R.J.W. Gartner, 1982. Whole body composition of heifers in relation to phosphorus status with particular reference to the skeleton. Journal of Agricultural Science, Cambridge, 98, 31-37.

Höller, H., Figge, A., Richter, J. and G. Breves, 1988. Nettoresorption von Calcium und von anorganischem Phosphat aus dem perfundierten Colon und Rectum von Schafen. Journal of Animal Physiology and Animal Nutrition 59, 9-15.

House, W.A. and A.W. Bell, 1993. Mineral accretion in the fetus and adnexa during late gestation in Holstein cows. Journal of Dairy Science 76, 2999-3010.

Hughes, M.R., Brumbaugh, P.F., Haussler, M.R., Wergedal, J.E. and D.J. Baylink, 1975. Regulation of serum $1\propto$, 25-dihydroxyvitamin D3 by calcium and phosphate in the rat. Science 190, 578-580.

Kaufmann, W. and A. Orth, 1966. Untersuchungen über Einflüsse des Futters und der Pansenfermentation auf die Speichelsekretion, Zeitschrift für Tierphysiologie, Tierernährung und Futtermittelkunde 21, 110-120.

Kaufmann, W., Pfeffer, E. and G. Dirksen, 1972. Untersuchungen zur Verdauungs-physiologie der Milchkuh mit der Umleitungstechnik am Duodenum. II. Messtechnik, Flussrate am Duodenum und Verdaulichkeit der Energie in Mägen und Darm. Fortschritte in der Tierphysiologie und Tierernährung 1, 33-37.

Kay, R.N.B., 1960a. The rate and composition of various salivary secretions in sheep and calves. Journal of Physiology 150, 515-537.

Kay, R.N.B., 1960b. The development of parotid salivary secretion in young goats. Journal of Physiology 150, 538-545.

Kay, R.N.B., 1969. Digestion of protein in the intestines of adult ruminants. Proceedings of the Nutrition Society 28, 140-151.

Kay, R.N.B. and E. Pfeffer, 1970. Movements of water and electrolytes into and from the intestine of the sheep. In: Phillipson, A.T. (ed.) Physiology of Digestion and Metabolism in the Ruminant, Oriel Press LTD., Newcastle upon Tyne, pp. 390-402.

Khorasani, G.R. and D.G. Armstrong, 1990. Effect of sodium and potassium level on the absorption of magnesium and other macro-minerals in sheep. Livestock Production Science 24, 223-235.

Khorasani, G.R. and D.G. Armstrong, 1992. Calcium, phosphorus, and magnesium absorption and secretion in the bovine digestive tract as influenced by dietary concentration of these elements. Livestock Production Science 31, 271-243.

Khorasani, G.R., Janzen, R.A., McGill, W.B. and J.J. Kennelly, 1997. Site and extent of mineral absorption in lactating cows fed whole-crop cereal grain silage or alfalfa silage. Journal of Animal Science 75, 239-248.

Kirk, D.J., Fontenot, J.P. and S. Rahnema, 1994. Effects of feeding Lasalocid and Monensin on digestive flow and partial absorption of minerals in sheep. Journal of Animal Science 72, 1029-1037.

Kleiber, M., Smith, A.H., Ralston, N.P. and A.L. Black, 1951. Radiophosphorus (P32) as tracer for measuring endogenous phosphorus in cows feces. Journal of Nutrition 45, 253-263.

Knowlton, K.F. and J.H. Herbein, 2002. Phosphorus partitioning during early lactation in dairy cows fed diets varying in phosphorus content. Journal of Dairy Science 85, 1227-1236.

Knowlton, K.F., Herbein, J.H., Meister-Weisbarth, M.A. and W.A. Wark, 2001. Nitrogen and phosphorus partitioning in lactating Holstein cows fed different sources of dietary protein and phosphorus. Journal of Dairy Science 84, 1210-1217.

Koddebusch, L. und E. Pfeffer, 1988. Untersuchungen zur Verwertbarkeit von Phosphor verschiedener Herkünfte an laktierenden Ziegen. Journal of Animal Physiology and Animal Nutrition 60, 269-275.

Law, F.M.K., Leaver, D.D., Martin, T.J., Selleck, K., Clarke, I.J. and P.J. Moate, 1994. Effect of treatment of dairy cows with slow release bovine somatotropin during the periparturient period on minerals in plasma and milk and on parathyroid hormone-released protein in milk. Journal of Dairy Science 77, 2242-2248.

Leibholz, J., 1974. The flow of calcium and phosphorus in the digestive tract of the sheep. Australian Journal of Agricultural Research 1974, 147-154.

Lofgreen, G.P., 1960. The availability of the phosphorus in dicalcium phosphate, bonemeal, soft phosphate and calcium phytate for mature wethers. Journal of Nutrition 70, 58-62.

Lofgreen, G.P. and M. Kleiber, 1953. The availability of the phosphorus in alfalfa hay. Journal of Animal Science 12, 366-371.

Lofgreen, G.P. and M. Kleiber, 1954. Further studies on the availability of the phosphorus in alfalfa hay. Journal of Animal Science 13, 258-264.

Lopez, H., Kanitz, F.D., Moreira, V.R., Wiltbank, M.C. and L.D. Satter, 2004a. Effect of dietary phosphorus on performance of lactating dairy cows: Milk production and cow health. Journal of Dairy Science 87, 139-145

Lopez, H., Kanitz, F.D., Moreira, V.R., Satter, L.D. and M.C. Wiltbank, 2004b. Reproductive performance of dairy cows fed two concentrations of phosphorus. Journal of Dairy Science 87, 146-157

Lueker, C.E. and G.P. Lofgreen, 1961. Effects of intake and calcium to phosphorus ratio on absorption of these elements by sheep. Journal of Nutrition 74, 233-238.

Manas-Almendros, M., Ross, R. and A.D. Care, 1982. Factors affecting the secretion of phosphate in parotid saliva in the sheep and goat. Quarterly Journal of experimental Physiology 67, 269-280.

Martz, F.A., Belo, A.T., Weiss, M.F., Belyea, R.L. and J.P. Goff, 1990. True absorption of calcium and phosphorus from alphalpha and corn silage when fed to lactating cows. Journal of Dairy Science 73, 1288-1295.

Maunder, E.M.W., Pillay, A.V. and A.D. Care, 1986. Hypophosphataemia and vitamin D metabolism in sheep. Quarterly Journal of Experimental Physiology 71, 391-399.

McDougall, E.L., 1948. Studies on ruminant saliva. I. The composition and output of sheep´s saliva. Biochemical Journal 43, 99-108.

Meyer, H., Flothow, C. and S. Radicke, 1997. Preileal digestibility of coconut fat and soybean oil in horses and their influence on metabolites of microbial origin of the proximal digestive tract. Archives of Animal Nutrition 50, 63-74.

Morse, D., Head, H.H., Wilcox, C.J., van Horn, H.H., Hissem, C.D. and B. Harris, 1992. Effects of concentration of dietary phosphorus on amount and route of excretion. Journal of Dairy Science 75, 3039-3049.

Müschen, H., Petri, P., Breves, G. and E. Pfeffer, 1988. Response of lactating goats to low phosphorus intake. 1. Milk yield and faecal excretion of P and Ca. Journal of Agricultural Science, Cambridge, 111, 255-263.

NRC [National Research Council], 2001. Nutrient Requirements of Dairy Cattle, 7th Revised Edition. National Academy Press, Washington, D.C., 381pp.

Perge, P., Hardebeck, H., Sommer, H. and E. Pfeffer, 1982. Untersuchungen zur Beeinflussung der Calcium- und Phosphorgehalte in Blutserum und Speichel von Hammeln durch die Versorgung. Zeitschrift für Tierphysiologie, Tierernährung und Futtermittelkunde 48, 113-121.

Pfeffer, E., 1989. Phosphorus requirements in goats. in: Recent progress on mineral nutrition and mineral requirements in ruminants. Proceedings of the International Meeting on mineral nutrition and mineral requirements in ruminants. Kyoto, Japan, 3-4 September, 1989, pp 42-46.

Pfeffer, E. and W. Kaufmann, 1972. Untersuchungen zur Verdauungsphysiologie der Milchkuh mit der Umleitungstechnik am Duodenum. 4. Netto-Bewegung einiger mineralischer Mengenelemente durch die Wand der Mägen und des Darms. Fortschritte in der Tierphysiologie und Tierernährung 1, 33-37.

Pfeffer, E. and M. Rodehutscord, 1998. Body chemical composition and utilization of dietary energy by male Saanen kids fed either milk to satiation or solid complete feeds with two proportions of straw. Journal of Agricultural Science, Cambridge, 131, 487-495.

Pfeffer, E., Thompson, A. and D.G. Armstrong, 1970. Studies on intestinal digestion in the sheep. 3. Net movement of certain inorganic elements in the digestive tract on rations containing different proportions of hay and rolled barley. British Journal of Nutrition 24, 197-204.

Pfeffer, E., Rodehutscord, M. and G. Breves, 1995. Effects of reducing dietary calcium and/or phosphorus on performance and body composition in male kids. Journal of Animal Physiology and Animal Nutrition 74, 243-252.

Pfeffer, E., Rodehutscord, M. and G. Breves, 1996. Effect of varying dietary calcium and phosphorus on growth and on composition of empty bodies and isolated bones of male kids. Archives of Animal Nutrition 49, 243-252.

Potthast, V., Moctar, M., Abdel Rahman, K. and E. Pfeffer, 1976. Untersuchungen an Hammeln über Sekretion und Absorption von Speichel-Phosphor. Zeitschrift für Tierphysiologie, Tierernährung und Futtermittelkunde 36, 334-340.

Preston, R.L. and W.H. Pfander, 1964. Phosphorus metabolism in lambs fed varying phosphorus intakes. Journal of Nutrition 83, 369-378.

Rahnema, S., Wu, Z., Ohajuruka, O.A., Weiss, W.P. and D.L. Palmquist, 1994. Site of mineral absorption in lactating cows fed high-fat diets. Journal of Animal Science 72, 229-235.

Rajaratne, A.A.J., Scott, D. and W. Buchan, 1994. Effects of a change in phosphorus requirement on phosphorus kinetics in the sheep. Research in Veterinary Science 56, 262-264.

Rajaratne, A. A. J., Buchan, W., Kyle, D. and D. Scott, 1996. Use of sheep maintained by intragastric infusion as a model for studying the endogenous loss of phosphorus in ruminants. Research in Veterinary Science 60, 92-93.

Reed, W.D.C., Elliott, R.C. and J.H. Topps, 1965. Phosphorus excretion of cattle fed on high-energy diets. Nature 208, 953-954.

Riad, F., Lefaivre, J. and J.-P. Barlet, 1987. 1,25-dihydroxycholecalciferol regulates salivary phosphate secretion in cattle. Journal of Endocrinology 112, 427-430.

Ribovich, M.L. and H.F. DeLuca, 1978. Effect of dietary calcium and phosphorus on intestinal calcium absorption and vitamin d metabolism. Archives of Biochemistry and Biophysics 188, 145-156.

Risley, C.R., Kornegay, E.T., Lindermann, M.D., Wood, C.M. and W.N. Eigel, 1992. Effect of feeding organic acids on selected intestinal content measurements at varying times postweaning in pigs. Journal of Animal Science 70, 196-206.

Rodehutscord, M. Haverkamp, R. and E. Pfeffer, 1998. Inevitable losses of phosphorus in pigs, estimated from balance data using diets deficient in phosphorus. Archives of Animal Nutrition 51, 27-38.

Rodehutscord, M., Heuvers, H. and E. Pfeffer, 2000. Effect of organic matter digestibility on obligatory faecal phosphorus loss in lactating goats, determined from balance data. Animal Science 70, 561-568.

Salih, Y., McDowell, L.R., Hentges, J.F., Mason, R.M. and C.J. Wilcox, 1987. Mineral content of milk, colostrums and serum as affected by physiological state and mineral supplementation. Journal of Dairy Science 70, 608-612.

Scarisbrick, R. and T.K. Ewer, 1951. The absorption of inorganic phosphate from the rumen of sheep. Biochemical Journal 49, 1xxix (Abstract).

Schröder, B., Käppner, H., Failing, K., Pfeffer, E. and G. Breves, 1995. Mechanisms of intestinal phosphate transport in small ruminants. British Journal of Nutrition 74, 635-648.

Schulz, E., Oslage, H. J. and R. Daenicke, 1974. Untersuchungen über die Zusammensetzung der Körpersubstanz sowie den Stoff- und Energieansatz bei wachsenden Mastbullen. Fortschritte in der Tierphysiologie und Tierernährung 4, 5-70.

Scott, D., 1978. The effects of intravenous phosphate loading on salivary phosphate secretionand plasma parathyroid hormone levels in the sheep. Quarterly Journal of Experimental Physiology 63, 147-156.

Scott, D. and W. Buchan, 1985. The effects of feeding either roughage or concentrate diets on salivary phosphorus absorption and urinary phosphorus excretion in the sheep. Quarterly Journal of Experimental Physiology 70, 365-375.

Scott, D. and W. Buchan, 1988. The effects of feeding pelleted diets made from either coarsely or finely ground hay on phosphorus balance and on the partition of phosphorus excretion between urine and faeces in sheep. Quarterly Journal of Experimental Physiology 73, 315-322.

Scott, D., McLean, A.F. and W. Buchan, 1984a. The effect of variation in phosphorus intake on net intestinal phosphorus absorption, salivary phosphorus secretion and pathway of excretion in sheep fed roughage diets. Quarterly Journal of Experimental Physiology 69, 439-452.

Scott, D., McLean, A.F. and W. Buchan, 1984b. The effects of intravenous phosphate loading on salivary phosphorus secretion, net intestinal phosphorus absorption and pathway of excretion in sheep fed roughage diets. Quarterly Journal of Experimental Physiology 69, 453-461.

Spiekers, H., Brintrup, R., Balmelli, M. and E. Pfeffer, 1993. Influence of dry matter intake on faecal phosphorus losses in dairy cows fed rations low in phosphorus. Journal of Animal Physiology and Animal Nutrition 69, 37-43.

Sutton, A. and D.K. Beede, 2003.. Feeding strategies to lower nitrogen and phosphorus in manure. In: Best Environmental Management Practices – Farm Animal Production. Publicaiton E-2822 (Michigan State University Extension), East Lansing, Michigan.

Ternouth, J.H., 1989. Endogenous losses of phosphorus by sheep. Journal of Agricultural Science, Cambridge, 113, 291-297.

Ternouth, J.H., Bortolussi, G., Coates, D.B., Hendricksen, R.E. and R.W. McLean, 1996. The phosphorus requirements of growing cattle consuming forage diets. Journal of Agricultural Science, Cambridge, 126, 503-510.

Théwis, A. and E. Francois, 1985. Intestinal absorption and secretion of total and lipid phosphorus in adult sheep fed chopped meadow hay. Reproduction, Nutrition, Développement 25, 389-397.

Théwis, A., Francois, E. and M.-F. Thielemans, 1978. Etude quantitative de l'absorption et de la sécrétion du phosphore total et du phosphore phospholipidique dans le tube digestif du mouton. Annales de Biologie animale, Biochimie, Biophysique 18, 1181-1195.

Tillman, A.D. and J.R. Brethour, 1958a. Dicalcium phosphate and phosphoric acid as phosphorus sources for beef cattle. Journal of Animal Science 17, 100-103.

Tillman, A.D. and J.R. Brethour, 1958b. Utilization of phytin phosphorus by sheep. Journal of Animal Science 17, 104-112.

Tillman, A.D. and J.R. Brethour, 1958c. Ruminant utilization of sodium meta-, ortho- and pyrophosphates. Journal of Animal Science 17, 792-796.

Tillman, A.D., Brethour, J.R. and S.L. Hansard, 1959. Comparative procedures for measuring the phosphorus requirement of cattle. Journal of Animal Science 18, 249-255.

Tomas, F.M., 1973. Parotid salivary secretion in sheep: its measurement and influence on phosphorus in rumen fluid. Quarterly Journal of Experimental Physiology 58, 131-138.

Tomas, F.M., Moir, R.J. and M. Somers, 1967. Phosphorus turnover in sheep. Australian Journal of Agricultural Research 18, 635-645.

Tribe, D.E. and L. Peel, 1963. Total salivation in grazing lambs. Australian Journal of Agricultural Research 14, 330-339.

Valdivia, R., Ammerman, C.B., Henry, P.R., Feaster, J.P. and C.J. Wilcox, 1982. Effect of dietary aluminium and phosphorus utilization and tissue mineral composition in sheep. Journal of Animal Science 55, 402-410.

Valk, H. and A.C. Beynen, 2002. Proposal for the assessment of phosphorus requirements of dairy cows. Livestock Production Science 79, 267-272.

Valk, H. and L.B.J. Sebek, 1999. Influence of long-term feeding of limited amounts of phosphorus on dry matter intake, milk production, and body weight of dairy cows. Journal of Dairy Science 82, 2157-2163.

Valk, H., Sebek, L.B.J. and A.C. Beynen, 2002. Influence of phosphorus intake on excretion and blood plasma and saliva concentrations of phosphorus in dairy cows. Journal of Dairy Science 85, 2642-2649.

van der Krogt, P., 2003. Elementymology & Elements Multidict: 15 Phosphorus. http://www.vanderkrogt. net/elements/p.html [Accessed 14 October 2003].

Walker, D.M. and S.J. Al-Ali, 1987. The endogenous phosphorus excretion of preruminant lambs. Australian Journal of Agricultural Research 38, 1061-1069.

Wildt, E., 1874. Ueber die Resorption und Secretion der Nahrungsbestandtheile im Verdauungskanal des Schafes. Journal für Landwirthschaft 22, 1-34.

Wright, R.D., Blair-West, J.R., Nelson, J.F. and G.W. Tregear, 1984. Handling of phosphate by a parotid gland (ovine.. American Journal of Physiology 246, F916-F926.

Wu, Z. and L.D. Satter, 2000. Milk production and reproductive performance of dairy cows fed two concentrations of phosphorus for two years. Journal of Dairy Science 83, 1052-1063.

Wu, Z., Satter, L.D. and R. Sojo, 2000. Milk production, reproductive performance, and fecal excretion of phosphorus by dairy cows fed three amounts of phosphorus. Journal of Dairy Science 83, 1028-1041.

Wu, Z., Satter, L.D., Biohowiak, A.J., Stauffacher, R.H. and J.H. Wilson, 2001. Milk production, estimated phosphorus excretion, and bone characteristics of dairy cows fed different amounts of phosphorus for two or three years. Journal of dairy Science 84, 1738-1748.

Wylie, M.J., Fontenot, J.P. and L.W. Greene, 1985. Absorption of magnesium and other macrominerals in sheep infused with potassium in different parts of the digestive tract. Journal of Animal Science 61, 1219–1229.

Young, V.R., Luick, J.R. and G.P. Lofgreen, 1966a. The influence of dietary phosphorus intake on the rate of bone metabolism in sheep. British Journal of Nutrition 20, 727-732.

Young, V.R., Richards, W.P.C., Lofgreen, G.P. and J.R. Luick, 1966b. Phosphorus depletion in sheep and the ratio of calcium to phosphorus in the diet with reference to calcium and phosphorus absorption. British Journal of Nutrition 20, 783-794.

Young, V.R., Lofgreen, G.P. and J.R. Luick, 1966c. The effects of phosphorus depletion, and of calcium and phosphorus intake, on the endogenous excretion of these elements by sheep. British Journal of Nutrition 20, 795-805.

Prevalence of subclinical hypocalcemia in U.S. dairy operations

R.L. Horst[1], J.P. Goff[1] and B.J. McCluskey[2]
[1]USDA-ARS-National Animal Disease Center, Ames, IA, USA; [2]Centers for Epidemiology and Animal Health, Fort Collins, CO, USA

Cows developing clinical hypocalcemia are known to experience biochemical and physiologic changes that predispose them to other diseases such as mastitis, retained placenta, displaced abomasum, and ketosis. For example, severe hypocalcemia results in higher plasma cortisol which may exacerbate the immunosuppression ordinarily present at calving, a greater decline in feed intake after calving exacerbating the negative energy balance, decreased secretion of insulin preventing tissue uptake of glucose, and increasing lipid mobilization. Although milk fever is the clinical manifestation of severe hypocalcemia, an emerging concern is subclinical hypocalcemia. Cows suffering from subclinical hypocalcemia have few overt clinical signs. However, they may be more susceptible to secondary problems than normocalcemic cows. The objective of this study was to determine the occurrence of subclinical hypocalcemia in U.S. dairy operations.

As part of the USDA's National Animal Health Monitoring System (NAHMS) Dairy 2002 study, blood samples were taken from 1,446 cows within 48 hours of parturition, representing 480 dairy operations in 21 states. Serum was harvested and frozen within 24 hours of collection. The samples were divided in three groups: 1st lactation (n=442); 2nd lactation (n=424); ≥3rd lactation (n=580). Subclinical hypocalcemia was defined as serum calcium <8.0 mg/dl. Subclinical hypocalcemia increased with advancing age and represented 25.3%, 43.9%, and 57.8% of 1st, 2nd and ≥3rd lactation cows, respectively.

Specifc management data was also evaluated for association with hypocalcemia. In this study, 38.7% of the animals were identifed as being on a DCAD program. Animals on the DCAD program suffered a signifcantly (P<0.01) lower incidence of subclinical hypocalcemia than those not being offered a DCAD program, with the biggest difference observed in the ≥3rd lactation group. Animals with calcium values of ≥8 mg/dl also had lower serum NEFAs than those that were <8 mg/dl, indicating that normocalcemic animals were in better energy status than those suffering from hypocalcemia. Subclinical hypocalcemia can induce some of the same secondary disease problems as clinical milk fever and should be viewed as an impediment to the health of a cow.

The effect of low phosphorus intake in early lactation on apparent digestibility of phosphorus and bone metabolism in dairy cows

A. Ekelund, R. Spörndly and K. Holtenius
Department of Animal Nutrition and Management. Swedish University of Agricultural Sciences, Kungsangen Research Center, S-753 23 Uppsala, Sweden

In intensive dairy farming systems fecal phosphorus excretion may cause phosphorus losses to the environment resulting in leaching to ground water and eutrophication. It is therefore essential to reduce dietary phosphorus surpluses. However phosphorus is involved in almost every aspect of metabolism and deficiency leads to decreased feed intake, lower milk production and reduced body weight[1]. The objective of this study was to attempt to improve phosphorus utilization by taking advantage of the naturally occurring bone resorption during early lactation. The hypothesis was that less phosphorus has to be fed in early lactation, when the cow is supported by phosphorus from bone tissue.

Twenty-two multiparous cows of the Swedish Red and White Breed were included in the experiment, which covered an entire lactation period plus the preceding and following dry period. Eleven cows received a dietary phosphorus concentration of 0.43% of DM during the whole lactation (NP), and eleven cows received a dietary P concentration of 0.32% of DM during the first four months of the lactation and 0.43% of DM during the rest of the lactation. Such a reduced phosphorus supply during the first four months of lactation was calculated to reduce the fecal phosphorus output into the environment by approximately 3.1 kg per animal.

The milk yield did not differ between the LP and NP cows at any period of the study. The average milk yield throughout the complete lactation was 30.4 and 30.5 kg energy corrected milk/day respectively. Total collection of feces was carried out during 5 different stages of lactation, for 5 consecutive days at each occasion. To evaluate bone metabolism, biochemical markers of bone formation and resorption respectively were monitored in blood plasma during the whole experiment. We measured osteocalcin, a specific product of the osteoblasts, to study bone formation and collagen type I fragments, specific products of the osteoclasts, as a measure of bone resorption.

At the first collection periods, 5 and 13 weeks after parturition, the apparent digestibility of phosphorus was significantly higher in the LP cows (52%), than in the NP cows (42%) when the LP cows received the low phosphorus diet. During the following collection periods, no difference in the apparent digestibility could be noted between the two groups of cows. It has been reported that cows offered a diet with 0.34 g phosphorus/kg DM were in a negative phosphorus balance in early lactation[2]. The cows in the present study were not in negative phosphorus balance in any of the 5 collection periods. However, the first collection was performed about 5 weeks after parturition, and with a collection period closer to parturition we might have observed a negative phosphorus balance. There were no significant differences in phosphorus retention between LP and NP cows at any collection period. The most pronounced retention was observed during late lactation and during the dry period. The profiles of the bone metabolism markers indicated a net resorption of bone during early lactation, but there were no differences in marker concentrations between the groups during early lactation, indicating that the LP diet did not induce a further elevated net bone resorption. Feed intake and milk production were not affected by treatment.

References

[1] Valk, H. and L.B. Šebek, 1999. Influence of long-term feeding of limited amounts of phosphorus on dry matter intake, milk production, and body weight of dairy cows. J. Dairy Sci. 82, 2157-2163.

[2] Knowlton, K.F. and H. Herbein, 2002. Phosphorus partitioning during early lactation in dairy cows fed diets varying in phosphorus content. J. Dairy Sci. 85, 1227-1236.

Bone metabolism of milk goat and sheep during pregnancy and lactation

A. Liesegang[1] and Juha Risteli[2]
[1]*Institute of Animal Nutrition, Vetsuisse Faculty University Zurich, Zurich, Switzerland;*
[2]*Department of Clinical Chemistry, University of Oulu, Oulu, Finland*

Substantial losses occur in skeletal metabolism during pregnancy and lactation. The goal of the present study was to follow these changes in pregnant and lactating goat and sheep and to investigate whether these two species show a different mobilization rate of calcium from bone. Blood samples were collected from 12 milk goat (Saanen breed) and 12 milk sheep (Ostfrisean breed) monthly during pregnancy, 2 to 3 days postpartum (pp), 2 weeks pp, 4 weeks pp, and then monthly during lactation until 6 months after parturition. Total bone mineral content (BMC) and total bone mineral density (BMD) were quantified monthly using peripheral quantitative computed tomography (pQCT). Bone resorption was measured in serum using a radioimmunoassay (RIA) and a one-step ELISA for two different domains of the carboxyterminal telopeptide of type I collagen (ICTP and crosslaps). Bone formation was quantified in serum with osteocalcin (OC) (RIA) and bone-specific alkaline phosphatase (bAP) (Enzyme immunoassay). In addition, Ca and 1,25-dihydroxy vitamin D (VITD) were determined in serum.

Mean ICTP and crosslaps concentrations of the two animal groups (goat and sheep) showed an increase in the first week after parturition. Already in the second week pp, the concentrations of both markers decreased again. Furthermore, the crosslaps concentrations were increased again in the 2nd month pp (=2nd month of lactation). In contrast, mean OC concentrations decreased slowly from the 3rd month of pregnancy until the first week pp. Already in the 2nd week pp, the mean concentrations started to increase again to reach levels as at the beginning of pregnancy. From the 3rd month pp on the concentrations started to decrease again until the end of the experiment (= 6 month after parturition). Also mean bAP activities showed a similar time course. Mean bAP activities decreased during pregnancy to reach a nadir in the first week pp in goat and 4 weeks pp in the sheep group. From the 1st week pp and the 4th week pp, mean bAP activities increased again in goat and sheep, respectively. In the second month and third month pp, respectively, the bAP activities reached a peek in goat and sheep, and decreased again in both animal species until the end of the experiment. VITD concentrations peaked in the first week pp and returned to prepartum values until 2 month pp in goat and 1 month pp in sheep, respectively. In the following months, the concentrations stayed at the same levels in sheep, but increased again in the 4th month pp in goats. Total bone mineral content and density decreased also from the 4th month of pregnancy until the first week pp in both groups. Afterwards, BMC increased until the first month pp in goat and the third month pp in sheep. BMD levels always stayed under the level of prepartum BMD in goat, whereas BMD levels of sheep returned to prepartum levels during lactation. Ca concentrations always stayed within the reference values for goat and sheep.

The increase of ICTP concentrations in both groups indicates that bone is substantially resorbed around parturition and at the beginning of lactation. This is consistent with the findings regarding BMC and BMD. At the same time, less Ca is embedded in bone, as the decrease of the OC concentrations and bAP activities indicates. During lactation BMD levels in goats remain at lower levels compared to prepartum values. In conclusion, goats and sheep resorb bone around parturition. This provides the lactating animals with adequate calcium. Increased bone remodeling during lactation may represent physiological mechanisms to help the maternal skeleton adapt to greatly increased requirements due to enormous calcium losses in milk. The mineral deficit in bone of these lactating animals seems to be in part reversible, since BMD and BMC in sheep and BMC in goat return to prepartum levels.

Influence of starvation on fermentation in bovine rumen fluid (*in vivo*)

M. Höltershinken, A. Höhling, N. Holsten and H. Scholz
Clinic for Cattle, School of Veterinary Medicine Hannover, Germany

Starvation in cattle is a frequent consequence of illness and further a consequence of longer transportation of the animal. As the consistent lack of energy is well documented, the impact on the fermentation in bovine rumen fluid is not known. The aim of this *in vivo* investigation was to look at the effects of starvation on the rumen fermentation.

Five cattle (Deutsch-Schwarz-Bunt) were investigated for nine days. After a control period of three days, while they were fed with hay and concentrate, the animals were starved for three days. Afterwards, they were again fed with hay and concentrate. Twice a day, the influence on ruminal pH, ammonia, concentrations of volatile fatty acids, potassium, sodium and chloride were investigated[1].

The following effects of starvation on the rumen fluid were noted: pH: +14.9%; ammonia −74.2%; volatile fatty acids: -65.4%; potassium: +672%; sodium: +105%; chloride: -29.5%. In summary it may be said that the fermentation in the bovine rumen is clearly decreased during starvation. Furthermore, one day after the end of the period of starvation, when hay and concentrate are fed again, the rumen of a healthy animal is able to compensate the three day long impact of starvation. In ill animals and after a longer period of starvation, however, it is rather to expect a delayed recovery of the ruminal fermentation. A working fermentation, though, is important to prevent after effects such as ketosis. Therefore it should be considered to support the recovery of the cattle by replacing rumen fluid from a healthy animal.

Further investigations on the impact of starvation on the rumen fermentation in ill cattle are carried out.

References
[1] Höltershinken M., V. Vlizlo, M. Mertens and H. Scholz, 1992. Investigations on rumen fluid characteristics, taken via stomache tube or via rumen fistula. Dtsch. Tierärztl. Wschr. 99, 228-230.

Serum mineral concentrations and periparturient disease in Holstein dairy cows

Robert J. Van Saun, Amy Todd and Gabriella Varga
Penn State University, University Park PA, USA

Serum mineral concentrations are very dynamic around the time of calving as homeostatic mechanisms are altered to facilitate transition into lactation. If homeostatic controls are unable to maintain normal physiologic mineral concentrations, a variety of periparturient metabolic diseases may occur. The objective of this study was to evaluate serum mineral concentration around the time of calving to animal health status and specific disease conditions.

A series of serum samples were obtained from 60 randomly selected cows that had participated in one of two feeding trials. Individual cow serum samples (n=8) represented a 4-wk collection period prior to and following calving and were analyzed for calcium (Ca), magnesium (Mg), potassium (K), sodium (Na), inorganic phosphorus (P), chloride (Cl), zinc (Zn), iron (Fe), copper (Cu) and selenium (Se). Veterinary disease diagnoses were recorded. Serum mineral concentrations and time-based regression coefficients were analyzed by ANOVA (repeated measures for weekly data) with time relative to calving, health status and their interaction as main effects and feeding trial as a covariate. Relative risk of postpartum disease events were determined using contingency tables of metabolite concentration categories and health status.

Healthy cows had higher Ca ($P<0.001$), Na ($P<0.0002$) and Cl ($P<0.004$) concentrations pre- and postpartum compared to disease cows. Diseased cows had lower Mg ($P<0.01$) concentrations postpartum. Sick cows had lower Ca ($P<0.05$) and Na ($P<0.0001$) concentrations at calving. Health status did not influence any mineral's change over time pre- or postpartum. Irrespective of time relative to calving, cows with serum Ca concentration below 2.0 mmol/l were 1.6 (1.1 to 2.4, 95% CI) times greater ($P<0.03$) risk for any postpartum disease. Cows with prepartum serum K concentrations greater than 4.8 mmol/l were 1.6 (1.0 to 2.4, 95% CI) times greater ($P<0.04$) risk for disease. Serum Na concentrations predicted disease risk both prepartum ($P<0.003$) and postpartum ($P<0.0003$). Cows with pre- or postpartum serum Na concentrations below 137 mmol/l were at 2.0 (1.2 to 3.3, 95% CI) or 2.7 (1.3 to 5.4, 95% CI) times greater risk for disease, respectively. Cows with mastitis had lower serum Cl ($P<0.01$) and higher Fe ($P<0.03$) concentrations compared to healthy cows across time periods. Mastitic cows also had lower ($P<0.002$) postpartum Mg concentrations. Metritis and retained placenta cows tended ($P=0.1$) to have overall lower serum Zn concentrations. Ketotic cows had lower serum Ca ($P<0.04$), Na ($P<0.001$) and Cl ($P<0.02$) concentrations compared to healthy cows across time periods. Cows with retained placenta had lower serum Ca ($P<0.006$) and Na ($P<0.03$). Serum Cl concentrations were influenced ($P<0.0001$) by and interaction between retained placenta presence and time relative to calving. Udder edema cows had lower Ca ($P<0.02$), Na ($P<0.01$) and Cl ($P<0.04$) and higher K ($P<0.05$) and Zn ($P=.1$) serum concentrations. Serum K ($P<0.05$) and Mg ($P<0.02$) concentrations were influenced by the presence of udder edema and time relative to calving.

The observed effects of disease on serum mineral concentrations, irrespective of time period relative to calving, suggest a possible diagnostic capacity to predict possible disease risk. Significance of Na, Cl and K concentrations with a number of periparturient diseases may suggest an important component of fluid regulation in periparturient disease pathogenesis.

Acknowledgements

Funded in part by the Schreyer Honors College, the College of Agricultural Sciences, and the Department of Dairy and Animal Science at The Pennsylvania State University.

The relevance of hypophosphataemia in cows

M. Fürll, T. Sattler and M. Hoops
Medizinische Tierklinik Leipzig, An den Tierkliniken 11, 04103 Leipzig, Germany

The points of view about the relevance of the since 1932 known hypophosphataemia are varying. On one hand it is regarded as meaningless, on the other hand it is taken as the cause of "hypophosphataemic downer cow syndrome". The inorganic phosphate (Pi) in blood serum of healthy cows differs between 1.55 and 2.29 mmol/l, peripartal the lower limit is 1.25mmol/l.

In controls of metabolism in cows 2 to 8 weeks post partum (W. p.p.) during 1983-1988 20%-, 1998 to 2002 22 to 56% of the cows were hypophosphataemic. Depletion and repletion of Pi leads to changes in serum. Five-day starvation does not decrease the Pi concentration significantly. Cows with hypocalcaemia have a chance of 42-64% for hypophosphataemia at the same time and of 10–33% for hypophosphataemia without hypocalcaemia. In chronical acidosis very often a hypo-phosphataemia is found.

At the time of diagnosis of DA, 12% of the cows had a hyper- and 18% a hypophosphataemia. In 94 cows with Pi concentrations lower than 1.25mmol/l during the course of disease, significant negative correlations to CK, bilirubin, glucose and creatinin, and significant positive correlations to calcium and cholesterol (?15) were to be found. The lowest Pi concentrations (median, mmol/l) were to be found in 15 cows with low calcium (0.60), high bilirubin (0.75), high AST (0.75), low cholesterol (0.78), low ph-value (0.78), low base excess (0.79), high chloride (0.79) and high CK (0.80), respectively. No significant correlations are existent between Pi and the liver function (GLDH, GGT, albumin, BHB, AST). There was a non significant correlation between low Pi and leucocytosis and high potassium. The correlations to calcium were significant but varying. There was no difference of Pi concentration to be found in cows with left or right DA. The significant lowest Pi concentrations were to be found in cows with DA and strong inflammation (endometritis, enteritis, panaritium (0.63)), followed by the group "ileus" (0.79). There was a significant difference of Pi concentration between the euthanized or dead cows (0.80) and the cured cows (0.99). The first had the lowest single values.

In 118 healthy cows of one farm the Pi concentrations were 10 days prepartum at 1.98 ± 0.41, at 3 days postpartum at 1.80 ± 0.44 and at 28 days postpartum at 1.87 ± 0.37mmol/l. The percentage of cows with Pi lower than 1.25, was 0, 10 and 6%. There was a significant correlation of Pi to be found with FFS (-0.27**), sodium (0.31**) and potassium (0.20*) concentrations, but not with the above mentioned parameters.

The clinical classification of hypophosphataemia shows, that it occurs mostly in diseases with microhaemolysis because of intoxication, especially puerperal septicaemia, and strongly disordered digestion. Its degree has connections to the prognosis.

Section I.
Micromineral nutrition, metabolism and homeostasis

Section I.

Trace element nutriture and immune function

Jerry W. Spears
Department of Animal Science, North Carolina State University, Raleigh, NC 27695-7621, USA

Abstract

Deficiencies of selenium, copper, zinc, iron, and cobalt have been shown to reduce various aspects of immunity in cattle and/or swine. The ability of neutrophils to kill ingested microorganisms is reduced by selenium, copper or cobalt deficiency. Severe zinc deficiency greatly impairs cell-mediated immunity, but there is little evidence that marginal zinc deficiency affects immune function in domestic animals. Deficiencies of certain trace elements, especially selenium, have been associated with increased disease susceptibility. Chromium supplementation of diets has increased cell-mediated and/or humoral immune response in a number of cattle studies, and has reduced incidence of respiratory disease in stressed cattle in some studies.

Keywords: trace minerals, immune response, selenium, copper

Introduction

An efficient host defense system is essential to animal health, because animals are exposed under practical conditions to a number of potential disease causing organisms. A number of trace minerals have been shown to affect various aspects of immunity in animals, and reduced disease resistance has been related to deficiencies of certain trace minerals. This paper will review the effects of trace minerals on immune responses and disease resistance in cattle and swine.

Selenium

It is impossible to discuss the role of selenium in immune function without also mentioning vitamin E. The synergistic relationship between selenium and vitamin E in regard to immune function is well documented (Stabel and Spears, 1993; Finch and Turner 1996). Although selenium and vitamin E have independent functions, one function of both nutrients relates to their involvement in the cellular antioxidant defense system. Selenium functions as a component of glutathione peroxidase, which destroys hydrogen peroxide and lipid hydroperoxides. Vitamin E scavenges free radicals and protects against lipid peroxidation. Therefore, vitamin E status of an animal can affect responses to dietary selenium, including immune response.

Selenium deficiency in cattle reduces the ability of blood and milk neutrophils to kill yeast and bacteria (Boyne and Arthur, 1979; Grasso *et al.*, 1990; Hogan *et al.*, 1990). Reduced neutrophil killing activity in selenium-deficient steers was associated with non-detectable glutathione peroxidase activity in neutrophils (Boyne and Arthur, 1979). In studies with cattle, selenium deficiency has not affected the ability of neutrophils to ingest or phagocytize yeast or bacteria. However, in sows selenium deficiency reduced both phagocytic and microbicidal activity of blood polymorphonuclear cells (Wuryastuti *et al.*, 1993).

Selenium deficiency in dairy cows resulted in a reduced response of isolated peripheral blood lymphocytes to stimulation with concanavalin A (Con A; Cao *et al.*, 1992). They suggested that the impaired response of lymphocytes from selenium-deficient cows may be related to altered arachidonic acid oxidation by lymphocytes via the 5-lipooxygenase pathway. Lymphocytes from selenium-deficient cows produced less products of arachidonic acid oxidation, specifically 5-hydroxyeicosatetraenoic and leukotriene B$_4$ when stimulated. Selenium deficiency in pigs also reduced blastogenic response of blood lymphocytes to phytohemagglutinin (PHA) stimulation (Larsen and Toillersrud, 1981). However, in other studies with swine (Wuryastyuti *et al.*, 1993) and ruminants (Stabel and Spears, 1993), low dietary selenium has not affected cell-mediated immunity.

Effects of dietary selenium on humoral immune response have also been variable (Stabel and Spears, 1993; Finch and Turner, 1996). Antibody responses have been most consistent when both selenium and vitamin E have been supplemented to diets (Finch and Turner, 1996).

Vitamin E and selenium status of dairy cows has been shown to affect the susceptibility of dairy cows to intramammary infections. Supplementing a diet low in selenium and vitamin E with 740 IU of vitamin E/day, throughout the dry period, reduced incidence of clinical mastitis at calving by 37% (Smith *et al.*, 1984). Injecting 0.1 mg of selenium per kg of body weight at 21 days prior to calving did not affect the incidence of clinical mastitis, but duration of clinical symptoms was reduced by 46% in this study (Smith *et al.*, 1984). Cows supplemented with both vitamin E and selenium had a shorter duration of clinical signs of mastitis than cows supplemented with either nutrient alone. Vitamin E and selenium also were related to rate of clinical mastitis and bulk tank milk somatic cell count in a survey study of dairy herds in Ohio (Weiss *et al.*, 1990). High serum selenium concentrations were associated with reduced rates of mastitis and lower bulk tank somatic cell counts, while high blood concentrations of vitamin E were associated with decrease rate of clinical mastitis. Experimental mastitis, induced by intramammary challenge with *E. coli* was more severe and of longer duration in cows receiving a selenium-deficient diet compared with those receiving supplemental selenium (Erskine *et al.*, 1989). However, the severity and duration of infection was not affected by selenium deficiency when mastitis was induced by intramammary challenge with *Staphylococcus aureus* (*S. aureus*; Erskine *et al.*, 1990). Vitamin E and selenium supplementation of dairy cow diets also has reduced incidence of retained placenta and severity of udder edema in some studies (Miller *et al.*, 1993).

A study with beef cows and their calves fed feedstuffs marginally deficient in selenium indicated that bimonthly selenium-vitamin E injections reduced calf death losses (4.2 vs 15.3%) from birth to weaning (Spears *et al.*, 1986). However, clinical signs and duration of infection were not affected by selenium deficiency in calves inoculated with infectious bovine rhinotracheitis virus (IBRV; Reffett *et al.*, 1988). Selenium deficiency also did not affect susceptibility of stressed steers to *Pasteuralla hemolytica* (*P. hemolytica*) challenge (Stabel *et al.*, 1989). Injecting stressed steers with selenium did not affect health status during the receiving period (Droke and Loerch, 1989).

Selenium addition to a basal diet deficient in selenium and vitamin E increased disease resistance in pigs inoculated with *Treponema hyodysenteriae* (Teige *et al.*, 1982). During a 22-day period following inoculation, selenium-deficient pigs had a higher mortality rate, lower feed intake, lower daily gain, and greater incidence of diarrhea compared with selenium-supplemented pigs.

Copper

Copper deficiency is a problem in ruminants primarily because of dietary antagonists that greatly reduce copper absorption. It is well documented that high dietary concentrations of molybdenum, sulfur, and iron reduce copper status in ruminants. In several immune studies, copper deficiency has been induced by feeding high concentrations of molybdenum or iron. In these studies, it is impossible to separate the effects of copper deficiency, *per se*, from possible direct effects of high dietary molybdenum or iron on immune variables.

Superoxide dismutase, a copper metalloenzyme, prevents oxidative tissue damage from superoxide radicals. Phagocytic cells, when stimulated by invading microorganisms, produce large amounts of superoxide radicals that must be detoxified by superoxide dismutase. Ceruloplasmin, another copper enzyme, exhibits anti-inflammatory activity and increases in the blood during infection and inflammation. In copper deficiency, superoxide dismutase and ceruloplasmin activities are reduced, and this may increase the likelihood of oxidative tissue damage resulting from infection and inflammation.

Neutrophils isolated from calves fed a diet severely deficient in copper had impaired ability to kill ingested *Candida albicans* (Boyne and Arthur, 1981). Copper deficiency did not affect the ability of phagocytic cells to phagocytize *C. albicans*. Neutrophils from calves born to cows fed a control diet marginally deficient in copper or the control diet supplemented with 2.5 mg Mo/kg tended to have depressed ability to kill *S. aureus* (Gengelbach *et al.*, 1997). Marginal copper deficiency in dairy heifers reduced neutrophil killing of *S. aureus* (Torre *et al.*, 1996). Copper deficiency induced by feeding cattle either 5 mg Mo or 500 mg Fe/kg diet depressed the ability of neutrophils to not only kill but also ingest *C. albicans* (Boyne and Authur, 1986).

A number of studies in rats and mice have indicated that both cell-mediated and humoral immunity is greatly reduced by copper deficiency (Prohaska and Failla, 1993). In young pigs, blastogenic response of blood mononuclear cells to stimulation with PHA and Con A was reduced by copper deficiency (Bala *et al.*, 1992). However, studies in cattle have not shown consistent effects of copper deficiency on either cell-mediated or humoral immune response (Minatel and Carfagnini, 2000; Spears, 2000).

In vitro lymphocyte blastogenesis has not been affected by copper deficiency in cattle (Stabel *et al.*, 1993; Arthington *et al.*, 1996; Ward *et al.*, 1997). *In vivo* cell-mediated immunity, assessed by measuring the response to an intradermal injection of PHA, has been reduced by copper deficiency (Ward *et al.*, 1997). Recently, Dorton *et al.* (2003) reported that cattle supplemented with 20 mg Cu/kg had a greater response to PHA administration than those supplemented with 10 mg Cu/kg diet.

There is evidence that dietary copper may affect cytokine production by immune cells in cattle. Mononuclear cells from lactating dairy cows receiving a marginal level of copper (6 to7 mg/kg diet) produced less interferon-γ (IFN-γ) when stimulated with Con A than cells isolated from cows fed adequate copper (Torre *et al.*, 1995). Interleukin-2 (IL-2) production by mononuclear cells was not affected in this study. Production of tumor necrosis factor-α (TNF-α) and IL-1 by isolated peripheral-blood monocytes following stimulation with *E. coli* lipopolysaccharide tended to be lower from monocytes obtained from calves fed a low-copper diet supplemented with 5 mg Mo/kg diet (Gengelbach and Spears, 1998). Copper deficient calves also had lower plasma TNF-α concentrations than control calves (Gengelbach et al., 1997).

Disease resistance has been studied experimentally in copper-deficient cattle by inoculating animals with pathogenic organisms. Clinical signs of respiratory disease following inoculation with IBRV and *P. hemolytica* were similar in copper-deficient and copper-adequate calves (Stabel et al., 1993). The rectal temperature response to intranasal inoculation with bovine herpervirus-1 was similar in heifers fed a diet supplemented with adequate copper and those fed the basal diet supplemented with molybdenum to achieve a dietary Mo:Cu ratio of 2.5:1 (Arthington et al., 1996). However, following the viral challenge, plasma fibrinogen (an acute-phase protein) increased in heifers fed the high molybdenum diet, but not in those fed the copper-adequate diet. It is unclear if copper deficiency affects disease susceptibility of cattle under practical conditions. Naturally occurring copper deficiency in lambs grazing improved pastures resulted in increased susceptibility to bacterial infections and greater mortality (Woolliams et al., 1986). Copper deficiency observed in this study was at least partly due to molybdenum, as pastures contained 1.2 to 3.1 mg Mo/kg DM.

Zinc

An array of research in humans and laboratory animals has indicated that zinc deficiency reduces immune responses and disease resistance (Chesters, 1997). A number of zinc dependent enzymes are involved in protein synthesis and cell division. The immune system depends on rapid cell proliferation when functioning properly; thus, it is not surprising that the immune system is adversely affected by zinc deficiency.

Zinc deficiency in young pigs drastically reduced thymus weight and increased leukocyte count (Miller et al., 1968). Expressed as a percentage of the total leukocyte count, lymphocytes were lower while neutrophils were higher in zinc-deficient pigs. Reduced antibody responses following vaccination with *S. hyodysenterial* have been reported in zinc-deficient pigs (Hall et al., 1993). Lymphocytes from zinc-deficient pigs also had reduced proliferative response to stimulation with pokeweed mitogen but not to PHA and Con A stimulation.

In cattle, surprisingly little research has been carried out to examine the relationship between dietary zinc and immune function. Controlled studies in steers (Spears and Kegley, 2002) and lambs (Droke and Spears, 1993; Droke et al., 1993) suggest that marginal zinc deficiency does not impair cell-mediated or humoral immune responses. This finding is in contrast with research in rats and humans in which even marginal zinc deficiency reduces immune responses (Fraker et al., 1984).

A genetic disorder (lethal trait A46) of zinc metabolism has been reported in Holstein and Shorthorn calves that results in severe zinc deficiency due to impaired zinc absorption. Thymus atrophy and reduced lymphocyte response to mitogen stimulation are seen in calves with lethal trait A46 (Perryman et al., 1989). Addition of 25 mg Zn/kg to a control diet containing 33 mg Zn/kg increased weight gain in growing steers by 10% (Spears and Kegley, 2002). However, zinc supplementation did not increase *in vitro* lymphocyte responses to mitogen stimulation, *in vivo* cellular response to intradermal PHA administration or antibody response following vaccination with IBRV. Lambs fed a semi-purified diet severely deficient in zinc showed a reduced blastogenic response to PHA (a T-cell mitogen) but an increased response to pokeweed mitogen (a T-dependent B-cell mitogen; Droke and Spears, 1993). Zinc deficient lambs had a lower percentage of lymphocytes and a higher percentage of neutrophils in blood. Immune responses in lambs fed a diet marginally deficient (8.7 mg Zn/kg) in zinc did not differ from those observed in lambs fed adequate zinc (44 mg Zn/kg; Droke and Spears, 1993). Marginal zinc deficiency in pregnant ewes also did not affect cellular immunity in their lambs (White *et al.*, 1994),

Despite the lack of effects of marginal zinc deficiency on immune function in ruminants, limited research suggests that the addition of zinc to practical diets may affect disease resistance. Increasing the level of supplemental zinc from 30 to 100 mg/kg slightly reduced morbidity from respiratory diseases in newly weaned calves that had been transported (Galyean *et al.*, 1995).

Chromium

It has been recognized for some time that chromium enhances insulin function. Considerable research has also indicated that chromium can affect immune response and disease resistance in cattle. Responses to supplemental chromium have been highly variable. Factors that may contribute to the inconsistent findings include differences between studies in: 1) amount of available chromium in the control diet, 2) form of chromium supplemented, 3) initial chromium status of the animals, and 4) type or degree of stress imposed on the animals. Concentrations of chromium in control diets have not been reported in many studies and furthermore, little is known regarding the bioavailability of chromium from various feeds.

Lymphocytes from dairy cows supplemented with 0.5 mg Cr/kg, as a chromium-amino acid chelate, from 6 weeks prepartum to 16 weeks postpartum had increased blastogenic responses to Con A stimulation (Burton *et al.*, 1993). Chromium supplementation prevented the decrease in blastogenic response that was observed in control cows 2 weeks prepartum. Addition of blood serum from chromium-supplemented cows to cultures of lymphocytes isolated from control cows also increased Con A-induced lymphocyte blastogenesis (Burton *et al.*, 1995). In stressed calves, chromium supplementation increased Con A-induced lymphocyte blastogenesis in calves showing signs of morbidity, but not in calves with normal body temperatures and no visual signs of sickness (Chang *et al.*, 1994). Other studies (Kegley and Spears, 1995; Kegley *et al.*, 1996; Arthington *et al.*, 1997) have found no effects of chromium supplementation on *in vitro* lymphocyte blastogenesis in cattle.

Effects of chromium on cell-mediated immunity have also been evaluated *in vivo* by measuring induration responses following percutaneous application of dinitrochlorobenzene or intradermal injection of PHA. Chromium supplementation did not affect response to dinitrochlorobenzene application (Moonsie-Shageer and Mowat, 1993; Kegley *et al.*, 1997). Addition of 0.4 mg Cr/kg diet, from either chromium chloride or a chromium-nicotinic acid complex, increased inflammatory responses to an intradermal injection of PHA in young calves receiving milk (Kegley *et al.*, 1996).

Mononuclear cells from dairy cows supplemented with a chromium-amino acid chelate produced lower concentrations of IL-2, IFN-γ and TNF-α following stimulation with Con A than cells from control cows (Burton *et al.*, 1996). However, plasma TNF-α concentrations were not affected by chromium supplementation, from high-chromium yeast, in calves before or following inoculation with bovine herpesvirus-1 (Arthington *et al.*, 1997).

Chromium does not appear to affect the ability of neutrophils to ingest (Chang *et al.*, 1996) or kill (Arthington *et al.*, 1997) foreign materials. Effects of chromium on humoral immunity have been variable (Spears, 2000).

Supplementation of stressed calves with chromium has reduced morbidity following transport in some studies (Moonsie-Shageer and Mowat, 1993; Lindell *et al.*, 1994), but not in other studies (Chang and Mowat, 1992; Chang *et al.*, 1995; Mathison and Engstrom, 1995). Chromium picolinate supplementation (3.5 mg Cr/day) of dairy cows during the last 9 weeks of pregnancy reduced incidence of retained placenta after parturition from 56 to 15% (Villalobos-F *et al.*, 1997).

Effects of dietary chromium on physiological responses of calves to an experimental disease challenge have also been evaluated. Supplementation of chromium was provided for 49 to 75 days before disease challenge in these studies. Supplementation with 0.4 mg Cr/kg diet (as either chromium-nicotinic acid complex or chromium chloride) tended to lower body temperatures in calves at certain points after intranasal inoculation with IBRV followed by *P. hymolytica* intratracheally 5 days later (Kegley *et al.*, 1996). Supplementing calves with 0.4 mg Cr/kg diet for 56 days before transport did not affect body temperature or feed intake responses to an IBRV challenge (Kegley *et al.*, 1997). Rectal temperature responses also were not affected by chromium in calves inoculated with bovine herpesvirus-1 (Arthington *et al.*, 1997).

In weanling pigs, chromium supplementation tended to improve blastogenic response of lymphocytes to mitogen stimulation (van Heugten and Spears, 1997). However, dietary chromium had little effects on the humoral immune response in sows or on transfer of antibodies to the neonate (van de Ligt *et al.*, 2002).

Iron

The relationship of iron status to immunity and disease resistance has been a controversial area for some time. A number of studies in humans and laboratory animals have shown that iron deficiency depresses phagocytic function as well as cell-mediated and humoral immune responses (Sherman and Spear, 1993). However, microorganisms require iron for growth, and in some studies, providing supplemental iron even to iron deficient subjects, has increased incidence of disease (Sherman and Spear, 1993). During an infection, iron is sequestered in response to cytokines released by immune cells.

Iron deficiency in cattle is unlikely because iron concentrations in practical feedstuffs generally exceed dietary requirements. Pigs are born with low iron stores and are generally given injections of iron shortly after birth to prevent anemia. Injecting pigs with 200 mg of iron from iron dextran increased growth and hematocrit compared with unsupplemented pigs (Hammerberg *et al.*, 1984). However, no differences in cell-mediated immunity were observed between unsupplemented and iron-supplemented pigs. Sells *et al.* (1991) found that iron deficiency in pigs suppressed natural killer cell cytotoxicity, but lymphocyte blastogenesis was not affected by iron deficiency. Oral administration of 200 mg of iron (from iron dextran at 30 minutes after birth) increased severity of diarrhea and incidence of mortality in pigs challenged with *E. coli* 6 hours after birth (Kadis *et al.*, 1984). Administration of 50 mg of iron orally did not increase mortality compared with unsupplemented pigs following *E coli* challenge. These results indicate that excess iron supplementation may increase bacterial growth under certain conditions, thus promoting disease.

Cobalt

Limited research indicates that dietary cobalt affects neutrophil function and resistance to parasitic infection in ruminants. Neutrophils isolated from calves deficient in cobalt had reduced ability to kill *C. albicans* (MacPherson *et al.*, 1987). Cobalt-deficient calves had a decreased prepatent period and increased fecal egg output following experimental infection with *Ostertagia ostertagi* (MacPherson *et al.*, 1987). Higher fecal egg counts were also observed in cobalt deficient lambs after natural infection with gastrointestinal nematodes (Vellema *et al.*, 1996).

References

Arthington, J.D., L.R. Corah and F. Blecha, 1996. The effect of molybdenum-induced copper deficiency on acute-phase protein concentrations, superoxide dismutase activity, leukocyte numbers, and lymphocyte proliferation in beef heifers inoculated with bovine herpesvirus-1. J. Anim. Sci. 74, 211-217.

Arthington, J.D., L.R. Corah, J.E. Minton, T.H. Elsasser and F. Blecha, 1997. Supplemental dietary chromium does not influence ACTH, cortisol, or immune responses in young calves inoculated with bovine herpesvirus-1. J. Anim. Sci. 75, 217-223.

Bala, S., J.K. Lunney and M.L. Failla, 1992. Effects of copper deficiency on T-cell mitogenic responsiveness and phenotypic profile of blood mononuclear cells from swine. Am. J. Vet. Res. 53, 1231-1235.

Boyne, R. and J.R. Arthur, 1979. Alterations of neutrophil function in selenium deficient cattle. J. Comp. Path. 89, 151-158.

Boyne, R. and J.R. Arthur, 1981. Effects of selenium and copper deficiency on neutrophil function in cattle. J. Comp. Path. 91, 271-276.

Boyne, R. and J.R. Arthur, 1986. Effects of molybdenum and iron induced copper deficiency on the viability and function of neutrophils from cattle. Res. Vet Sci. 41, 417-419.

Burton, J.L., B.A. Mallard and D.N. Mowat, 1993. Effects of supplemental chromium on immune responses of periparturient and early lactation dairy cows. J. Anim. Sci. 71, 1532-1539.

Burton, J.L., B.J. Nonnecke, P L. Dubeski, T.H. Elsasser and B.A. Mallard, 1996. Effects of supplemental chromium on production of cytokines by mitogen-stimulated bovine peripheral blood mononuclear cells. J. Dairy Sci. 79, 2237-2246.

Burton, J.L., B.J. Nonnecke, T.H. Elsasser B.A. Mallard, W. Z. Yang, and D. N. Mowat, 1995. Immunomodulatory activity of blood serum from chromium-supplemented periparturient dairy cows. Vet. Immuno. Immunopathol. 49, 29-38.

Cao, Y., J.F. Maddox, A.M. Mastro, R.W. Scholz, H. Hildenbrandt and C.C. Reddy, 1992. Selenium deficiency alters the lipoxygenase pathway and mitogenic response in bovine lymphocytes. J. Nutr. 122, 2121-2127.

Chang, X. and D.N. Mowat, 1992. Supplemental chromium for stressed and growing feeder calves. J. Anim. Sci. 70, 559-565.

Chang, X., B.A. Mallard and D.N. Mowat, 1994. Proliferation of peripheral blood lymphocytes of feeder calves in response to chromium. Nutr. Res. 14, 851-864.

Chang, X., B.A. Mallard and D.N. Mowat, 1996. Effects of chromium on health status, blood neutrophil phagocytosis and *in vitro* lymphocyte blastogenesis of dairy cows. Vet. Immuno. Immunopathol. 52, 37-52.

Chang, X., D.N. Mowat and B.A. Mallard, 1995. Supplemental chromium and niacin for stressed feeder calves. Can. J. Anim. Sci. 75, 351-358.

Chesters, J.K., 1997. Zinc. In: Handbook of Nutritionally Essential Mineral Elements. B.L. O'Dell and R.A. Sunde, ed. Marcel Dekker Inc., New York, pp 185-230.

Dorton, K.L., T.E. Engle, D.W. Hamar, P.D. Siciliano and R.S. Yemm, 2003. Effects of copper source and concentration on copper status and immune function in growing and finishing steers. Anim. Feed Sci. Technol. 110, 31-44.

Droke, E.A. and S.C. Loerch, 1989. Effects of parenteral selenium and vitamin E on performance, health and humoral immune response of steers new to the feedlot environment. J. Anim. Sci. 67, 1350-1359.

Droke, E.A. and J.W. Spears, 1993. *In vitro* and *in vivo* immunological measurements in growing lambs fed diets deficient, marginal or adequate in zinc. J. Nutr. Immunol. 2, 71-90.

Droke, E.A., J.W. Spears, T.T. Brown, Jr. and M.A. Qureshi, 1993. Influence of dietary zinc and dexamethasone on immune responses and resistance to *Pasteurella hemolytica* challenge in growing lambs. Nutr. Res. 13, 1213-1226.

Erskine, R.J., R.J. Eberhart, P.J. Grasso and R.W. Scholz, 1989. Introduction of *Escherichia coli* mastitis in cows fed selenium-deficient or selenium-supplemented diets. Am. J. Vet. Res. 51, 2093-2100.

Erskine, R.J., R.J. Eberhart and R.W. Scholz, 1990. Experimentally induced *Staphylococcus aureus* mastitis in selenium-deficient and selenium-supplemented dairy cows. Am J. Vet. Res. 51, 1107-1111.

Finch, J.M. and R.J. Turner. 1996, Effects of selenium and vitamin E on the immune responses of domestic animals. Res. Vet. Sci. 60, 97-106.

Fraker, P.J., K. Hildebrandt and R.W. Lueck, 1984. Alteration of antibody-mediated responses of suckling mice to T-cell dependent and independent antigens by maternal marginal zinc deficiency: Restoration of responsively by nutritional repletion. J. Nutr. 114, 170-179.

Gengelbach, G.P. and J.W. Spears, 1998. Effects of dietary copper and molybdenum on copper status, cytokine production, and humoral immune response of calves. J. Dairy Sci. 8, 3286-3292.

Gengelbach, G.P., J.D. Ward, J.W. Spears and T.T. Brown, Jr., 1997. Effects of copper deficiency and copper deficiency coupled with high dietary iron or molybdenum on phagocytic cell function and response of calves to a respiratory disease challenge. J. Anim. Sci. 75, 1112-1118.

Glayean, M.L., K.J. Malcolm-Callis, S.A. Gunter and R.A. Berrie. 1995, Effects of zinc source and level and added copper lysine in the receiving diet on performance by growing and finishing steers. Prof. Anim. Sci. 11, 139-148.

Grasso, P.J., R.W. Scholz, R.J. Erskine and R.J. Eberhart, 1990. Phagocytosis, bactericidal activity, and oxidative metabolism of milk neutrophils from dairy cows fed selenium-supplemented and selenium-deficient diets. Am. J. Vet. Res. 51, 269-274.

Hall, V.L., R.C. Ewan and M.J. Wannemuehler, 1993. Effect of zinc deficiency and zinc source on performance and immune response in young pigs. J. Anim. Sci. 71(Suppl. 1), 173. (Abstr.)

Hammerberg, C., E.T. Kornegay, H.P. Veit and G. Schurig, 1984. The effect of iron supplementation on the cell-mediated immunity of iron deficiency piglets. Nutr. Reports Int. 30, 347-353.

Hogan, J.S., K.L. Smith, W.P. Weiss, D.A. Todhunter and W.L. Schockey, 1990. Relationship among vitamin E, selenium and bovine neutrophils. J. Dairy Sci. 73, 2372-2378.

Kadis, S., F.A. Udeze, J. Polanco and C.W. Dressen, 1984. Relationship of iron administration to susceptibility of newborn pigs to enterotoxic colibacillosis. Am. J. Vet. Res. 45, 255-259.

Kegley, E.B. and J.W. Spears, 1995. Immune response, glucose metabolism, and performance of stressed feeder calves fed inorganic or organic chromium. J. Anim. Sci. 73, 2721-2726.

Kegley, E.B., J.W. Spears and T.T. Brown, Jr., 1996. Immune response and disease resistance of calves fed chromium nicotinic acid complex or chromium chloride. J. Dairy Sci. 79, 1278-1283.

Kegley, E.B., J.W. Spears and T.T. Brown, Jr., 1997. Effect of shipping and chromium supplementation on performance, immune response, and disease resistance of steers. J. Anim. Sci. 75, 1956-1964.

Larsen, H.J. and S. Tollersrud, 1981. Effect of dietary vitamin E and selenium on the phytohaemagglutinin response of pig lymphocytes. Res. Vet. Sci. 31, 301-305.

Lindell, S.A., R.T. Brandt, Jr., J.E. Milton, F. Blecha, G.L. Stokka and C.T. Milton, 1994. Supplemental Cr and revaccination effects on performance and health of newly weaned calves. J. Anim. Sci. 72(Suppl. 1), 133. (Abstr.)

MacPherson, A., D. Gray, G.B.B. Mitchell and C.N. Taylor, 1987. Ostertagia infection and neutrophil function in cobalt-deficient and cobalt-supplemented cattle. Br. Vet. J. 143, 348-353.

Mathison, G.M. and D.F. Engstrom, 1995. Chromium and protein supplements for growing-finishing beef steers fed barley-based diets. Can. J. Anim. Sci. 75, 549-558.

Miller, E.R., R.W. Luecke, D.E. Ullrey, B.V. Baltzer, B.L. Bradley and J.A. Hoefer, 1968. Biochemical, skeletal, and allometric changes due to zinc deficiency in the baby pig. J. Nutr. 95, 278-286.

Miller, J.K., E. Brzezinska-Slebodzinska and F.C. Madsen, 1993. Oxidant stress, antioxidants, and animal function. J. Dairy Sci. 76, 2812-2823.

Minatel, L. and J.C. Carfagnini, 2000. Copper deficiency and immune response in ruminants. Nutr. Res. 20, 1519-1529.

Moonsie-Shageer, S. and D.N. Mowat, 1993. Effect of level of supplemental chromium on performance, serum constituents, and immune status of stressed feeder calves. J. Anim. Sci. 71, 232-238.

Perryman, L.E., D.R. Leach, W.C. Davis, W.D. Mickelson, S.R. Ochs, J.A. Ellis and E. Brummerstedt, 1989. Lymphocyte alterations in zinc-deficient calves with lethal trait A46. Vet. Immuno. Immunopath. 21, 239-248.

Prohaska, J.R., and M.L. Failla, 1993. Copper and immunity. In: Human Nutrition - A Comprehensive Treatise, Vol. 8., D.M. Klurfeld, ed. Plenum Press, New York, pp. 309-332.

Reffett, J.K., J.W. Spears and T.T. Brown, Jr., 1988. Effect of dietary selenium on the primary and secondary immune response in calves challenged with infectious bovine rhinotracheitis virus. J. Nutr. 118, 229-235.

Sells, J.S., J. J. McGlone, R.J. Hurst and E.A. Lumpkin-Miller, 1991. Interactive effects of supplemental iron, and dietary fat on piglet performance, immunity and hemoglobin concentration. Texas Tech. Univ. Agric. Sci. Tech. Rep. No. T-5-283. p. 74-75.

Sherman, A.R. and A.T. Spear, 1993. Iron and immunity. In: Human Nutrition - A Comprehensive Treatise, Vol. 8., D.M. Klurfeld, ed. Plenum Press, New York, pp 285-307.

Smith, L.K., J.H. Harrison, D.D. Hancock, D.A. Todhunter and H.R. Conrad, 1984. Effect of vitamin E and selenium supplementation on incidence of clinical mastitis and duration of clinical symptoms. J. Dairy Sci. 67, 1293-1300.

Spears, J.W., 2000. Micronutrients and immune function in cattle. Proc. Nutr. Soc. 59, 587-594.

Spears, J.W. and E.B. Kegley, 2002. Effect of zinc source (zinc oxide vs zinc proteinate) and level on performance, carcass characteristics, and immune response of growing and finishing steers. J. Anim. Sci. 80, 2747-2752.

Spears, J.W., R.W. Harvey and E.C. Segerson, 1986. Effects of marginal selenium-deficiency and winter protein supplementation on growth, reproduction, and selenium status of beef cattle. J. Anim. Sci. 63, 586-594.

Stabel, J.R. and J.W. Spears, 1993. Role of selenium in immune responsiveness and disease resistance. In: Human Nutrition - A Comprehensive Treatise, Vol. 8. D.M. Klurfeld, ed. Plenum Press, New York, pp. 333-356.

Stabel, J.R., J.W. Spears and T.T. Brown, Jr., 1993. Effect of copper deficiency on tissue, blood characteristics, and immune function of calves challenged with infectious bovine rhinotracheitis virus and *Pasteurella hemolytica*. J. Anim. Sci. 71, 1247-1355.

Stabel, J.R., J.W. Spears, T.T. Brown, Jr. and J. Brake, 1989. Selenium effects on glutathione peroxidase and the immune response of stressed calves challenged with *Pasteurella hemolytica*. J. Anim. Sci. 67, 557-564.

Teige, J., S. Tollersrud, A. Lund and H.J. Larsen, 1982. Swine dysentery: the influence of dietary vitamin E and selenium on the clinical and pathological effects of *Treponema hyodyserteriae* infection in pigs. Res. Vet. Sci. 32, 95-100.

Torre, P.M., R.J. Harmon, R.W. Hemken, T.W. Clark, D.S. Trammell and B.A. Smith, 1996. Mild dietary copper insufficiency depresses blood neutrophil function in dairy cattle. J. Nutr. Immunol. 4, 3-24.

Torre, P.M., R.J. Harmon, L.M. Sordillo, G.A. Boissonneault, R.W. Hemken, D.S. Trammell and T.W. Clark, 1995. Modulation of bovine mononuclear cell proliferation and cytokine production by dietary copper insufficiency. J. Nutr. Immunol. 3, 3-20.

Villalobos-F, J.A., C. Romero-R, R. Tarrago-C,and A. Rosado, 1997. Supplementation with chromium picolinate reduces the incidence of placental retention in dairy cows. Can. J. Anim. Sci. 77, 329-330.

van de Ligt, J.L.G., M.D. Lindemann, R.J. Harman, H.J. Monegue and G.L Cromwell, 2002. Effect of chromium tripicolinate supplementation on porcine immune response during the postweaning period. J. Anim. Sci. 80, 456-466.

van Heugten, E. and J.W. Spears, 1997. Immune response and growth of stressed weanling pigs fed diets supplemented with organic or inorganic forms of chromium. J. Anim. Sci. 75, 409-416.

Vellema, P., V.P. M.G. Rutten, A. Hoek, L. Moll and G.H. Wentink, 1996. The effect of cobalt supplementation on the immune response in vitamin B_{12} deficient Texel lambs. Vet. Immunol. Immunopath. 55, 151-161.

Ward, J.D., G.P. Gengelbach,and J.W. Spears, 1997. The effects of copper deficiency with or without high dietary iron or molybdenum on immune function of cattle. J. Anim. Sci. 75, 1400-1408.

Weiss, W.P., J.S. Hogan, K.L. Smith and K.H., Hoblet, 1990. Relationships among selenium, vitamin E, and mammary gland health in commercial dairy herds. J. Dairy Sci. 73, 381-390.

White, C.L., H.J. Kyme and M.J. Barnes, 1994. Marginal zinc deficiency in pregnant Merino ewes. Proc. Nutr. Soc. Aust. 18, 85.

Woolliams, C., N.F. Suttle, J.A. Woolliams, D.G. Jones and G. Wiener, 1986. Studies on lambs genetically selected for low and high copper status. 1. Differences in mortality. Anim. Prod. 43, 293-301.

Wuryastuti, H., H.D. Stowe, R.W. Bull, and E.R. Miller, 1993. Effects of vitamin E and selenium on immune responses of peripheral blood, colostrum, and milk leukocytes of sows. J. Anim. Sci. 71, 2464-2472.

Expected changes of the selenium content in foods of animal origin at a change from inorganic to organic selenium compounds for supplementation of the diet of farm animals

B.G. Pehrson
Department of Animal Environment and Health, Swedish, University of Agricultural Sciences, P.O. Box 234, S-53223 Skara, Sweden

It is well known that organic selenium compounds are retained to a higher degree in tissues than inorganic compounds[1]. If the diet of animals will be supplemented with organic instead of inorganic selenium, it can be expected that an optimal selenium status and a secured immune capacity will follow, and that the most common of the selenium deficiency diseases, nutritional muscular degeneration, will be eliminated[2]. Moreover, it will significantly increase the selenium content in different foods of animal origin, and, consequently, increase the daily selenium intake in human beings[3, 13].

Hitherto, no one has tried to quantify the degree of the increased selenium status in man which will follow a change from inorganic to organic selenium compounds for supplemention of the diet of farm animals. However, studies of the literature reveal that it can be expected that - at an equal dosage level of selenium - the content of milk will increase by at least 100%[3,4,5,6], of meat from beef and swine by about 75%[5,7,8], of chicken meet by about 100%[9], of eggs by at least 100%[10], and of liver by about 25%[5,8]. It can also be presumed that the selenium content in cheese will increase even to a significantly higher degree than in milk. However, there are no data in the literature about cheese, but an ongoing study at our department will make it possible to present such data at the conference.

In Sweden, the average intake of human beings is 33 µg per day, which is lower than the officially recommended intake of 45 µg[11]. In many other European countries the average intake is equal to that in Sweden, but the figures for recommended intake are always higher (up to 75 µg)[12]. The situation in USA varies significantly between different regions, but in many states it is comparable to that in Sweden[12].

With the eating habits of the Swedish population, a change to organic selenium in the mineral feeds for animals will increase the intake of selenium by at least 50%[3]. A similar effect can be expected in many other countries, although its degree will vary with the eating habits in each particular country. As a conclusion it can be assumed that a change from inorganic selenium (as selenite or selenate) to organic selenium (as adequately produced selenium yeast) for supplementation of the diets of farm animals should be recommended, at least in countries where the selenium status of the animals and/or man is insufficient. In fact, such a situation exists in most European countries as well as in many states of USA[12].

References

[1] Wagner, P.D. and J.A. Butler, 1988. Effects of various dietary levels of selenium as selenite or selenomethionine on tissue selenium levels and glutathione peroxidase activity in rats. J. Nutr. 118, 846-852.

[2] Pehrson, B., K. Ortman, N. Madjid and U. Trafikowska, 1999. The influence of dietary selenium as selenium yeast or sodium selenite on the concentration of selenium in the milk of suckler cows and on the selenium status of their calves. J. Anim. Sci. 77, 3371-3376.

[3] Pehrson, B. and A. Arnesson, 2003. Tillförsel av oorganiskt och organiskt selen till mjölkkor (Supplementation with inorganic and organic selenium to dairy cows). Svensk Vet. Tidn. 55, 17-23.

[4] Ortman, K. and B. Pehrson B., 1999. Effect of selenate as feed complement to dairy cows in comparison to selenite and selenium yeast. J. Anim. Sci. 77, 3365-3370.

[5] Ortman, K. and B. Pehrson, 1997. Selenite and selenium yeast as feed supplements for dairy cows. J. Vet. Med. A 44, 373-380.

[6] Knowles, S., N.D. Grace, K. Wurms and J. Lee, 1999. Significance of amounts and form of dietary selenium on blood, milk, and casein selenium concentrations in grazing cows. J. Dairy Sci. 82, 429-437.

[7] Mahan, D.C. and N.A. Parett, 1996. Evaluating the efficacy of selenium-enriched yeast and sodium selenite on tissue selenium retention and serum glutathione peroxidase activity in grower and finisher swine. J. Anim. Sci. 74, 2967-2974.

[8] Mahan, D.C., T.R. Cline and B. Richert, 1999. Effects of dietary levels of selenium-enriched yeast and sodium selenite as selenium sources fed to growing-finishing pigs on performance, tissue selenium, glutathione peroxidase activity, carcass characteristics, and loin quality. J. Anim. Sci. 77, 2172-2179.

[9] Downs, K.M., J.B. Hess and S.F. Bilgili, 2000. Selenium source effect on broiler carcass characteristics, meat quality and drip loss. J. Appl. Anim. Res. 18, 61-72.

[10] Surrai, P., 2000. In: Natural Antioxidants in Avian Nutrition and Reproduction. Nottingham University Press, Nottingham UK, pp. 234-236, 254-259.

[11] Lindmark-Månsson, H., 2000. Bioproteins in Bovine Milk. Thesis. Center for Chemistry and Chemical Engineering. University of Lund, Sweden.

[12] Rayman, M.P., 2000. The importance of selenium to human health. Lancet 356, 233-241.

[13] Givens D.I., R. Allison, B. Cottrill and J.S. Blake, 2004. Enhancing the selenium content of bovine milk through alteration of the form and concentration of selenium in the diet of the dairy cow. J. Sci. Food & Agric. 84, 811-817.

Long acting injectable remedies to prevent cobalt and selenium deficiencies in grazing lambs and calves

N.D. Grace and S.O. Knowles
AgResearch Limited, Grasslands Research Centre, Private Bag 11008, Palmerston North, New Zealand

The unique geochemistry of New Zealand soils means cobalt (Co; vitamin B_{12}) and selenium (Se) deficiencies are widespread among grazing lambs and calves, leading to poor growth rates and occasional deaths. Deficient pastures typically contain <0.08 mg Co/kg DM (ppm) and <0.03 ppm Se. Lambs grazing those pastures will have serum vitamin B_{12} concentrations <220 pmol/l, while lambs and calves will have blood Se concentrations <130 nmol/l. Remedies, effective for varying intervals, include supplementation of the animals directly via injections and drenches or indirectly through trace element amended fertilizers. Two recently developed long-acting injectables based on controlled release technology can prevent Co and Se deficiencies in lambs and calves for many months.

A subcutaneous injection in the anterior neck region of 3-5 week old Romney lambs, at the time of tailing, with 3 mg of microencapsulated vitamin B_{12} in 0.5 ml SMARTShot B_{12}™ (Stockguard Laboratories Ltd, Hamilton, NZ) significantly increased (P<0.01) serum (246 *vs*. 143 pmol/l) and liver (158 *vs*. <70 nmol/kg fresh tissue) vitamin B_{12} concentrations for at least 160 and 125 days respectively, compared to untreated control animals (n=10/group). The pasture was very Co deficient (0.05 ppm Co); 68% of the untreated lambs lost weight and were removed from the trial for welfare reasons. After 240 days the mean liveweight of the supplemented lambs was 45 kg (n=30) compared to 28 kg for the remaining controls (n=16)[1].

Treatment of Friesian heifer calves of mean liveweight 110 kg with a single injection of SMARTShot B_{12} at 0.24, 0.12 or 0 mg vitamin B_{12}/kg liveweight significantly (P<0.05) increased serum (383 *vs*. 283 *vs*. 225 pmol/l) and liver (1047 *vs*. 828 *vs*. 592 nmol/kg) vitamin B_{12} for at least 110 days compared to controls (n= 10/group). In this study pasture Co was considered adequate at 0.16 ppm Co and no weight gain response was observed[2].

SMARTShot B_{12} plus Se™ contains 3 mg of microencapsulated vitamin B_{12} plus 12 mg Se as barium selenate per ml. Subcutaneous injection of 4 week old Romney x Suffock lambs with 1 ml significantly (P<0.01) increased serum vitamin B_{12} (706 *vs*. 426 pmol/l) and blood Se (791 *vs*. 151 nmol/l) for at least 190 days compared to controls (n=10). The treatment also improved (P<0.05) growth rates from 125 ± 4 to 136 ± 4 g/day over the full 246 day trial (n=25). This flock grazed pastures containing 0.07 ppm Co and 0.02 ppm Se.

Treating Angus-cross weaner calves (n=10) of liveweight 220 kg with SMARTShot B_{12} plus Se™ (0.11 mg vitamin B_{12}/kg plus 0.43 mg Se/kg liveweight) significantly (P<0.05) increased growth rates from 560 ± 30 to 710 ± 30 g/day over 295 days when pastures were 0.09 ppm Co and 0.02 ppm Se. Compared to controls, serum and liver vitamin B_{12} were increased 1.4- and 4.1-fold, respectively, for up to 105 days while blood and liver Se were 4.8- and 2.6-fold greater for up to 266 days.

Administering new long-acting slow-release forms of vitamin B_{12} and Se to young animals prevents Co deficiency for up to 4 and 8 months in calves and lambs respectively, and Se deficiency for up 12 months. These new products offer an efficacious, long term and low labor input approach to trace element problems in grazing animals.

References

[1] Grace N.D., S.O.Knowles, G.R. Sinclair and J. Lee, 2003. Growth response to increasing doses of microencapsulated vitamin B12 and related changes in tissue vitamin B12 in cobalt deficient lambs. New Zealand Veterinary Journal 51, 89-92.

[2] Grace N.D. and D.M.West, 2000. Effect of an injectable microencapsulated vitamin B12 on serum and liver vitamin B12 concentrations in calves. New Zealand Veterinary Journal 48, 70-73.

Parenteral and oral selenium supplementation of weaned beef calves

W.S. Swecker, Jr.[1], R.K. Shanklin[2], K.H. Hunter[2], C.L. Pickworth[2], G. Scaglia[2], and J.P. Fontenot[2]
[1]Department of Large Animal Clinical Sciences; [2]Department of Animal and Poultry Sciences, Virginia Tech, Blacksburg, VA, USA

Selenium (Se) is an essential trace element for cattle. When native forages are deficient in Se, it can be supplemented through supplemental feeds, mineral mixes or parenteral injections. The objective of this trial was to compare single injection of Se at arrival to oral supplementation with sodium selenite in weaned beef calves.

Thirty-six Angus and Angus-cross calves (180-250 kg) were purchased at a regional stockyard and backgrounded for 42 d on endophyte-free fescue. Calves were allotted to one of 3 treatments: Control, Se injection (0.05 mg Se/kg BW) SC on d 0, or oral Se (Na selenite mixed with ground corn fed at 1% of body weight, 0.25 ppm Se). The control and injection groups were fed corn (0.08 ppm Se) at the same rate without added Se. Three calves were placed in each paddock. Each treatment group was replicated 4 times across previous pasture treatments for a total of 12 paddocks. Response variables included body weight, rectal temperature, blood Se, serum Se, plasma lipid peroxides, lymphocyte count and neutrophil count. Blood and serum Se were measured by atomic absorption spectrophotometry. Leukocytes were counted on an automated blood analyzer (Celldyne). Response variables were measured on d 0, 7, 14, 28, and 42 except for blood Se which was not measured on d 42. Data was analyzed by repeated measures in time analysis of variance.

Serum Se tended to be higher in calves injected with Se (P<0.07) as compared to control calves and oral Se calves. Rectal temperature was higher for calves injected with selenium (P<0.03) as compared to control and oral Se calves. Treatment differences were not detected in body weight, blood Se, plasma lipoperoxides, lymphocyte count, or neutrophil count. Selenium supplementation of calves prior to enrollment in the study was unknown. However, the Mean ± SEM blood Se (81.7 ± 5.7 ppb) and serum Se (40.2 ppb ± 2.1) on d 0 would suggest that many calves were marginal to deficient in Se status. Calf performance was acceptable with an average daily gain of 1.4 kg/d for the 42 d-period, however some of the gain may be considered recovery of shrink from the marketing process.

In conclusion, Se supplementation by injection marginally increased serum Se concentrations. However, other parenteral doses, sources of oral Se, or amounts of oral Se should be investigated to determine effective measures to rapidly restore Se status in weaned beef calves.

Selenium yeast prevents nutritional muscular degeneration (NMD) in nursing calves

B.G. Pehrson
Department of Animal Environment and Health, Swedish University of Agricultural Sciences,
P.O.Box 234, S-53223 Skara, Sweden

In most western countries mineral feeds supplemented with inorganic selenium as sodium selenite are routinely given to suckler cows. As a consequence, the incidence of NMD in their calves has been reduced. However, many practitionars report that NMD still is a common disease among fast growing calves, particularly during their first months of life. Our own trials strongly indicate that the reason is the insufficient capacity of inorganic selenium compounds to increase the selenium content in cows´ milk, resulting in a suboptimal selenium concentration in the blood of their calves[1].

After changing the source of the selenium supplement from selenite to a selenium yeast product at an equal selenium dosage level, all calves reached a blood selenium value exceeding 100 µg/l, which by most researchers is considered to be high enough, both for prevention of NMD and for securing an optimal immune competence. On the contrary, when the cows were fed a conventional selenite-supplemented mineral feed, all their calves had blood selenium values lower than 100 µg/l, which is considered suboptimal for achieving a maximal immune competence. Some of the calves even had values lower than 50 µg/l, which is considered to imply obvious risks of clinical manifest NMD.

Later on, our results have been verified by researchers in the USA. A reasonable conclusion from these trials is that NMD can effectively be prevented if selenite in mineral feeds to suckler cows is replaced by organic selenium in the form of a selenium yeast product, on the assumption that that product is produced in an adequate way.

References
[1] Pehrson, B., K. Ortman, N. Madjid, *et al.*, 1999. The influence of dietary selenium as selenium yeast or sodium selenite on the concentration of selenium in milk of suckler cows and on the selenium status of their calves. J. Anim. Sci. 77, 3371-3376.

Section J.
Role of specific fatty acids in animal health,
reproduction, and performance

Using conjugated linoleic acids for healthier animal products and as a management tool

M.A. McGuire[1], E.E. Mosley[1], S.A. Mosley[1], A. Nudda[2] and A. Corato[3]
[1]University of Idaho, Department of Animal & Veterinary Sciences, Moscow, USA 83844-2330;
[2]University of Sassari, Department of Animal Science, 07100 Sassari, Italy; [3]University of Padova, Department of Animal Science, Legnaro (PD), I-35020, Italy

Abstract

Conjugated linoleic acids (CLA) are fatty acids found mostly in ruminant tissues that may be beneficial to human health. The primary isomer of CLA is *c9,t11*-CLA. These fatty acids are formed during ruminal biohydrogenation and via synthesis from vaccenic acid by action of the Δ^9-desaturase enzyme in mammary tissue. Another CLA isomer, *t10,c12*-CLA, is produced in the rumen when fermentation is altered and is a potent inhibitor of milk fat synthesis. Milk fat depression can occur through feeding rumen-protected CLA mixtures containing *t10,c12*-CLA. Altering energy output by reducing milk fat production may enhance milk yield in early lactation, improve profitability and could also improve reproduction. Enriching muscle of beef cattle with CLA through alterations in rumen biohydrogenation or feeding of rumen protected CLA is less responsive than milk. Feeding of CLA mixtures to growing pigs reduces back fat significantly while increasing concentrations of CLA in edible tissues. Enhancing the content of CLA in animal products improves their quality for the consumer.

Keywords: fatty acids, milk fat depression, rumen biohydrogenation, CLA

Introduction

Conjugated linoleic acid (CLA) is a collective term to describe positional and geometric isomers of linoleic acid (*c9, c12* 18:2) which contain a conjugated double-bond system. Thus, CLAs have 18 carbons with 2 double bonds separated by a single bond. Many beneficial effects of CLA mixtures in cancer, atherosclerosis, diabetes and immunity have been demonstrated in animal models (see McGuire and McGuire, 2000; Parodi, 2002). The first beneficial effect for CLA determined was that topical application from extracts of ground beef reduced skin cancer in mice (Pariza and Hargraves, 1985). Clement Ip (Roswell Park Cancer Institute, Buffalo, NY) then led a series of studies to demonstrate that laboratory-produced CLA when fed to rats reduced the incidence and number of mammary tumors (Ip *et al.*, 1991, 1994, 1995, 1999). Conjugated linoleic acid is unique among the naturally occurring anticarcinogens in that it is potent at extremely low levels and present in foods from ruminant animals. Butter containing CLA naturally was able to reduce mammary cancer similar to synthetic CLA (Ip *et al.*, 1999). The primary sources of CLA in the human diet are dairy products and beef (Ritzenthaler *et al.*, 2001). Data suggest that humans consume anywhere from negligible amounts of CLA to about 500 mg/d from natural sources (McGuire *et al.*, 1999). Many CLA isomers have been identified in ruminant tissues and products with the most abundant form being *c9,t11*-CLA (Sehat *et al.*, 1999). Mixtures of CLA used in research typically contain similar concentrations of *c9,t11* and *t10,c12*-CLA. Research has been conducted to enrich animal products with CLA to improve the healthfulness for the consumer. This information has led to a greater understanding of the control of milk fat synthesis.

CLA in bovine milk

Parodi (1976) confirmed the presence of CLA in bovine milk fat after indirect methods had for years noted that milk fat contained conjugated dienes. Parodi (1976) identified the principle CLA isomer of bovine milk fat, butter and adipose tissue as c9,t11-CLA, but other positional and geometric isomers were present. The c9,t11 isomer represents about 80% of the total CLA isomers found in bovine milk fat with the t7,c9-CLA the second most abundant (Yurawecz et al., 1998). Concentrations of total CLA in bovine milk fat range from 3 to 40 mg/g fat (Jahreis et al., 1999). Typical concentrations from cows fed a total mixed ration are approximately 5 mg/g fat.

Biohydrogenation leads to CLA

Efforts to enrich animal products with CLA have focused on either altering the rumen environment or by direct addition of CLA to the diet (Bauman et al., 2000). Rumen bacteria have specific enzymes that perform the steps of biohydrogenation. Ruminal biohydrogenation of linoleic acid (18:2) leads to the formation of c9,t11-CLA as an intermediate (Figure 1). Trans-vaccenic acid (t11 18:1; VA) is then formed from CLA before complete hydrogenation to stearic acid. Biohydrogenation of linolenic acid (18:3) does not lead to c9,t11-CLA formation but does produce VA. The biohydrogenation of oleic acid also produces t18:1 intermediates but little VA (Mosley et al., 2002).

Efforts to enhance ruminal production have identified inhibition of the final reduction to stearic acid and greater substrate (primarily linoleic acid but also linolenic acid) as two key control points (Figure 2). By inhibiting the final reduction to stearic acid, biohydrogenation yields mostly VA. Griinari et al. (2000) demonstrated that the Δ^9 desaturase enzyme could convert VA to c9,t11-CLA. The Δ^9 desaturase enzyme is present in various tissues, but primarily found in mammary

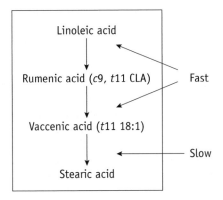

Figure 1. The pathway of biohydrogenation of linoleic acid by bacteria in the rumen.

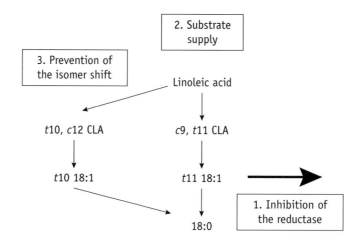

Figure 2. Three key control points to enhance milk with CLA.

and adipose in ruminants. Sixty to 90% of $c9,t11$-CLA is estimated to arise from the desaturation of VA (Corl et al., 2002; Griinari et al., 2000; Kay et al., 2004). Thus, by preventing the final reduction to stearic acid, VA is available for conversion to $c9,t11$-CLA (Figure 3). The $t7,c9$-CLA is also produced via action of the Δ^9 desaturase enzyme (Corl et al., 2002).

Concentrations of CLA in milk fat have been shown to vary according to seasonal or feeding management (Agenäs et al., 2002; Dhiman et al., 1999; Precht and Molkentin, 1997). Cows fed pasture alone had $c9,t11$-CLA concentrations in milk fat ranging from 7 to 22 mg/g (Agenäs et al., 2002; Dhiman et al., 1999; Precht and Molkentin, 1997). Concentrations of $c9,t11$-CLA were greatest when pasture was lush and immature and declined as the pasture became more mature. Further, cutting the pasture and drying the grass to hay reduced the concentrations of $c9,t11$-CLA. Along with the increases in CLA on lush pasture is also an increase in VA content. Thus, something in lush pasture must block the final reduction of VA to stearic acid.

The feeding of fish oil consistently increases the concentration of both VA (21 to 41 mg/g fat) and $c9,t11$-CLA (10 to 21 mg/g fat) in milk fat similar to that in pasture fed cows (AbuGhazaleh et al., 2002, 2003a, b, 2004; Whitlock et al., 2002). Thus, it appears that some component of fish oil can block the final reduction of VA to stearic acid leading to an increase in VA outflow from the rumen as the content of linoleic and linolenic acid is low in fish oil. AbuGhazaleh and Jenkins (2004) demonstrated that the addition docosahexaenoic acid (DHA) to mixed ruminal cultures promoted VA accumulation. Further, Franklin et al. (1999) increased $c9,t11$-CLA content of milk fat to nearly 25 mg/g by feeding algae enriched in DHA. Thus, DHA is an active inhibitor of the final step to complete biohydrogenation of linoleic acid. Control point 2 is also important as amount of substrate for biohydrogenation ultimately impacts how much VA could leave the rumen. The addition of DHA and soybean oil to ruminal cultures confirmed that both level of substrate and inhibition of the final reduction are critical to high concentrations of VA being available to the cow for $c9,t11$-CLA synthesis (AbuGhazaleh and Jenkins, 2004). Feeding of fish oil with high linoleic acid sunflower seeds yielded the greatest concentrations of both VA (41 mg/g) and $c9,t11$-CLA (17 mg/g) compared to cows fed fish oil with either stearic, oleic or linolenic acid (AbuGhazaleh et al., 2003b).

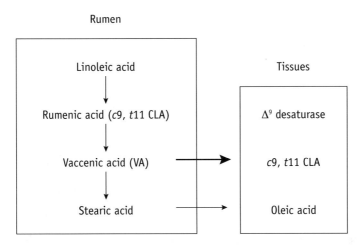

Figure 3. The role of the desaturase enzyme in bovine tissues in the synthesis of CLA and oleic acid. Production of CLA by the tissues is enhanced by increasing the output of VA from the rumen with a reduction in oleic acid.

The *t*11 to *t*10 shift and milk fat depression

It is also important when enriching tissues with *c*9,*t*11-CLA to prevent a shift in biohydrogenation through *t*10,*c*12-CLA and *t*10 18:1 (Bauman and Griinari, 2003). The shift in biohydrogenation to *t*10,*c*12-CLA and *t*10 18:1 not only limits the production of *c*9,*t*11-CLA but leads to milk fat depression. The effect of *trans* fatty acids on milk fat synthesis has been known for many years (see Bauman and Griinari, 2003). The idea that only certain *trans* fatty acids, those containing a *t*10 double bond, were capable of causing milk fat depression was established by Griinari *et al.* (1998). This led to the biohydrogenation theory of milk fat depression that proposed that under certain dietary conditions, typical pathways of rumen biohydrogenation are altered to produce unique fatty acids inhibit milk fat synthesis (Bauman and Griinari, 2001). This theory amended the *trans* fatty acid theory to include these concepts. Further work (Chouinard *et al.*, 1999) showed that milk fat depression occurred with abomasal infusion of a CLA mixture. The specific CLA isomer that caused milk fat depression was *t*10,*c*12-CLA while *c*9,*t*11-CLA did not affect milk fat (Baumgard *et al.*, 2000). Abomasal infusion of as little as 2.5 g/d of *t*10,*c*12-CLA caused milk fat depression while 10 g/d of *c*9,*t*11-CLA did not alter milk fat percentage demonstrating that the shift in biohydrogenation to *t*10,*c*12-CLA is the mechanism for milk fat depression. Changes in rumen conditions leading to milk fat depression are related to a shift toward production of *t*10,*c*12-CLA rather than *c*9,*t*11-CLA (Figure 2).

Feeding of protected CLA

Direct feeding of CLA to ruminants is ineffective without protection from rumen biohydrogenation. Calcium salts of CLA mixtures (50 to 100 g/d) have been fed to lactating cows and have increased concentrations of CLA and caused milk fat depression (Figure 4). The response to dietary CLA is

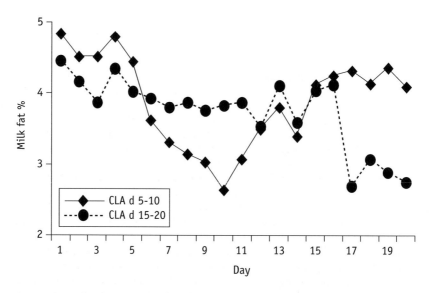

Figure 4. The response of milk fat to dietary supplementation with calcium salts of a mixture of conjugated linoleic acid (CLA). Cows (n=6) were fed 50 g CLA per day for days 5-10 (circles) or days 15-20 (diamonds). Cows were fed an equivalent amount of calcium salts of palm oil during the non-CLA feeding periods (Hanson and McGuire, unpublished data).

rapid and reversible. Altering energy output by reducing milk fat production may enhance milk yield in early lactation and could also improve reproduction (Table 1). Feeding rumen protected CLA would be very beneficial in regions with a milk fat quota system.

Table 1. Long-term studies with rumen protected CLA.

Study	CLA (g/d)	DIM at start	Length (d)	Feed type[1]	Fat yield (%)	Milk yield (%)	Protein yield (%)
Giesy et al., 1999[2]	50	13	67	TMR	-23.5	+5.8	+6.5
Medeiros et al., 2000	59	28	56	P+C	-24.5	+4.5	+16.0
Bernal-Santos et al., 2004	30	0	140	TMR	-6.6	+7.6	+4.1
Perfield et al., 2002	30	-[3]	140	TMR	-23.0	+1.0	+1.0
Selberg et al., 2004	225	-28	77	TMR	-14.4	+1.2	0

[1] TMR = total mixed ration; P+C= pasture plus concentrate.
[2] Yield results are pooled means of samples collected from 35 to 80 DIM.
[3] Treatment began 200 d prepartum in pregnant lactating cows and continued until dry off at 60 d prepartum.

CLA in beef

Beef contains a number of different CLA isomers. Fritsche *et al.* (2000) using sensitive techniques found 14 isomers of CLA in samples of the *longissimus* muscle. As with milk, *c*9,*t*11-CLA was the most abundant isomer (1.95 ± 0.54 mg/g fat) found in beef, followed by *t*7,*c*9-CLA (0.19 ± 0.04 mg/g fat). Total concentrations of CLA averaged 3.52 ± 0.88 mg/g fat. Strategies to enhance the content of *c*9,*t*11-CLA in beef have used similar methods to those for enriching milk. Most studies have examined how additional substrate (*i.e.*, linoleic acid) could improve CLA. Responses to added oil or oilseeds have been quite variable (0 to 300%) with lower responses most typical (Beaulieu *et al.*, 2002; Garcia *et al.*, 2003; Griswold *et al.*, 2003; Madron *et al.*, 2002; Mir *et al.*, 2002). When the final reduction of linoleic acid is blocked by feeding fresh pasture compared to preserved forage, *c*9,*t*11-CLA concentrations have increased (French *et al.*, 2000). However, the effect of fish oil is not as dramatic as the response in milk fat (Enser *et al.*, 1999). One reason for the poor response to either added substrate or by blocking the final reduction to stearic acid is that the *t*10/*t*11 shift predominates in cattle fed diets high in concentrates, as the *t*10 pathway is the primary path of biohydrogenation in diets low in forage. Factors impacting the response include tissue sampled, length of treatment, amount of forage to concentrate, and fresh versus preserved forage. Thus, other factors regulating *c*9,*t*11-CLA concentrations are involved. Further, consuming beef with enhanced CLA concentrations may not deliver any greater CLA as total lipid content also drives CLA intake. For example, although concentrations of CLA are greater in tissues from pasture fed animals, total lipid content is typically less. Therefore, one must look at both lipid content and fatty acid concentration to examine benefits to the consumer.

Direct feeding of a rumen protected CLA has been tested (Gillis *et al.*, 2004a, b). Finishing heifers were fed diets (2-2.5% of diet dry matter) delivering about 60 g/d of rumen protected CLA for 56 d. Concentrations of *c*9,*t*11 and *t*10,*c*12-CLA were increased in adipose tissue compared to the control (Gillis *et al.*, 2004a). No measure of performance (feed intake, average daily gain, gain: feed, or carcass composition) was altered by feeding of the rumen protected CLA (Gillis *et al.*, 2004b). It does not appear that beef respond as quickly or to the extent of lactating animals in the enrichment of CLA or in response to feeding of a rumen protected CLA.

CLA in other animals

Feeding of CLA mixtures to other growing animals may alter growth. Pigs fed mixtures of CLA (0.5 to 2.5% of diet DM) had reduced back fat while increasing concentrations of CLA in edible tissues (Dugan *et al.*, 1997; Ostrowka *et al.*, 1999, 2003; Wiegand *et al.*, 2002). Broilers fed mixtures of CLA also have enhanced CLA concentrations with some reductions in internal fat deposition (Simon *et al.*, 2000; Szymczyk *et al.*, 2001). Addition of CLA to diets did not affect the whole body composition in Atlantic salmon (Berge *et al.*, 2004), yellow perch (Twibell *et al*, 2001) or channel catfish (Twibell and Wilson, 2003), but did decrease intraperitoneal fat and increase muscle moisture in hybrid striped bass (Twibell *et al.*, 2000). Feeding fish with CLA enriched CLA concentrations in whole body and muscle lipids (Berge *et al.*, 2004; Twibell *et al*, 2001).

Content of CLA (22 mg/g fat) and VA (45 mg/g fat) in sheep milk is greater than in cow milk when sheep are in the early grazing season (Nudda *et al.*, 2003). Concentrations of both decline as summer begins. These changes are probably related to the nutrient composition of the grass consumed. Studies have also shown that the CLA in the meat of sheep can be altered through similar nutritional means as found with lactating cows (Demirel *et al.*, 2004; Mir *et al.*, 2000; Wachira *et al.*, 2002).

Supplementation of humans with CLA

Supplements of CLA for humans contain approximately equal concentrations of *t*10,*c*12 and *c*9,*t*11-CLA. These CLA supplements are being marketed based primarily upon potential impacts on body weight and composition. Several studies (Blankson *et al.*, 2000; Kreider *et al.*, 2002; Medina *et al.*, 2000; Mougious *et al.*, 2000; Risérus *et al.*, 2001; Thom *et al.*, 2001; von Loeffelholz *et al.*, 2003; Zambell *et al.*, 2000) have been published on the effects of CLA supplementation on nutrient partitioning in humans. Doses of CLA ranged from 0.7 to 9 g/d, substantially greater than the range of CLA intakes from normal diets. Results remain difficult to interpret and suggest that the relationship between CLA intake and nutrient partitioning is complex. A recent long-term (1 yr) intervention study (Gaullier *et al.*, 2004) detected a 5% reduction in body fat mass in overweight (25-30 body mass index) when they received 4.5 g/d of a 75-80% CLA mixture compare to an olive oil placebo. No adverse effects were reported in this long-term trial. Impacts of CLA supplementation on other indices of human health such as cholesterol concentrations and immune function are equivocal at this time.

Conclusion

Milk and meat from ruminant animals provide humans with an important source of beneficial fatty acids including CLA and VA. CLA can also be dietary tool for animal production that can reduce milk fat synthesis and adipose tissue deposition in growing animals. Further, CLA improves the healthiness of animal products for the consumer.

References

AbuGhazaleh, A.A. and T.J. Jenkins, 2004. Docosahexaenoic acid promotes vaccenic acid accumulation in mixed ruminal cultures when incubated with linoleic acid. J. Dairy Sci. 87, 1047-1050.

AbuGhazaleh, A.A., D.J. Schingoethe, A.R. Hippen and K.F. Kalscheur, 2002. Feeding fish meal and extruded soybeans enhances the conjugated linoleic acid (CLA) content of milk. J. Dairy Sci. 85, 624-631.

AbuGhazaleh, A.A., D.J. Schingoethe, A.R. Hippen and K.F. Kalscheur, 2003a. Milk conjugated linoleic acid response to fish oil supplementation of diets differing in fatty acid profiles. J. Dairy Sci. 86, 944-953.

AbuGhazaleh, A.A., D.J. Schingoethe, A.R. Hippen and K.F. Kalscheur, 2003b. Conjugated linoleic acid and vaccenic acid in rumen, plasma, and milk of cows fed fish oil and fats differing in saturation of 18 carbon fatty acids. J. Dairy Sci. 86, 3648-3660.

AbuGhazaleh, A.A., D.J. Schingoethe, A.R. Hippen and K.F. Kalscheur, 2004. Conjugated linoleic acid increases in milk when cows fed fish meal and extruded soybeans for an extended period of time. J. Dairy Sci. 87, 1758-1766.

Agenäs, S., K. Holtenius, M. Griinari and E. Burstedt, 2002. Effects of turnout to pasture and dietary supplementation on milk fat composition and conjugated linoleic acid in dairy cows. Acta Agric. Scand. 52, 25-33.

Bauman, D.E. and J.M. Griinari, 2001. Regulation and nutritional manipulation of milk fat: low-fat milk syndrome. Livest. Prod. Sci. 70, 15-29.

Bauman, D.E. and J.M. Griinari, 2003. Nutritional regulation of milk fat synthesis. Annu. Rev. Nutr. 23, 203-227.

Bauman, D.E., L.H. Baumgard, B.A. Corl and J.M. Griinari, 2000. Biosynthesis of conjugated linoleic acid in ruminants. Proc. Amer. Soc. Anim. Sci., 1999. Available at: http://www.asas.org/jas/symposia/proceedings/0937.pdf. Accessed July 10, 2000.

Baumgard, L.H., B.A. Corl, D.A. Dwyer, A. Sæbø and D.E. Bauman, 2000. Identification of the conjugated linoleic acid isomer that inhibits milk fat synthesis. Amer. J. Physiol. 278, R179-R184.

Beaulieu, A.D., J.K. Drackley and N.R. Merchen, 2002. Concentrations of conjugated linoleic acid (cis-9, trans-11-octadecadienoic acid) are not increased in tissue lipids of cattle fed a high-concentrated diet supplemented with soybean oil. J. Anim. Sci. 80, 847-861.

Berge, G.M., B. Ruyter and T. Åsgård, 2004. Conjugated linoleic acid in diets for juvenile Atlantic salmon (Salmo salar); effects on fish performance, proximate composition, fatty acid end mineral content. Aquaculture 237, 365-380.

Bernal-Santos, G., J.W. Perfield, D.M. Barbano, D.E. Bauman and T.R. Overton, 2003. Production responses of dairy cows to dietary supplementation with conjugated linoleic acid (CLA) during the transition period and early lactation. J. Dairy Sci. 86, 3218-3228.

Blankson, H., J.A. Stakkestad, H. Fagertun, E. Thom, J. Wadstein and O. Gudmundsen, 2000. Conjugated linoleic acid reduces body fat mass in overweight and obese humans. J. Nutr. 130, 2943-2948.

Chouinard, P.Y., L. Corneau, D.M. Barbano, L.E. Metzger and D.E. Bauman, 1999. Conjugated linoleic acids alter milk fatty acid composition and inhibit milk fat secretion in dairy cows. J. Nutr. 129, 1579-1584.

Corl, B.A., L.H. Baumgard, J.M. Griinari, P. Delmonte, K.M. Morehouse, M.P. Yurawecz and D.E. Bauman, 2002. Trans-7, cis-9 CLA is synthesized endogenously by Δ^9-desaturase in dairy cows. Lipids 37, 681-688.

Demirel, F., J.D. Wood and M. Enser, 2004. Conjugated linoleic acid content of the lamb muscle and liver fed different supplements. Small Ruminant Res. 53, 23-28.

Dhiman, T.R., G.R. Anand, L.D. Satter and M.W. Pariza, 1999. Conjugated linoleic acid content of milk from cows fed different diets. J. Dairy Sci. 82, 2146-2156.

Dugan, M.E.R., J.L. Aalhus, A.L., Schaefer and J.K.G. Kramer 1997. The effect of conjugated linoleic acid on fat to lean repartitioning and feed conversion in pigs. Can. J. Anim. Sci. 77, 723-725.

Enser, M., N.D. Scollan, N.J. Choi, E. Kurt, K. Hallett and J.D. Wood, 1999. Effect of dietary lipid on the content of conjugated linoleic acid (CLA) in beef muscle. Anim. Sci. 69, 143-146.

Franklin, S.T., K.R. Martin, R.J. Baer, D.J. Schingoethe and A.R. Hippen, 1999. Dietary marine algae (Schizochytrium sp.) increases concentrations of conjugated linoleic, docosahexaenoic and transvaccenic acids in milk of dairy cows. J. Nutr. 129, 2048-2052.

French, P., C. Stanton, F. Lawless, E.G. O'Riordan, F.J. Monahan, P.J. Caffrey and A.P. Moloney, 2000. Fatty acid composition, including conjugated linoleic acid, of intramuscular fat from steers offered grazed grass, grass silage, or concentrate-based diets. J. Anim. Sci. 78, 2849-2855.

Fritsche, J., S. Fritsche, M.B. Solomon, M.M. Mossoba, M.P. Yurawecz, K. Morehouse, and Y. Ku. 2000. Quantitative determination of conjugated linoleic acid isomers in beef fat. Eur. J. Lipid Sci. Technol. 102, 667-672.

Garcia, M.R., M. Amstalden, C.D. Morrison, D.H. Keisler and G.L. Williams, 2003. Age at puberty, total fat and conjugated linoleic acid content of carcass, and circulating metabolic hormones in beef heifers fed a diet high in linoleic acid beginning at four months of age. J. Anim. Sci. 81, 261-268.

Gaullier, J.-M., J. Halse, K. Høye, K. Kristiansen, H. Fagertun, H. Vik and O. Gudmundsen, 2004. Conjugated linoleic acid supplementation for 1 y reduces body fat mass in healthy overweight humans. Amer. J. Clin. Nutr. 79, 1118-1125.

Giesy, J.G., T.W. Hanson, M. Robison, H.C. Hafliger, S. Viswanatha, M.A. McGuire, C.H. Skarie and A. Vinci, 1999. Effects of calcium salts of conjugated linoleic acid (CLA) on milk yield, fat and CLA content of milk fat in Holstein cows early in lactation. J. Dairy Sci. 82(Suppl. 1), 83-84.

Gillis, M.H., S.K. Duckett and J.R. Sackmann, 2004a. Effects of supplemental rumen-protected conjugated linoleic acid or corn oil on fatty acid composition of adipose tissue in beef cattle. J. Anim. Sci. 82, 1419-1427.

Gillis, M.H., S.K. Duckett, J.R. Sackmann, C.E. Realini, D.H. Keisler and T.D. Pringle, 2004b. Effects of supplemental rumen-protected conjugated linoleic acid or linoleic acid on feedlot performance, carcass quality, and leptin concentrations in beef cattle. J. Anim. Sci. 82, 851-859.

Griinari, J.M., B.A. Corl, S.H. Lacy, P.Y. Chouinard, K.V.V. Nurmela and D.E. Bauman, 2000. Conjugated linoleic acid is synthesized endogenously in lactating dairy cows by Δ^9-desaturase. J. Nutr. 130, 2285-2291.

Griinari, J.M., D.A. Dwyer, M.A. McGuire, D.E. Bauman, D.L. Palmquist and K.V.V. Nurmela, 1998. *Trans*-octadecenoic acids and milk fat depression. J. Dairy Sci. 81, 1251-1261.

Griswold, K.E., G.A. Apgar, R.A. Robinson, B.N. Jacobson, D. Johnson and H.D. Woody, 2003. Effectiveness of short-term feeding strategies for altering conjugated linoleic acid content of beef. J. Anim. Sci. 81, 1862-1871.

Ip, C., S. Banni, E. Angioni, G. Carta, J. McGinley, H.J. Thompson, D. Barbano and D. Bauman, 1999. Alterations in rat mammary gland leading to a reduction in cancer risk by conjugated linoleic acid (CLA)-enriched butter fat. J. Nutr. 129, 2135-2142.

Ip, C., S.F. Chin, J.A. Scimeca and M.W. Pariza, 1991. Mammary cancer prevention by conjugated dienoic derivative of linoleic acid. Cancer Res. 51, 6118-6124.

Ip, C., J.A. Scimeca and H. Thompson, 1995. Effect of timing and duration of dietary conjugated linoleic acid on mammary cancer prevention. Nutr. Cancer 24, 241-247.

Ip, C., M. Singh, H.J. Thompson and J.A. Scimeca, 1994. Conjugated linoleic acid suppresses mammary carcinogenesis and proliferative activity of the mammary gland in the rat. Cancer Res. 54, 1212-1215.

Jahreis, G., J. Fritsche, P. Möckel, F. Schöne, U. Möller and H. Steinhart, 1999. The potential anticarcinogenic conjugated linoleic acid, *cis*-9, *trans*-11 C18:2, in milk of different species: cow, goat, ewe, sow, mare, woman. Nutr. Res. 19, 1541-1549.

Kay, J.K., T.R. Mackle, M.J. Auldist, N.A. Thomson and D.E. Bauman, 2004. Endogenous synthesis of *cis*-9, *trans*-11 conjugated linoleic acid in dairy cows fed fresh pasture. J. Dairy Sci. 87, 369-378.

Kreider, R.B., M.P. Ferreira, M. Greenwood, M. Wilson and A.L. Almada, 2002. Effects on conjugated linoleic acid supplementation during resistance training on body composition, body density, strength and selected hematologic markers. J. Strength Conditioning Res. 16, 325-334.

Madron, M.S., D.G. Peterson, D.A. Dwyer, B.A. Corl, L.H. Baumgard, D.H. Beermann and D.E. Bauman, 2002. Effect of extruded full-fat soybeans on conjugated linoleic acid content of intramuscular, intermuscular, and subcutaneous fat in beef steers. J. Anim. Sci. 80, 1135-1143.

McGuire, M.A. and M.K. McGuire, 2000. Conjugated linoleic acid (CLA): A ruminant fatty acid with beneficial effects on human health. (Invited Review). Proc. Am. Soc. Anim. Sci., 1999. Available at: http://www.asas.org/jas/symposia/proceedings/0938.pdf. Accessed September 6, 2000.

McGuire, M.K., M.A. McGuire, K. Ritzenthaler and T.D. Shultz, 1999. Dietary sources and intakes of conjugated linoleic acid intake in humans. In: Advances in Conjugated Linoleic Acid Research, Volume 1. (Eds. Yurawecz, M.P., M.M. Mossoba, J.K.G. Kramer, M.W. Pariza and G.J. Nelson). AOCS Press, Champaign, IL. pp. 369-377.

Medeiros, S.R., D.E. Oliveira, L.J.M. Aroeira, M.A. McGuire, D.E. Bauman and D.P.D. Lanna, 2000. The effect of long-term supplementation of conjugated linoleic acid (CLA) to dairy cows grazing tropical pasture. J. Anim. Sci. 78(Suppl. 1)/J. Dairy Sci. 83(Suppl. 1), 169.

Medina, E.A., W.F. Horn, N.L. Keim, P.J. Havel, P. Benito, D.S. Kelley, G.J. Nelson and K.L. Erickson, 2000. Conjugated linoleic acid supplementation in humans: effects on circulating leptin concentrations and appetite. Lipids 35, 783-788.

Mir, Z., M.L. Rushfeldt, P.S. Mir, L.J. Paterson and R.J. Weselake, 2000. Effect of dietary supplementation with either conjugated linoleic acid (CLA) or linoleic acid rich oil on the CLA content of lamb tissues. Small Rum. Res. 36, 25-31.

Mir, P.S., Z. Mir, P.S. Kuber, C.T. Gaskins, E.L. Martin, M.V. Dodson, J.A. Elias Calles, K.A. Johnson, J.R. Busboom, A.J. Wood, G.J. Pittenger and J.J. Reeves, 2002. Growth, carcass characteristics, muscle conjugated linoleic acid (CLA) content, and response to intravenous glucose challenge in high percentage Wagyu, Wagyu X Limousin, and Limousin steers fed sunflower oil-containing diets. J. Anim. Sci. 80, 2996-3004.

Mosley, E.E., G.L. Powell, M.B. Riley and T.C. Jenkins, 2002. Microbial biohydrogenation of oleic acid to trans isomers in vitro. J. Lipid Res. 43, 290-296.

Mougious, V., A. Matsakas, A. Petridou, S. Ring, A. Sagredos, A. Mslissopoulou, N. Tsigilis and M. Nikolaidis, 2001. Effect of supplementation with conjugated linoleic acid on human serum lipids and body fat. J. Nutr. Biochem. 12, 585-594.

Nudda, A., M. Mele, G. Battacone, M.F. Usai and N.P.P. Macciotta, 2003. Comparison of conjugated linoleic acid (CLA) content in milk of ewes and goats with the same dietary regimen. Proc. Natl. Congr. ASPA. Italy. pp.515-517.

Ostrowska, E., M. Muralitharan, R.F. Cross, D.E. Bauman and F.R. Dunshea, 1999. Dietary conjugated linoleic acids increase lean tissue and decrease fat deposition in growing pigs. J. Nutr. 129, 2037-2042.

Ostrowska, E., D. Suster, B.J. Leury and F.R. Dunshea, 2003. Conjugated linoleic acid decreases fat accretion in pigs: evaluation by dual-energy X-ray absorptiometry. Brit. J. Nutr. 89, 219-229.

Pariza, P.W. and W.A. Hargraves, 1985. A beef-derived mutagenesis modulator inhibits initiation of mouse epidermal tumors by 7,12-dimethylbenz[a]anthracene. Carcinogenesis 6, 591-593.

Parodi, P.W., 1976. Distribution of isomeric octadecenoic fatty acids in milk fat. J. Dairy Sci. 59, 1870-1873.

Parodi, P.W., 2002. Health benefits of conjugated linoleic acid. Food Industry J. 5, 222-259.

Perfield, J.W., G. Bernal-Santos, T.R. Overton and D.E. Bauman, 2002. Effects of dietary supplementation of rumen-protected conjugated linoleic acid in dairy cows during established lactation. J. Dairy Sci. 85, 2609-2617.

Precht, D. and J. Molkentin, 1997. Effect of feeding on conjugated cisΔ9, transΔ11- octadecadienoic acid and other isomers of linoleic acid in bovine milk fats. Nahrung 41, 330-335.

Risérus, U., L. Berglund and B. Vessby, 2000. Conjugated linoleic acid (CLA) reduced abdominal adipose tissue in obese middle-aged men with signs of the metabolic syndrome: a randomized controlled trial. Inter. J. Obesity 25, 1129-1135.

Ritzenthaler, K., M.K. McGuire, R. Falen, T.D. Shultz and M.A. McGuire, 2001. Estimation of conjugated linoleic acid intake by written dietary assessment methodologies underestimates actual intake evaluated by food duplicate methodology. J. Nutr. 131, 1548-1554.

Sehat, N., J.K.G. Kramer, M.M. Mossoba, M.P. Yurawecz, J.A.G. Roach, K. Eulitz, K.M. Morehouse and Y. Ku, 1999. Identification of conjugated linoleic acid isomers in cheese by gas chromatography, silver ion high performance liquid chromatography and mass spectral reconstructed ion profiles. Comparison of chromatographic elution sequences. Lipids 33, 963-971.

Selberg, K.T., A.C. Lowe, C.R. Staples, N.D. Luchini and L. Badinga, 2004. Production and metabolic responses of periparturient Holstein cows to dietary conjugated linoleic acid and trans-octadecenoic acids. J. Dairy Sci. 87, 158-168.

Simon, O., K. Männer, K. Schäfer, A. Sagredos and K. Eder, 2000. Effects of conjugated linoleic acids on protein to fat proportions, fatty acids, and plasma lipids in broilers. Eur. J. Lipid Sci. Technol. 102, 402-410.

Szymczyk, B., P.M. Pisulewski, W. Szczurek and P. Hanczakowski, 2001. Effects of conjugated linoleic acid on growth performance, feed conversion efficiency, and subsequent carcass quality in broiler chickens. Brit. J. Nutr. 85, 465-473.

Thom, E., J. Wadstein, and O. Gudmundsen. 2001. Conjugated linoleic acid reduces body fat in healthy exercising humans. J. Inter. Med. Res. 29, 392-396.

Twibell, R.G., and R.P. Wilson. 2003. Effect of dietary conjugated linoleic acids and total dietary lipid concentrations on growth responses of juvenile channel catfish, *Ictalus punctatus*. Aquaculture 221, 621-628.

Twibell, R.G., B.A. Watkins, and P.B. Brown. 2001. Dietary conjugated linoleic acids and lipid source alter fatty acid composition of juvenile yellow perch, *Perca flavescens*. J. Nutr. 131, 2322-2328.

Twibell, R.G., B.A. Watkins, L. Rogers, and P.B. Brown. 2000. Effect of dietary conjugated linoleic acid on hepatic and muscle lipids in hybrid striped bass. Lipids 35, 155-161.

von Loeffelholz, C., J. Kratzsch, and G. Jahreis. 2003. Influence of conjugated linoleic acids on body composition and selected serum and endocrine parameters in resistance-trained athletes. Eur. J. Lipid Sci. Technol. 105, 251-259.

Wachira, A.M., L.A. Sinclair, R.G. Wilkinson, M. Enser, J.D. Wood, and A.V. Fisher. 2002. Effects of dietary fat source and breed on the carcass composition, *n*-3 polyunsaturated fatty acid and conjugated linoleic acid content of sheep meat and adipose tissue. Brit. J. Nutr. 88, 697-709.

Wiegand, B.R., J.C. Sparks, F.C. Parrish, Jr., and D.R. Zimmerman. 2002. Duration of feeding conjugated linoleic acid influences growth performance, carcass traits, and meat quality of finishing barrows. J. Anim. Sci. 80, 637-643.

Whitlock, L., D.J. Schingoethe, A.R. Hippen, K.F. Kalscheur, R.J. Baer, N. Ramaswamy, and K.M. Kasperson. 2002. Fish oil and extruded soybeans fed in combination increase CLA in milk of dairy cows more that when fed separately. J. Dairy Sci. 85, 234-243.

Yurawecz, M.P., J.A.G. Roach, N. Sehat, M.M. Mossoba, J.K.G. Kramer, J. Fritsche, H. Steinhart, and Y. Ku. 1998. A new conjugated linoleic acid isomer, 7 *trans*, 9 *cis*-octadecadienoic acid, in cow milk, cheese, beef and human milk and adipose tissue. Lipids 33, 803-809.

Zambell, K.L., N.L. Keim, M.D. Van Loan, B. Gale, P. Benito, D.S. Kelly, and G.J. Nelson. 2000. Conjugated linoleic acid supplementation in humans: effects on body composition and energy expenditure. Lipids 35, 777-782.

Fatty acid composition of phospholipids and levels of alpha-tocopherol, total antioxidative capacity and malondialdehyde in liver and muscular tissue after dietary supplementation of various fats in cattle

J. Rehage[1], O. Portmann[1], R. Berning[1], M. Kaske[1], W. Kehler[1], M. Hoeltershinken[1], R. Duehlmeier[2], M. Coenen[3] and H.-P. Sallmann[2]
[1]Clinic for Cattle; [2]Department of Physiological Chemistry; [3]Department of Animal Nutrition, School of Veterinary Medicine Hannover, Bischofsholer Damm 15, D–30173 Hannover, Germany

Aim of the study was to investigate the effect of dietary fat supplements on fatty acid composition of phospholipids in liver and muscular tissue including the assessment of the wash-out period, and the ascertainment of effects on tocopherol, total antioxidative capacity and malondialdehyde concentrations.

In a Latin Square design, six non-lactating and non-pregnant rumen fistulated German HF cows were used. Dietary fat supplements (5% of DM) were palmstearin fat (C16:0: 57%, C18:0: 35%, C18:1n9: 6%), linseed (C16:0: 4%, C18:0: 3%, C18:3n3: 62%, C:18:2n6: 14%, C:18:1n9: 16%) and sunflower oil (C16:0: 4%, C18:2n6: 69%, C18:1n9:25%), resp.. Fats were supplemented four weeks each. Thrice daily, palmstearin was administered intraruminally, linseed and sunflower oil intraabomasally to avoid bio hydrogenation. Rations were isocaloric, calculated for maintenance and based on hay and concentrate. Muscle and liver biopsies were obtained between, before and after each supplementation period and analyzed for fatty acid composition in the phospholipid fraction by gas chromatography. In same intervals alpha-tocopherol was measured by HPLC and malondialdehyde (TBAs) and total antioxidative capacity (FRAP) was assessed photometrically in plasma and liver and muscular tissue.

Muscular phospholipid mean fatty acid composition did not change significantly after supplementation of palmstearin compared to baseline values, whereas significant changes in fatty acid composition were induced by linseed (mean ± sem before and after supplementation: C18:3n3: 2.3±0.3 *vs.* 9.7±0.4%; C18:2n6: 14.2±0.5 *vs.* 19.9±1.2%; C18:1n9: 19.1±0.7 *vs.* 12.2±0.2%;) and sunflower oil (C18:2n6: 14.2±0.5 *vs.* 32.9±1.91; C18:1n9: 19.0±0.7 *vs.* 9.9±1.1) supplements. Similarly, also hepatic phospholipid fatty acid composition was significantly altered by linseed and sunflower oil supplements. Whereas in muscular tissue baseline values of the fatty acid composition of phospholipids (wash-out period) were not reached before 150 days after finishing fat supplementation, in hepatic tissue baseline values were observed already after 45 days. In plasma mean alpha-tocopherol concentrations increased significantly by approximately 80% after linseed and sunflower oil supplements, in contrast in liver tissue alpha-tocopherol levels decreased by 20% (n.s.). Similar results were obtained for mean FRAP levels. Only after linseed supplementation in average a significant increase of TBAs – levels by about 100% was observed in plasma and studied tissues, whereas after sunflower oil and palmstearin supplements mean TBAs concentrations remained almost unchanged.

Results indicate a) that the fatty acid composition of phospholipids of various tissues is altered considerably by the source of supplemented fats, and b) the wash-out period after finishing the fat supplementation can vary substantially between different tissues. Additionally, the source of different fat supplements may affect the antioxidative capacity of the treated animal. However, in particular after supplementation of fats rich in poly-unsaturated fatty acids, such as linseed oil, the risk of lipoperoxidative processes appears to be increased.

A field sample study investigating possible indicators of undernutrition in cattle

S. Agenäs[1,2], M.F. Heath[1], R.M. Nixon[3], J.M. Wilkinson[1] and C.J.C. Phillips[1,4]
[1]*Farm Animal Epidemiology and Informatics Unit, Department of Clinical Veterinary Medicine, University of Cambridge, Madingley Road, Cambridge CB3 0ES, UK;* [2]*Swedish University of Agricultural Sciences, Department of Animal Nutrition and Management, Kungsängen Research Centre, 753 23 Uppsala, Sweden;* [3]*MRC Biostatistics Unit, Institute of Public Health, University Forvie Site, Robinson Way, Cambridge CB2 2SR, UK;* [4]*School of Veterinary Science, University of Queensland, Gatton 4343, QLD, Australia*

The objective of the study was to investigate potential biochemical indicators of long-term undernutrition in cattle, which could be used objectively, reliably and routinely by veterinary practitioners. This was addressed in a field sample study that included analysis of metabolites in fresh samples from adequately nourished suckler cows and stored samples from suckler cows in severe undernutrition. Evidence of suffering in cases of undernutrition in farm animals is currently a matter of subjective veterinary judgement, often based on body condition scoring. In addition to this, metabolic profiling is often used. However, metabolic data for cattle in long-term undernutrition is very limited and the precision of metabolic profiling is generally low for individual cows. Objective measures of undernutrition are needed to assist in the determination of possible infringements of legislation by livestock farmers.

The study included samples obtained in 2003 from nine English herds that were considered, by the attending veterinary surgeon, to be receiving adequate nourishment. Also, 27 frozen serum samples from two herds (A and B) were made available to us by the Royal Society for the Protection of Cruelty to Animals. These samples were collected in the spring of 1998 in connection with suspected severe undernutrition in two herds of beef suckler cows. The herds comprised mature animals, some with calves at foot, that had been offered a restricted allocation of a total mixed ration, and were emaciated, according to the attendant veterinarian.

All samples were analysed in 2003 for serum albumin, total protein, urea, BHB, NEFA, creatinine, fructosamine and globulin. For 20 animals (Herd B) the samples had also been analysed when fresh in 1998, for albumin, BHB, total protein and urea. Statistical analyses were carried out to determine whether any of the analytes had significantly different average values in either of the under-nourished herds when compared to adequately-nourished animals. The two under-nourished herds were each compared to the combined adequately-nourished animals by the Mann-Whitney U test. The 1998 values for Herd B were also compared with the adequately-nourished animals by the Mann-Whitney U test. The 1998 and 2003 analytical results were compared by paired t-tests. Furthermore, tests were carried out to determine whether any of the analytes provided a useful diagnostic test for undernutrition. The sensitivity and specificity of each analyte as a test for undernutrition was assessed by relative operating characteristic (ROC) analysis.

Significant regressions were found only for albumin (regression coefficient: 1.11 ± 0.35, P=0.0014), for creatinine (6.65 ± 2.25, P=0.0033) and for fructosamine/albumin ratio (-0.27 ± 0.09, P=0.0042). For albumin and creatinine, both under-nourished herds had values significantly lower than the adequately-nourished animals (P<0.001), while both were significantly higher for fructosamine (P<0.001) and for fructosamine/albumin ratio (P<0.001). With the condition for undernutrition set at ≥ 10.75 µmol/g, the fructosamine/albumin ratio gave sensitivity and specificity of 100%. The apparently satisfactory sensitivity of the fructosamine/albumin ratio to detect significant differences between undernourished and adequately nourished cattle offers some promise, but further confirmatory work, on fresh samples from animals in different herds is required.

Section K.
Mastitis

Activation of immune cells in bovine mammary gland secretions by zymosan treated bovine serum

K. Kimura and Jesse P. Goff
National Animal Disease Center, USDA–ARS, Ames, IA, USA

Bacterial infection of the mammary gland remains a major disease of dairy cattle. The greatest risk of mammary infection occurs at the end of lactation and at initiation of the next lactation when the cow calves. The bacteria causing mastitis during these two periods are similar. If we can produce memory/effector T-cells that recognize bacteria in mammary glands at dry off, we may be able to reduce mastitis at dry off and prevent mastitis in early lactation. Memory cells are easily activated by encounter with Ags to which they are sensitive, and transform to effector cells, which then attack the bacteria. Other workers have successfully recruited more neutrophils (PMN) into the mammary gland at dry off using inflammatory substances such as LPS, colchicine, oyster glycogen, *etc.* These studies focused on acceleration of mammary gland involution and recruitment of PMN to eliminate mastitis at dry off period. The few studies examining the long-term effect of these modulators have not shown a beneficial effect on mastitis in early lactation. These studies did not examine the effect the treatments might have on lymphocytes, or the production of effector/memory cells.

Treating serum with zymosan causes complement cleavage allowing serum to be used as a source of C5a, a potent activator and chemoattractant of immune cells. C5a not only recruits/activates PMN, but also activates antigen presenting cells and T-cells. We hypothesized that intra-mammary infusion of zymosan treated serum (ZTS) just after cessation of lactation can recruit/activate PMN and generate memory/effector lymphocytes. In order to test our hypothesis, two of the four quarters of the mammary gland of eight dairy cows were infused with ZTS (12.5 ml/quarter). The other two quarters were treated with physiological saline (control). Mammary secretions were collected throughout the dry period (0, 1, 8, and 15 d after treatment) and the first two weeks of the next lactation (-1wk, 0 d, 1 & 2 wk after calving). Activation of lymphocytes and PMN in those secretions was evaluated based on intracellular IFN-g and IL-8 up-regulation as assessed by flow cytometry using anti-human IFN-g and IL-8 mAbs.

Results are shown in Table 1. ZTS infusion greatly increased % PMN & SCC in mammary secretions at 1 d after treatment only. The % IFN-g positive lymphocytes and PMN, and the % IL-8 positive PMN exhibited a sustained increase in secretions from ZTS treated quarters through the first two weeks of lactation compared to control (P<0.05). ZTS treatment showed temporarily high fever at 7 h after treatment (40.6 C) in one cow and it returned to normal by 24 hours after treatment. Milk production in the ZTS treated and control half udders were 195.7 ± 37.8 and 187.1 ± 48.7 kg/head/15 days, respectively, and combined total milk yield was higher than former lactation (first lactation, 253.3 ± 36.8 kg/head/15 days). ZTS can stimulate innate and acquired immune defense mechanisms in the mammary gland. Further studies will be required to determine if this treatment provides a useful means of preventing mastitis.

Table 1. Evaluation of mammary secretions.

Group	Parameter	Pre treatment	1d after treatment	8d after treatment	Day of calving	2wk after calving
Control	SCC	4.3 ± 1.0	4.2 ± 0.8	4.6 ± 0.6	0.49 ± 0.1	0.05 ± 0.02
ZTS	(x106/ml)	5.0 ± 1.1	10.8 ± 2.3	4.5 ± 0.7	0.43 ± 0.1	0.02 ± 0.007
Control	% PMN	21.8 ± 9.2	15.9 ± 8.0	7.9 ± 1.7	0.8 ± 0.19	4.7 ± 4.4
ZTS		23.6 ± 9.2	46.8 ± 10.4	10.8 ± 2.0	1.7 ± 0.7	8.1 ± 4.4
Control	% IFN+ in	6.2 ± 3.4	6.4 ± 4.2	2.4 ± 0.4	16.8 ± 5.7	8.7 ± 2.5
ZTS	lymphocytes	6.1 ± 2.4	17.3 ± 5.7	4.8 ± 0.8	27.2 ± 10.8	10.4 ± 2.8
Control	% IFN+ in	7.3 ± 1.6	4.3 ± 0.8	6.2 ± 2.7	14.9 ± 5.4	7.4 ± 3.2
ZTS	PMN	8.5 ± 1.7	11.7 ± 3.3	17.2 ± 7.4	27.0 ± 11.4	19.7 ± 8.8
Control	% IL-8+ in	24.7 ± 7.4	6.9 ± 4.6	4.1 ± 2.2	46.0 ± 15.1	23.9 ± 10.3
ZTS	PMN	16.3 ± 6.9	26.3 ± 10.7	15.2 ± 3.9	65.9 ± 13.9	45.4 ± 12.8

The effect of milk yield at dry-off on the likelihood of intramammary infection at calving

P.J. Rajala-Schultz[1], K.L. Smith and J.S. Hogan[2]
[1]Dept. of Veterinary Preventive Medicine; [2]Dept. of Animal Sciences, The Ohio State University, Columbus OH, USA

The importance of the dry period regarding udder and overall health and productivity of dairy cows is well documented and recently reviewed[1-4]. The effect of milk yield level at dry-off on the rate of intramammary infections (IMI) at calving, however, has not been intensively studied. A recent study reported a positive association between milk production on the day before dry-off and the risk of new IMI in the dry period[5]. The objective of this study was to evaluate the effect of milk yield at dry-off on the probability of a cow having an IMI at calving.

The study was conducted in the Krauss Dairy Research Herd at the Ohio Agricultural Research and Development Center (OARDC) in Wooster, OH. All cows in the herd are routinely treated with antibiotics after the last milking at the end of lactation. Duplicate quarter milk samples were obtained within 3 days of calving from each cow in the herd (173 lactations in total, 116 from 2nd lactation or older cows) between January 2001 and April 2002. Daily milk yield data were available for all cows. Cows were classified as infected with major environmental pathogens or with coagulase negative staphylococci (CNS), if one or more quarters were infected at calving with the respective pathogen. If infected with both types of organisms, cows were considered infected with the major pathogen. Data were also analyzed using infection status of a quarter as the outcome. Data were analyzed using generalized estimation equations (PROC GENMOD) in SAS® to account for the correlated data structure. Cow's lactation number, calving season, somatic cell count prior to dry-off, dry period length, and milk yield on the day of drying off were considered as explanatory variables in the model.

The results indicated that increasing milk yield at dry-off significantly increased the risk of a cow (or an individual quarter) being infected at calving with Gram-negative organisms or esculin-positive streptococci. Season of calving, lactation number of a cow, somatic cell count prior to dry-off and dry period length were not statistically significantly associated with the probability of a cow being infected with environmental organisms at calving. When infection status with CNS was modeled, only lactation number of the cow was significantly associated with the probability of infection at calving. The association between increased risk of IMI caused by environmental organisms at calving and increased milk yield at dry-off may be related to milk leakage from quarters as well as failure of teat-canal closure.

References

[1] Berry E.A., 2003. Recent evaluations of dry cow strategies. 42nd Annual Meeting of National Mastitis Council, pp. 31-41.

[2] Dingwell R.T., D.F. Kelton and K.E. Leslie, 2003. Management of the dry cow in control of peripartum disease and mastitis. Vet. Clin. Food Anim. 19, 235-265.

[3] Ekman T. and O. Østerås, 2003. Mastitis control and dry cow therapy in the Nordic countries. 42nd Annual Meeting of National Mastitis Council, pp. 18-30.

[4] Leslie K.E. and R.T. Dingwell, 2003. Background to dry cow therapy: what, where, why - is it still relevant? 42nd National Mastitis Council Annual Meeting, pp. 5-17.

[5] Dingwell R.T., K.E. Leslie, T.F. Duffield, *et al.*, 2003. Efficacy of intramammary tilmicosin and risk factors for cure of Staphylococcus aureus infection in the dry period. J. Dairy Sci. 86, 159-168.

Mastitis therapy for persistent *Escherichia coli* on a large dairy

P.M. Sears[1], C.E. Ackerman[1], K.D. Crandall[1] and W.M. Guterbock[2]
[1]College of Veterinary Medicine, Michigan State University, East Lansing, Michigan, USA, sears@ cvm.msu.edu; [2]Den Dulk Dairy, LLC, Ravenna, MI, USA

In a previous clinical mastitis study[1], 39% of *E. coli* intramammary infections persisted for more than 21 days. Since *E. coli* is isolated from less than 20% of the clinical cases, persistent *E. coli* infections make up less than 10% of the clinical mastitis cases. A protocol that includes the treatment of all cows to address these persistent infection is neither practical nor cost-effective. Currently, many large farms are not treating *E. coli* mastitis with antibiotic therapy. However, in herds where milk from clinical mastitis cases is cultured and pathogens identified before treatment, the use of antibiotics in *E. coli* infections may be both practical and cost-effective. In this study, we evaluated two antibiotic treatments for their effect on persistent *E. coli* infections. Milk from clinical mastitis cases were cultured at the farm and antibiotic treatment was withheld for 18-24 hours until the pathogen was identified. *E. coli* cases (n=30) were assigned to a treatment group by randomized block table. The treatment groups included: 1) no antibiotic treatment; 2) 30cc of intramuscular tetracycline once daily for 3 days; or 3) intramammary infusion of 200 mg ceftiofur once daily for 3 days. Milk was observed for clinical response at each milking and milk was cultured between 7-14 days and 21-28 days after treatment. Data was analyzed for *E. coli* isolation, quarters lost and cows removed from the dairy herd for mastitis.

The mean time to reculture was 14 days for tetracycline treated cases and 9 days for ceftiofur treated cases. When clinical quarters were recultured and antibiotic treatments combined, there were no difference in the recovery of persistent *E. coli* from treated and untreated cases. In the 30 clinical cases of *E. coli*, the cow treated with antibiotics had a 33% (7/21) persistent *E. coli* which was the same as the non-treated cases, 33% (3/9). The persistence was lower for ceftiofur (20%) than tetracycline (45%), but the number of quarters and cows lost to production was not different (38% for treatment cases and 33% in the non-treated cases). However, cows with persistent infections made up the majority of cows and quarters lost. Cows were recultured between 6 and 14 days before re-entering the milk herd. Isolation of *E. coli* at 21-28 days were the same as 7-14 days post-treatment. Infections that cleared early after treatment remained clear, while *E. coli* presence at 7 days persisted past 21 days.

Treating *E. coli* cases with antibiotic therapy did not significantly lower the persistence of *E. coli* infections. Treatment had no effect on the number of cows or quarters lost to production. Because the low incidence of persistent *E. coli* infections in clinical mastitis, the treatment of these cases with antibiotics did not produce an economic benefit by reducing their persistence in the herd or reducing the number of cows and quarters lost to the infection.

References
[1] Ackerman, C.E., C.T. Wehreman, P.A. Busman, L.M. Neuder and P.M. Sears, 2003. Identifying persistent infection from clinical cases of coliform mastitis. Proc. Am. Assoc. Bovine Pract. 2003, pp. 170.

Diagnostic key data for lactate dehydrogenase activity measurements in raw milk for the identification of subclinical mastitis in dairy cows

A. Neu-Zahren, U. Müller, S. Hiss and H. Sauerwein
Institute of Physiology, Biochemistry and Animal Hygiene, University of Bonn, Germany

For decades, determination of lactate dehydrogenase (LDH) activity in milk alone or in combination with other soluble factors has been proposed as diagnostic marker for mastitis. However, most of the data available are related to clinical mastitis and are limited to off-line measurements, *i.e.* laboratory analyses of stabilized or frozen and thawed samples. Besides parenchymal sources, LDH is contained in leukocytes and if somatic cells are lysed during sample centrifugation and/or storage, the actual LDH activity determined in row versus treated milk might differ. In order to develop a cow-site testing system in particular for the detection of subclinical mastitis, we initially validated a reflectrophotometrical assay system (Ortho Diagnostik Systems GmbH, Neckargemünd, Germany) for raw fresh milk. We then applied the system to record LDH activity in quarter foremilk samples freshly obtained from 30 German Holstein cows that showed no signs of clinical mastitis from a dairy farm using an automated milking system (Leonardo, WestfaliaSurgeGmbH). Sampling was repeated 3–6 times in weekly intervals and a total number of 408 samples was evaluated. Somatic cell counts (SCC) and presence of mastitis pathogens were determined in parallel. To relate the data to udder health, the definitions of the International Dairy Federation (1967) and the German Veterinary Society DVG (1994) were used for orientation.

Median values are given. From healthy quarters average LDH activities of 88 U/l were recorded, in quarters affected with subclinical mastitis (SCC above 130,000 cells/mL and bacteriologically positive findings in 2 out of 3 testings) LDH activity was increased ($P<0,001$) to 131 U/l. On the basis of the bacteriological findings the positive quarter samples were subdivided in two classes being affected either with minor pathogens (minP, *e.g.* C. bovis and coagulase-negative Staphylococcus, n = 254) or with major pathogens (majP, *e.g.* Staph. aureus and E. coli n = 77). In majP samples LDH activities were higher than in minP samples (94 U/l versus 168 U/l, $P<0,001$). SCC was 38,000 in samples from healthy quarters; in mastitic samples SCC values were 100,000 in minP and 1,600,000 in majP, respectively. The correlation coefficient between LDH activity and SCC determined on log data was r = 0.72 ($P=0.01$). For the determination of diagnostic key numbers, we tried upper threshold range between 90 to 110 U/l for LDH-activity in negative samples. Using 100 U/l we obtained a classification of mastitis with a sensitivity of 88% and a specificity of 86%. A threshold of 110 U/l resulted in a sensitivity of 70% and a specificity of 94%.

Our results demonstrate that LDH activity can successfully be used as diagnostic tool for mastitis. By assigning defined ranges of LDH activity to different levels of alertness, a simple and practicable traffic-light system could be established in which green is healthy, yellow means jeopardized and red stands for mastitis positive findings. Moreover, the applicability of the assay in raw milk opens up new vistas of on-line identification of diseased quarters or cows, respectively.

Acknowledgements
We thank WestfaliaSurge GmbH, Research and Design (Germany) for financial support.

Evaluation of leukocyte subset for occurrence of mastitis on dairy herd

H. Ohtsuka[1], M. Kohiruimaki[2], T. Hayashi[3], K. Katsuda[3], R. Abe[3] and S. Kawamura[1]
[1]Veterinary Internal Medicine, School of Veterinary Medicine and Animal Sciences, Kitasato University, Towada, Aomori 034 8628, Japan; [2]Kohiruimaki Large Animal Medical Center, Touhoku, Aomori 039 2683, Japan; [3]Department of Immunology, National Institute of Animal Health, Tsukuba, Ibaraki 305 0856, Japan; [4]Division of Immunobiology, Research Institute for Biological Sciences, Tokyo University of Science, Noda, Chiba 278 0022, Japan

To clarify the cellular immune condition in the dairy herd with occurrence of astitis, leukocyte population was analyzed in two mastitis herd groups (A and B herd), and one other healthy herd group (C herd). There was a 40% occurrence of mastitis in the herd A, and mastitis occurred 30% in the herd B. All herds were evaluated for five milking terms.

In feeding contents, the rate of total digestible nutrients (TDN) and crude protein (CP) were lower in A and B groups than those in C group. Levels of serum total cholesterol and blood urea nitrogen were lower in A and B herd than those in the C herd. The number of neutrophil in A and B herd increased compared to the C herd, while the number of CD3+, CD4+ T-cells and CD14 MHC class_+ cells were lower in the A and B herd than in the C herd.

These results suggest that there was a relationship between the immune condition in herds and sporadically occurring mastitis following to malnutrition about herd entirely.

Section K.

Antimicrobial treatment strategies for Streptococcal and Staphylococcal mastitis

K.D. Crandall[1], Philip M. Sears[1] and Walter M. Guterbock[2]
[1]Department of Large Animal Clinical Science, Michigan State University, East Lansing, Michigan; [2]Den Dulk Dairy LLC, Ravenna, MI. USA

Mastitis is one of the most costly diseases encountered by dairymen. Traditional therapy aimed at curing clinical mastitis cases includes intramammary (IMM) antimicrobial therapy. In spite of multiple available IMM antimicrobial products, cure rates for clinical mastitis run about 46% for *Streptococcus* spp, 21% for *Staphylococcus* spp, and 9% for *Staphylococcus aureus* mastitis[1]. This study investigated the use of systemic antimicrobial therapy (ampicillin) in conjunction with IMM antimicrobial therapy for *Streptococcus* spp, *Staphylococcus* spp., and *S. aureus*.

Milk from lactating Holstein cows with clinical mastitis from a 3,000-cow dairy herd was cultured on sheep blood agar. All cows with a *Streptococcus* spp. (n=80), *Staphylococcus* spp. (n=60), or *Staph aureus* (n=25) positive culture result were enrolled in the trial and assigned one of three treatments based on a randomized six-block table. For *Streptococcus* spp., cows were grouped as not treated, treated with IMM amoximast once a day for five days, or treated with IMM amoximast plus 30 cc ampicillin IM once a day for 5 days. For *Staphylococcus* spp. and *S. aureus* cows were grouped as not treated, treated with IMM pirsue once a day for five days, or treated with IMM pirsue plus 30 cc amplicillin IM once a day for 5 days. One milk sample was collected and cultured from the affected quarters 21-28 days after completion of antimicrobial therapy. A cow was considered cured if there was no growth at the 21-28 day culture. Days until clinical cure, days in hospital, somatic cell count (SCC), milk production, previous mastitis events, lost quarters, and whether the cow was sold/died were all recorded.

For cows with *Streptococcus* spp. mastitis in the no treatment group, 6/15 cows were cured, 4 cows lost a quarter and 10 cows were sold/died. Within the IMM amoximast group, 11/20 cows were cured, 4 cows lost a quarter and three cows were sold/died. Within the IMM amoximast plus IM ampicillin group, 16/20 cows were cured, three cows lost a quarter, and none were sold/died. For cows with *Staphylococcus* spp. mastitis in the no treatment group, 3/8 cows were cured, 1 cow lost a quarter and 2 cows were sold/died. Within the IMM pirsue group, 5/9 cows were cured, 3 cows lost a quarter, and 1 cow was sold/died. Within the IMM pirsue plus IM amplicillin group, 4/14 cows were cured, 2 cows lost a quarter, and no cows were sold/died. For cows with *S. aureus* mastitis in the no treatment group, 0/5 cows were cured, 5 cows lost a quarter and 2 cows were sold/died. Within the IMM pirsue group, 0/9 cows were cured, 1 cow lost a quarter, and 1 cow was sold/died. Within the IMM pirsue plus IM ampicillin group, 0/8 cows were cured and no cows lost a quarter or were sold/died.

Treatment with systemic ampicillin in conjuction with IMM therapy proved more effective in eliminating *Streptococcus* spp. organisms from the mammary gland and kept more cows in production over the traditional IMM therapy. The same is not true for *Staphylococcus* spp. or *S. aureus*. Further investigation is needed to develop effective treatment strategies for *Staphylococcus* spp. and *S. aureus* mastitis clinical cases. A producer should rely on culture based treatment protocols as an effective and economical means to reduce mastitis in the overall herd.

References

[1] Wilson, D.J., *et al.*, 1996. Efficacy of Florfenicol versus clinical and subclinical cases of bovine mastitis. *NMC proceedings* 1996, pp. 164-165.

Subclinical mastitis in dairy cows in farms with organic and with integrated production: prevalence, risk factors and udder pathogens

M. Roesch[1], E. Homfeld[1], M.G. Doherr[2], W. Schaeren[3], M. Schällibaum[3], J.W. Blum[1]
[1]Division of Nutrition and Physiology; [2]Division of Clinical Research, Vetsuisse Faculty, University of Berne, Berne; [3]Swiss Federal Dairy Research Station, Liebefeld-Berne, Switzerland

Subclinical mastitis is one of the most costly diseases in dairy production and is more frequent in farms with organic production (OP) than in farms with integrated production (IP). Epidemiological studies comparing chronic mastitis in OP and IP farms are lacking.

We have therefore investigated 970 cows in 60 randomly selected OP farms and 60 comparable IP farms in the Canton of Berne, Switzerland to test the hypothesis that prevalence, risk factors and patterns of udder-pathogenic microbes in OP and IP cows are different. California Mastitis Tests (CMT) were performed at 21-43 days p.p. (period I) in 483 OP and 487 IP cows and at 72-135 days p.p. (period II) in 419 OP and 421 IP cows. Cows with CMT =1+ in one quarter, but without clinical signs of mastitis, were considered to have subclinical mastitis. Of quarters with CMT =2+ a milk sample was taken for bacteriological investigation. Somatic cell counts (SCC) were available from breeder organisations.

Prevalences of subclinical mastitis per cow in periods I and II were 39 and 40% in OP farms and 34 and 35% in IP farms, respectively. On a quarter level prevalences of mastitis were 15 and 18% in OP farms and 12 and 15% in IP farms, respectively. The SCC (cells x 10^3/mL milk) in periods I and II were 163 and 152 in OP cows and 136 and 119 in IP cows, respectively. Prevalences of *Staphylococcus aureus* and other Staphylococci were higher in IP than in OP farms, whereas prevalences of *Streptococci* other than *Streptococcus agalactiae* and *Corynebacterium bovis* were higher in OP than in IP farms. Breed, teat or udder injuries, use of antibiotic dry cow therapy, and SCC before drying off were risk factors for the development of subclinical mastitis. In conclusion, there was a higher number of subclinical mastitis in the first 100 days of lactation in OP than in IP farms. Some of the found risk factors were strongly related to the two different farming systems.

Evaluation of a novel on-farm test for antibiotic susceptibility determination in mastitis pathogens

B.C. Love[1] and P. Rajala-Schultz[2]
[1]Department of Veterinary Science, Pennsylvania State University, University Park, Pennsylvania;
[2]Department of Veterinary Preventive Medicine, The Ohio State University, Columbus, OH, USA

MASTik[TM] (ImmuCell, Portland, ME) is marketed as a rapid, cost effective, on-farm test for determining antibiotic susceptibility profiles of mastitis causing bacteria in milk samples, to be used in place of, or as an adjunct to, reference laboratory bacterial culture and susceptibility testing services. Concerns about increasing antibiotic resistance in bacteria, coupled with continual economic difficulties for farmers has increased the need for a test such as this. If this test proved accurate it would provide a rational basis for choosing which antibiotic should be used for mastitis treatment in a timely fashion. The purpose of this study was to compare culture and antimicrobial susceptibility results from standard laboratory methods (Sensititre™, Trek Diagnostics, Westlake, OH) with results from an on-farm antibiotic susceptibility test (MASTik™, ImmuCell Corporation, Portland, ME).

All milk samples used in the study had been submitted to the Pennsylvania State University Animal Diagnostic Laboratory (PSU ADL) for culture and antimicrobial susceptibility testing. A total of 102 samples were processed using both the laboratory culture and sensitivity protocol, done in accordance with National Mastitis Council guidelines and the MASTik™ protocol, according to manufacturer's instructions. Of these, bacterial culture resulted in 28 samples with No Growth, 16 with No Significant Growth, 7 with Contamination, 3 with Yeast and 1 with *Arcanobacterium pyogenes*. Antimicrobial susceptibility testing was not performed on these 55 samples. Of the remaining 47 samples, CNS (n=14), S.aureus (6), E.coli (6), Klebsiella (6) and streptococcal species (9) were most commonly isolated.

Susceptibility testing was performed on a subset of 28 isolates derived from the samples with significant bacterial growth. These isolates included *E. coli* (5), *Klebsiella pneumoniae* (5), *Enterobacter* spp. (1), *Staphylococcus aureus* (4), coagulase-negative staphylococci (4), *Streptococcus dysgalactiae* (4), *Streptococcus uberis* (2), non-agalactiae streptococci (2), and *Enterococcus* spp. (1).

Of the 102 samples, 70 (68%) had a valid internal positive control and 92 (89%) had a valid internal negative control on the MASTik™ kit. The internal controls performed best when gram-positive organisms were recovered or when no bacterial growth occurred on culture. Performance of internal controls more often failed when growth on culture was considered not significant or contamination, or when gram-negative organisms were recovered. There were 21 samples for which both a susceptibility test was performed in the laboratory and the MASTik™ kit had valid internal controls. Comparison of the interpretation of susceptibility results between the two methods is ongoing.

Section L.
Rumen digestion and metabolism

Physically effective fiber and regulation of ruminal pH: more than just chewing

M.S. Allen, J.A. Voelker and M. Oba
Michigan State University, Department of Animal Science, East Lansing, MI 48824 USA

Abstract

Ruminal acidosis is caused by an imbalance between the production of fermentation acids by microbes in the rumen and the absorption, passage, neutralization, and buffering of those acids. The production rate of fermentation acids is highly variable across diets and increases greatly with highly fermentable starch sources. Hydrogen ions are removed primarily by absorption from the rumen, and the concentration gradient across the ruminal epithelium is likely the major factor affecting absorption of acids from the rumen. Coarse forage fiber affects ruminal pH by retaining digesta in the rumen which provides the buffering capacity inherent in feedstuffs, increases salivary buffer flow through stimulation of rumination, and increases the concentration gradient through stimulation of ruminal motility. Selecting optimal diet fermentability and maintaining an adequate digesta pool are likely key factors in preventing subacute ruminal acidosis.

Keywords: subacute rumen acidosis, rumination, motility, concentration gradient, buffering

Introduction

A slightly acidic ruminal pH is desirable to maximize milk yield of dairy cattle because diet digestibility and yield of microbial protein produced in the rumen are maximized when highly fermentable diets are fed. Ruminal pH near or above 7.0 can be thought of as lost opportunity; energy intake and microbial protein production will be sub-maximal, limiting milk yield and (or) decreasing diet cost. However, as ruminal pH decreases, appetite (Shinozaki, 1959), ruminal motility (Ash, 1959; Shinozaki, 1959), microbial yield (Hoover, 1986), and fiber digestion (Hoover, 1986; Terry *et al.*, 1969) are reduced. The optimal ruminal pH is dependent on within-day variation determined by diet composition, production level, and feeding systems. Ruminal pH measured at one time point or averaged throughout the day has little biological significance without considering the pattern of ruminal pH within a day; even minimum ruminal pH means little if the duration at this pH is very short. Both the time spent at low pH and the extent to which pH is depressed should be considered (Mackie and Gilcrist, 1979). Within-day variation in ruminal pH varies greatly, increasing with increased dietary starch concentration (Oba and Allen, 2000, 2003), and is expected to differ among feeding systems. Minimizing fluctuations in ruminal pH is expected to allow greater energy intake, microbial protein production and improved animal health. The objective of this paper is to identify and discuss factors affecting ruminal pH as a dynamic system, so that feeding systems can be managed to limit subacute ruminal acidosis.

Ruminal acid production

Ruminal hydrogen ion concentration (described by pH) is determined by the balance between fermentation acid production and hydrogen ion removal by absorption, neutralization, buffering, and passage. Ruminal acids are end products of microbial fermentation and include acetic, propionic, and butyric acids. They are mild acids with pKa's around pH 4.8 and are the primary acids involved in subacute ruminal acidosis. Lactic acid also is produced by ruminal microbes (*e.g., Streptococcus bovis, Lactobacilli*), and its acid dissociation constant is 10-fold higher than the major VFA, so it has a much greater influence on ruminal pH below pH 5. Although lactic acid is clearly involved in acute acidosis, it is not likely a common contributor to subacute ruminal acidosis (see Figure 1B) because it is normally only a minor fermentation endproduct that is metabolized rapidly in the rumen (Gill *et al.*, 1986).

Both the amounts and the patterns of acid production and removal determine the physiological response to ruminal pH. Total daily production of fermentation acids is determined primarily by organic matter (OM) intake and proportion of ruminally degraded organic matter (RDOM) in the diet, and both OM intake and RDOM vary widely. Allen (1997) reported that RDOM content of diets for lactating dairy cattle averaged 50% and ranged from 29 to 67% of the total OM for experiments using duodenally canulated cows (48 treatment means). The amount of RDOM averaged 9.8 kg/d and ranged from 5.7 to 15.4 kg/d. Because approximately 7.4 equivalents of acid are produced per kg ruminally degraded organic matter (RDOM) when microbial cell yield from hexose fermented is 0.33 (Allen, 1997), approximate fermentation acid production in these experiments varied from 42 to 114 Eq/d. In addition to this wide range of potential acid production, both the pattern of acid production and the extent and pattern of acid neutralization

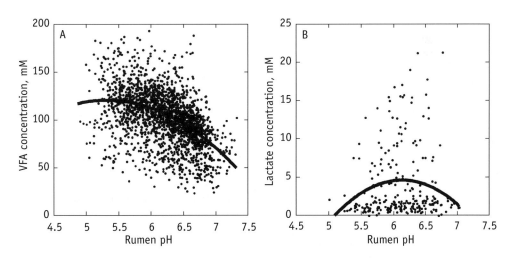

Figure 1. Panel A shows the relationship between ruminal concentration of VFA and ruminal pH for rumen liquid sampled every 20 minutes for 24 h periods in a 2x2 factorial experiment with replicated 4x4 Latin squares (Oba and Allen, 2003). Treatments were conservation method of corn grain (high moisture vs. dry) and diet starch content (21 to 32%). The relationship is quadratic (P<0.0001). Panel B shows ruminal lactate concentration for samples from the same experiment (quadratic, P < 0.01).

and removal contribute to physiological response to ruminal pH. The production of hydrogen ions per second for a cow consuming 20 kg of 50 rumen-degraded OM is 0.86 meq/sec, which is more than 10 times the free ruminal hydrogen ion pool size (0.08 meq at pH 6.0 in 80 L). Therefore, understanding effects of diet and management on the moment-by-moment turnover of hydrogen ions is very important for limiting subacute ruminal acidosis.

The amount of fermentable OM in the rumen at any moment, and thus the acid production at any moment, depend upon DMI, meal patterns, flow rate of OM from the rumen, and rate of fermentation. Feeds vary in the fraction of OM available for fermentation and in rates of digestion and passage of the available fraction from the rumen. The same ruminal digestibility can be attained for a feed with a slow rate of digestion as for a feed with a very fast rate of digestion if it is retained in the rumen long enough. Also, feeds with very fast rates of fermentation such as ground high-moisture corn and steam-rolled barley cause greater variation in ruminal pH than feeds with more moderate rates of fermentation. Fiber generally ferments more slowly than starch and usually has a longer retention time. Therefore, digestible fiber provides a consistent supply of energy to microbes and to the animal over time. In contrast, finely ground grain ferments and passes from the rumen quickly, providing pulses of energy to rumen microbes and VFA to the animal. Rapid energy availability within a day or across diets can result in variation in the efficiency of microbial cell yield from hexose because of the uncoupling of digestion and cell growth (Voelker and Allen, 2003), further contributing to within-day variation in ruminal pH.

Feed intake amount and feed characteristics contribute to variation in both amount and pattern of fermentation acid production. Simultaneously, variation in the amount and pattern of acid removal may either stabilize or increase variation in ruminal acid concentration.

Hydrogen ion removal

A model of hydrogen ion production and removal from the rumen indicated that most (> 50%) acid is removed from the rumen by absorption across the rumen wall (Allen, 1997). Other routes of removal were identified as incorporation into water (28%) via carbonic acid from salivary bicarbonate, and flow from the rumen as dihydrogen phosphate (10%), VFA (~3%), ammonium (~2%), and particulate matter (<2%).

Conversion to water

A large fraction of hydrogen ions is removed by the carbonate buffer system; hydrogen ions combine with bicarbonate to carbonic acid (H_2CO_3) which is rapidly converted to H_2O and CO_2. Although the pKa for bicarbonate is 6.1, the effective pKa is much higher (~7.0) because the rumen is an open system; CO_2 is constantly lost by eructation with a loss of bicarbonate and hydrogen ions from the system.

Passage

The phosphate buffer system is different from the carbonate system because phosphate is removed by passage only (except for minor incorporation into microbial cells) and the pKa of hydrogen phosphate is higher at 7.2. At pH 6.0, approximately 94% of the hydrogen phosphate secreted is complexed as dihydrogen phosphate, which is removed by passage with the liquid fraction. Hydrogen ions also pass from the rumen associated with VFAs (~3%), ammonia (~2%), and feed residues (~2%; Allen, 1997). Therefore, passage rates of both solid and liquid digesta can affect ruminal pH.

Saliva secretion

Ruminal pH often decreases following meals and often increases during bouts of rumination (Figure 2). The decrease following meals is because of the production of fermentation acids from the OM consumed while the increase during rumination is usually attributed to the secretion of buffers in saliva. Ruminal pH was highly responsive in the individual cow represented in the figure; the reponsiveness of ruminal pH to meals and ruminating bouts depends on the presence of existing buffer reserves and therefore is not always as dramatic. Saliva contains bicarbonate and phosphate that neutralize and buffer acids in the rumen. The bicarbonate and phosphate concentrations of saliva have been reported as 126 and 26 meq/l, respectively (Bailey and Balch, 1961) and saliva composition is relatively constant and not greatly affected by diet or feed intake (Erdman, 1988). Therefore differences in the flow of salivary buffers into the rumen are a function of saliva flow. Total chewing time per day and the fraction of total chewing time spent ruminating were highly related to forage NDF content of the diet and the particle length of the forage (Allen, 1997), confirming the role of forage NDF in stimulating chewing. A meta-analysis of treatment means from the literature revealed that concentration of forage NDF in diets was positively related to ruminal pH ($r^2 = 0.63$, $P < 0.001$) while concentration of NDF from all sources was not related, showing the importance of effectiveness of forage NDF to maintain ruminal pH (Allen, 1997). Although this correlation is partly the result of increased saliva flow during rumination, there are likely additional effects of effective fiber. Using saliva flow rates reported in the literature for eating, ruminating, and idle activities, Allen (1997) calculated that while increasing dietary forage NDF concentration can increase total chewing time by 200 min/d, salivary buffer flow only increased 2.3 eq/d, or about 5%. This suggests that either the methods used to measure saliva flow drastically underestimated the true difference between idle and chewing saliva flow rates, or additional mechanisms determine the relationship between forage NDF and ruminal pH. It is possible that the effect of forage NDF on ruminal pH is the result of increased ruminal motility and maintenance of ruminal digesta pool in addition to increased salivary buffer flow.

Buffering of hydrogen ions

Basal fermentation of digesta in the rumen provides a fairly consistent flux of acids but the total flux is pulsatile with a frequency and amplitude determined by meal patterns and fermentability of the diet. Portal appearance of VFA, and therefore removal of hydrogen ions by absorption, is also pulsatile and is linked to meal patterns (Benson *et al.*, 2002). Each meal might provide substrate for the production of over 7,000 meq of acid over time for cows consuming 10 meals/d. Removal of hydrogen ions by neutralization is also expected to be pulsatile because saliva secretion increases during rumination and cows ruminate in bouts between meals throughout the day (Figure 2).

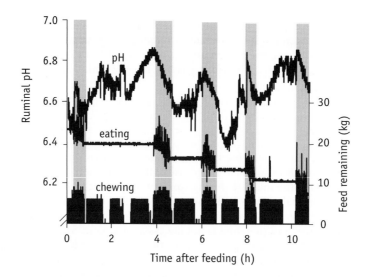

Figure 2. The relationship among ruminal pH, meals, and chewing activity for one cow fed a 35% NDF diet twice daily. Ruminal pH is represented by the top line. The weight of the feed remaining was measured by a manger suspended from a load cell and is represented by the middle line. Meals are represented by the vertical bars. Increases in feed remaining that were recorded during eating bouts were due to downward pressure applied by the cow on the manger. Chewing activity is represented by the bottom line. Because many points are represented, chewing activity appears as blocks of eating and ruminating bouts. Ruminal pH decreased rapidly following meals and increased rapidly during rumination (from Allen, 1997).

Diurnal variation in ruminal hydrogen ion concentration varies depending upon the link between meal patterns and rumination patterns. The buffering capacity of ruminal contents is a factor whose effect on the free hydrogen ion pool and therefore on ruminal pH is poorly understood. Buffers in the rumen include undigested feed residues and microbial cells, salivary buffers, and VFA. Hydrogen phosphate is less of a buffer in the rumen than it is an alkalizer because it is nearly completely complexed with hydrogen ions below pH 6. Fermentation acids and salivary buffers might be expected to have a greater effect on ruminal pH when pH is relatively high (*i.e.*, 6.5), because pH is a log-scale and hydrogen concentration is an order of magnitude lower at pH 6.5 than at pH 5.5. However, pH at 5.5 is determined by the concentration of VFA, buffering of digesta, and the removal of hydrogen ions through the bicarbonate system. This introduces great variation in the response of ruminal pH to VFA concentration (Figure 1A). The reserve bicarbonate pool in the rumen is expected to be depleted quickly as pH declines because CO_2 is lost from the system by eructation. Therefore, the lack of relationship between ruminal VFA concentration and ruminal pH at pH less than 6.0 shown in Figure 1A is probably because of buffering by digesta and previous removal of hydrogen ions by the bicarbonate system and hydrogen phosphate, and less the result of actual buffering by saliva. As ruminal pH declines below 5.0, it is very resistant to change in either direction because of the high hydrogen ion concentration and the great buffering capacity of VFA and digesta; any additional acid produced is slight relative to the existing pool and is also more likely to be buffered.

The buffering capacity (BC) of feedstuffs varies considerably. Cereal grains have low BC, low protein feeds and grass forages have intermediate BC, and legume forages and high protein feeds have high BC (Jasaitis *et al.*, 1987). Among forages, BC tends to increase with maturity (Jasaitis *et al.*, 1987) and with ensiling (Erdman, 1988). Normally only a fraction of the BC of feeds is used as most BC is at pH lower than normal ruminal pH (Allen, 1997). However, as pH decreases below 5.5, an increasingly greater fraction of the total BC of digesta is used. Therefore, the BC of feeds and ruminal digesta are very important for stabilizing ruminal pH. Maintaining ruminal digesta pool will likely be of special importance during the periparturient period and other occurrences of low feed intake so that the buffering capacity of digesta is maintained.

Absorption

The rumen is the major site of VFA absorption and absorbed molecules are predominately in the undissociated form (Ash and Dobson, 1963). Therefore VFA absorption results in the net removal of hydrogen ions from the rumen. Rate of absorption of VFA is dependent on the absorptive surface area, pH, and the concentration gradient across the rumen epithelium. Absorptive surface area is a function of rumen size, degree of fill (Dijkstra *et al.*, 1993), and papillae surface area (Dirksen *et al.*, 1985). In addition, parakeratosis reduces effective surface area for absorption; acidosis can have cumulative effects, diminishing surface area for absorption and increasing risk for future bouts of acidosis. The adaptive responses of ruminal papillae size to diets varying in ruminally fermented OM has been proposed as an important factor affecting the susceptibility of some animals to ruminal acidosis (Dirksen *et al.*, 1985). Surface area of ruminal papillae is affected by diet (Dirksen *et al.*, 1985; Xu and Allen, 1999) and offering high roughage diets during the dry period might decrease papillae surface area enough to decrease rate of VFA absorption. However, in practice, papillae surface area might not limit rate of VFA absorption of transition cows. Fermentability of diets commonly recommended for dry cows, particularly those prior to parturition, is likely adequate to maintain absorptive surface area. Also, absorption rate displays a diminishing response to surface area as surface area increases (Xu and Allen, 1999). In addition, relationships observed between papillae surface area and rate of VFA absorption through the transition period is confounded by other factors such as concentration gradient and ruminal pH, making it difficult to assess the importance of adapting ruminal papillae prior to parturition.

Effects of pH on fractional absorption rates of acetic, propionic, and butyric acids were reported by Dijkstra *et al.* (1993) using evacuated and washed rumens. Rates of absorption were similar among the VFA at pH 7.2, but as pH was reduced to 4.5, rate of absorption nearly tripled for butyrate, nearly doubled for propionate, and was not affected for acetate. Because they have similar acid dissociation constants with similar fractions in the unionized form, the differences in absorption rates are probably due to different concentration gradients cause by metabolism in the ruminal epithelium (Dijkstra *et al.*, 1993). Butyrate is extensively metabolized by the ruminal epithelium, followed by propionate, but acetate is metabolized to a much lesser extent (Sander *et al.*, 1959). Fermentation acids are absorbed by passive diffusion from high concentrations inside the rumen to lower concentrations in the epithelial cells; therefore factors that increase VFA concentration at the epithelial border inside the rumen and decrease the concentration inside the cells are expected to increase the rate of absorption. Concentration gradient might be a major factor affecting rate of VFA absorption from the rumen and therefore ruminal pH.

Because VFA are absorbed primarily in the undissociated state, and because a larger proportion of VFA exist in that form at lower pH, it is expected that fractional rate of absorption increases as pH decreases. While this has been shown using evacuated and washed rumens (Dijkstra *et al.*, 1993), it might not be the case in practice. The washed rumen technique maximizes the concentration gradient, decreasing the intracellular concentration of VFA because of removal of the rumen contents and increasing the concentration at the epithelial border inside the rumen because the solutions of VFA added are easily mixed. We developed a method to measure relative rate of VFA absorption that does not require rumen evacuation, by pulse dosing valerate and a liquid marker (Co EDTA) and calculating rate of absorption as the rate of change of the ratio of valerate concentration to cobalt concentration (Allen *et al.*, 2000). Using this technique, we found that rate of absorption of valerate ranged from ~20 to 60%/h (Voelker and Allen, 2003) and was the primary factor related to VFA concentration in the rumen which ranged from ~125 to 155 mM. As expected, ruminal VFA concentration was negatively correlated with both valerate absorption rate ($r^2 = 0.48$; $P<0.001$) and liquid passage rate ($r^2 = 0.24$; $P<0.01$); 72% of variation in ruminal VFA concentration was accounted for by the combined rates of valerate absorption and liquid passage.

We also recently reported that rate of VFA absorption was slower, not higher, under lower ruminal pH (Voelker and Allen, 2003); this contradicts data obtained using washed rumens. Although rumen motility was not measured in our study, valerate absorption rate increased with greater passage rate of indigestible NDF, and tended to increase with greater liquid passage rate, both of which might be indicators of rumen motility. Furthermore, rate of indigestible NDF passage decreased as ruminal pH decreased. Slower valerate absorption at lower pH was verified in another recent study from our laboratory (Taylor and Allen, unpublished). This is additional evidence that concentration gradient might be a dominant factor affecting rate of VFA absorption and therefore ruminal pH. Ruminal motility affects rate of absorption because constant mixing of ruminal contents is needed to maintain VFA concentrations at the ruminal epithelium where absorption occurs. Ruminal contractions are inhibited at low ruminal pH (Shinozaki, 1959), decreasing mixing and rate of VFA absorption and decreasing pH further. Stimulation of ruminal contractions by physically effective fiber is expected to increase mixing and rate of VFA absorption. Diets that provide more consistent rumen fill over time in animals prone to drastically reduced feed intake (periparturient or early lactation animals) might reduce the occurrence of subacute acidosis upon refeeding by continuing to stimulate ruminal motility in spite of low feed intake and by maintaining buffering capacity of ruminal digesta. Another factor that might be related to concentration gradient is milk yield; Voelker and Allen (2003) reported a positive relationship between valerate absorption rate and FCM yield of cows ($R = 0.49$, $P<0.01$). Cows with high milk yield have greater blood flow and a greater ability to metabolize absorbed VFA, which will decrease their concentrations in the blood, increasing the concentration gradient from the rumen to the blood. Greater ruminal motility is also expected to increase blood flow to the rumen and utilization of VFA within ruminal tissue, further increasing the concentration gradient and absorption rate for VFA.

Because more than 50% of acid produced in the rumen is removed by absorption across the rumen wall (Allen, 1997), maximizing and maintaining VFA absorption rate should be a primary strategy for stabilizing ruminal pH. Promoting ruminal motility and maintaining the concentration gradient across the rumen wall should contribute to this process. Physically effective fiber, especially from forages, should prevent acidosis and reduce fluctuations in pH not only by stimulating chewing and salivary buffer flow but also by maintaining the ruminal digesta pool and promoting ruminal motility. Forage fiber promotes mat formation and feed particle retention, which maintains a more constant supply of VFA and should reduce fluctuations in ruminal pH. However, the need for fiber to promote rumination, ruminal motility, and digesta retention must be balanced against its potential to limit feed intake due to ruminal filling effects (Allen, 2000). Using forage fiber to maintain ruminal digesta pool will be especially important when depressed feed intake is expected and when feed intake is unlikely to be limited by filling effects, such as immediately before and after calving.

Conclusions

Ruminal pH is a function of the rates of VFA production and absorption, and of hydrogen ion passage, neutralization, and buffering. Diets and feeding systems should be designed to provide highly fermentable diets while limiting variation in ruminal pH. Important diet characteristics affecting acid production, buffering, and absorption are fermentability and coarseness of fiber; highly fermentable feeds should be limited while adequate effective fiber should be provided. The concentration gradient across the ruminal epithelium probably has a great effect on rate of VFA absorption and is affected by rumen motility, ruminal pH, blood flow, and possibly by milk yield. This emphasizes the importance of physically effective fiber to maintain digesta pool size in order to provide adequate buffering by ruminal digesta, stimulate rumination and salivary buffer flow, and promote motility to enhance VFA absorption.

References

Allen, M.S., 1997. Relationship between ruminal fermentation and the requirement for physically effective fiber. J. Dairy Sci. 80, 1447-1462.

Allen, M.S., 2000. Effects of diet on short-term regulation of feed intake by lactating dairy cattle. J. Dairy Sci. 83, 1598-1624.

Allen, M.S., L.E. Armentano, M.N. Pereira and Y. Ying, 2000. Method to measure fractional rate of volatile fatty acid absorption from the rumen. Abstracts Conference on Rumen Function 25:24, Chicago, IL, Nov. 14-16, 2000, Department of Animal Science, Michigan State University, East Lansing, http://www.msu.edu/user/rumen/index.htm

Ash, R.W., 1959. Inhibition and excitation of reticulo-rumen contractions following the introduction of acids into the rumen and abomasum. J. Physiol. 169, 39.

Ash, R.W. and A. Dobson, 1963. The effect of absorption on the acidity of rumen contents. J. Phsiol. 169, 39.

Bailey, C.B. and C.C. Balch, 1961. Saliva secretion and its relation to feeding in cattle. 2. The composition and rate of secretion of mixed saliva in the cow during rest. Br. J. Nutr. 15, 383.

Benson, J.A., Reynolds, C.K., Aikman, P.C., Lupoli, B. and D.E. Beever, 2002. Effects of abomasal vegetable oil infusion on splanchnic nutrient metabolism in lactating dairy cows. J. Dairy Sci. 85, 1804-1814.

Dijkstra, J., H. Boer, J. van Bruchem, M. Bruining and S. Tamminga, 1993. Absorption of volatile fatty acids from the rumen of lactating dairy cows as influenced by volatile fatty acid concentration, pH, and rumen liquid volume. Br. J. Nutr. 69, 385.

Dirksen, G.U., H.G. Liebich and E. Mayer, 1985. Adaptive changes of the ruminal mucosa and their functional and clinical significance. Bovine Pract. 20, 116.

Erdman, R.A., 1988. Dietary buffering requirements of the lactating dairy cow: a review. J. Dairy Sci., 71, 3246.

Gill, M., R.C. Siddons and D.E. Beever, 1986. Metabolism of lactic acid isomers in the rumen of silage-fed sheep. 55, 399.

Hoover, W.H., 1986. Chemical factors involved in ruminal fiber digestion. J. Dairy Sci., 69, 2755.

Jasaitis, D.K., J.E. Wohlt and J.L. Evans, 1987. Influence of feed ion content on buffering capacity of ruminant feedstuffs *in vitro*. J. Dairy Sci. 70, 1391.

Mackie, R.I. and F.M.C. Gilchrist, 1979. Change in lactate producing and lactate utilizing bacteria in relation to pH in the rumen of sheep during stepwise adaptation to a high concentrate diet. Applied and Environ. Micro. 38, 422.

Oba, M. and M.S. Allen, 2000. Effects of brown midrib 3 mutation in corn silage on productivity of dairy cows fed two concentrations of dietary neutral detergent fiber: 1. Feeding behavior and nutrient utilization. J. Dairy Sci 83, 1333-1341.

Oba, M. and M.S. Allen, 2003. Effects of corn grain conservation method on feeding behavior and productivity of lactating dairy cows at two dietary starch concentrations. J. Dairy Sci. 86, 174-183.

Sander, E.G., R.G. Warner, H.N. Harrison and J.K. Loosli, 1959. The stimulatory effect of sodium butyrate and sodium propionate on the development of the rumen mucosa in the young calf. J. Dairy Sci. 42, 1600-1605.

Shinozaki, K., 1959. Studies on experimental bloat in ruminants. 5. Effects of various volatile fatty acids introduced into the rumen on the rumen motility. Tohoku J. Agric. Res. 9, 237.

Terry, R.A., J.M.A. Tilley and G.E. Outen 1969. Effect of pH on the cellulose digestion under *in vitro* conditions. J. Sci. Food Agric. 20, 317.

Voelker, J.A. and M.S. Allen, 2003. Pelleted beet pulp substituted for high-moisture corn: 3. Effects on ruminal fermentation, pH, and microbial protein efficiency in lactating dairy cows. J. Dairy Sci. 86, 3562-3570.

Xu, J. and M.S. Allen, 1999. Effects of dietary lactose compared with ground corn grain on the growth rate of ruminal papillae and rate of valerate absorption from the rumen. S. Afr. J. Anim. Sci. 29(ISRP2).

High dietary cation difference induces a state of pseudohypoparathyroidism in dairy cows resulting in hypocalcemia and milk fever

J.P. Goff and R.L. Horst
National Animal Disease Center, USDA-ARS, Ames IA, USA

How adjusting dietary cation-anion difference (DCAD) effectively reduces milk fever incidence remains debatable. Our hypothesis is that DCAD affects acid-base status of the cow impairing the action of parathyroid hormone (PTH) on its target tissues, primarily bone and kidney. PTH normally stimulates renal synthesis of 1,25-dihydroxyvitamin D (1,25-vit D) and the effects of 1,25-vit D on intestine, and PTH on bone should cause a rapid rise in plasma Ca concentration. To test this hypothesis, Jersey cows in the last month of gestation were fed either a Low DCAD (N=8) or a High DCAD diet (N=8) for 2 wks. Urine pH averaged 5.7 in Low DCAD and 8.3 in High DCAD cows indicating the diets did affect acid-base status of the cows. Baseline blood samples were collected from each cow. Then 0.5 mg PTH 1-34 was administered IM, with additional injections of 3.3 mg PTH administered every 3 h up to 48 h. Blood samples were obtained from each cow every 3 h during PTH treatment and plasma Ca and 1,25-vit D concentrations were determined. Repeated measures ANOVA demonstrated significant time by diet interactions on both plasma Ca and 1,25-vit D concentrations.

Cows fed the Low DCAD diet exhibited a significant increase in plasma Ca concentration from baseline at 6 h of PTH treatment and plasma Ca continued to increase throughout the experiment. In High DCAD cows plasma Ca concentration did not significantly increase from baseline until 21 h of PTH treatment, and continued to increase slowly thereafter. Plasma 1,25-vit D was increased above baseline in Low DCAD cows at 6 h of PTH treatment and continued to increase, peaking at 24 h of PTH treatment and remaining elevated thereafter. In high DCAD cows there was a small but significant spike in plasma 1,25-vit D concentration at 6 h of PTH treatment but values at 9 and 12 h of PTH treatment were not significantly elevated from baseline. Plasma 1,25-vitD concentration rose above baseline at 15, 24, 30 and 33 h of PTH treatment but were similar to baseline at 18, 21 and 27 h of PTH treatment. These data demonstrate that High DCAD causes a pseudo-hypo-parathyroid state in which tissues are less responsive to PTH stimulation, which impairs Ca homeostasis to cause milk fever. Low DCAD diets restore tissue PTH sensitivity to prevent milk fever.

Effects of two different dry-off strategies on metabolism in dairy cows

M. Odensten[1], K. Person Waller[2] and K. Holtenius[3,1]
[1]Department of Animal Nutrition and Management, Swedish University of Agricultural Sciences, Uppsala, Sweden; [2]Department of Ruminant and Porcine Diseases, National Veterinary Institute, and Department of Obstetrics and Gynaecology, Swedish University of Agricultural Sciences, Uppsala, Sweden; [3]Department of Animal Nutrition and Management, Swedish University of Agricultural Sciences, Uppsala, Sweden.

Dry-off (DO) can be very stressful due to abrupt reduction of the energy supply and cessation of milking in late-term pregnancy. Until now most attention has been paid at prevention of mastitis[1]. However it is likely that DO also might create metabolic problems especially in cows with a high level of milk production prior to DO. The objective of this study was to investigate the effects of two different diets during DO on the metabolism and health in high yielding dairy cows.

Twenty-one cows of the Swedish Red and White Breed with an average milk yield of 19.2 kg/day prior to DO were allotted to one of two DO treatments. One group of cows were fed straw *ad libitum* (STRAW) during the 5 day DO period. The other group was fed straw *ad libitum* and also 4 kg DM silage per day (SILAGE). All cows were milked at two occasions during the DO period; the mornings of day 3 and day 5. All cows were dry on day 6. In order to monitor the metabolic responses to the two different DO treatments daily samples of non esterified fatty acids (NEFA), beta-hydroxybutyrate, urea, glucose, insulin and cortisol were analysed.

The plasma concentration of NEFA was markedly elevated during the DO period in STRAW cows, but in SILAGE cows the increase was less pronounced. The glucose level in plasma increased in SILAGE cows, but was not affected in the STRAW cows. The insulin concentration was markedly reduced in both groups during DO. Both the beta-hydroxybutyrate and urea concentrations in plasma were significantly reduced during DO. These reductions were most pronounced in SILAGE cows. The diurnal profile of cortisol, by means of blood samples collected every second hour for 24 hours, was monitored prior to DO and during the third day of DO. The cortisol level was significanly elevated during DO in STRAW cows but not in SILAGE cows.

We also measured serum amyloid A (SAA), an acute phase protein, during the experiment. SAA can act as non-specific indicator of inflammatory reactions and a help to identify sub-clinical disease. SAA varied markedly between cows before DO. A rise in SAA during DO in both groups indicated an inflammatory response. There was no significant difference between the two treatments in SAA.

The results of this study indicate that DO markedly affected the metabolism of the cows, and that those cows that were offered only straw during the DO were most affected. There was furthermore an indication of stress in the STRAW cows since the plasma level of cortisol was consistently elevated during DO compared to the level immediately prior to DO.

The cows responded with a manifold increase in NEFA during DO, indicating a negative energy balance and a substantial mobilisation of adipose tissue. On the other hand, the level of beta-hydroxybutyrate was reduced indicating that the carbohydrate status was not negatively affected by DO.

References
[1] Dingwell R.T., D.F. Kelton and K.E. Leslie, 2003. Management of the dry cow in control of peripartum disease and mastitis. Veterinary Clinics of North America. Food Animal Practice 19, 235-265.

Ruminal pH, concentrations and post-feeding pattern of VFA and organic acids in cows experiencing subacute ruminal acidosis

G.R. Oetzel and K.M. Krause
School of Veterinary Medicine, University of Wisconsin-Madison, WI USA

The objective of this study was to describe the changes in ruminal pH and concentration of VFA and other organic acids in dairy cows induced with subacute ruminal acidosis (SARA).

Eleven lactating cows in two studies were induced with SARA by submitting the cows to one day of 50% restricted feeding of TMR followed by the addition of 4 to 5 kg of barley/wheat pellets to the TMR fed *ad libitum* the next day. Ruminal pH was measured continuously using indwelling electrodes and rumen fluid was collected every half hour for a 24 h period following feeding of TMR and pellets. Rumen pH values were averaged across half hour intervals to match the rumen fluid sampling. Rumen fluid was analyzed for the major VFA and for lactate, succinate, formate, 2,3-butanediol, and ethanol using HPLC.

The following description is based on mean values from the eleven cows. Ruminal pH was 7.07 prior to feeding, but started declining immediately after feeding and declined over the next 7.5 h, where nadir of 5.47 was reached. Ruminal pH stayed low (between 5.47 and 5.70) for the next several hours and started increasing again 16 h post feeding. However, pH did not return to pre-feeding level. Acetate concentration increased immediately after feeding from a mean of 40 mM to 70 mM at 5.5 h post feeding. Acetate concentration stayed at this level until approximately 15 h post feeding and then declined slowly. Propionate and butyrate concentrations were less variable and averaged 18 and 12 mM, respectively, for the 24 h period of sampling. A small peak in lactate concentration (8 mM) appeared at 1.5 h post feeding, but the concentration quickly decreased to 2-3 mM. At 8 h post feeding lactate concentration started to increase and peaked at 15 mM 10 h post feeding. Lactate concentration remained at this level for four hours and then declined to approximately zero at 19 h post feeding. The concentration of 2,3-butanediol followed the same biphasic pattern as lactate, with the first peak reaching 0.6 mM 1 h post feeding and the second more prolonged peak reaching 0.5 mM at 9.5 h post feeding. Ethanol concentration increased gradually after feeding until 8 h post feeding (1.3 mM), and then increased substantially to 3.4 mM at 10 h post feeding. Ethanol concentration remained elevated until 20 h post feeding and then declined to the pre-feeding level close to zero. Concentration of succinate remained close to zero until 8 h post feeding where levels increased and peaked at 3 mM 14.5 h post feeding. No clear pattern in formate concentration was observed for the 24 h period. Concentration of formate was unchanged and close to zero for most cows, but did increase to levels between 1 and 2 mM for some cows. Ruminal pH was negatively correlated ($P<0.0001$) with acetate (r = -0.51), butyrate (r = -0.39), lactate (r = -0.43) and ethanol (r = -0.45) concentrations. Although one cow had ruminal lactate concentrations above 60 mM, acetate appeared to be the fermentation product mostly responsible for the observed decrease in ruminal pH.

These data indicate that SARA causes a disruption of the normal ruminal fermentation pattern, resulting in the appearance of otherwise rare fermentation products.

Characterization of the Na^+/Mg^{2+} exchanger as a major Mg^{2+} transporter in isolated ovine ruminal epithelial cells

M. Schweigel[1,2], B. Etschmann[2], F. Buschmann[2], H.S. Park[2] and H. Martens[2]
[1]*Research Institute for the Biology of Farm Animals, Nutritional Physiology Research Unit;*
[2]*Department of Veterinary-Physiology, Free University Berlin, D-14163 Berlin, Germany*

The rumen is the most important site of net Mg^{2+} absorption in the gastrointestinal tract of sheep and maintains Mg^{2+} homeostasis. The passive, paracellular permeability of the rumen epithelium is low and therefore, the transfer of the ion is mainly (85-95%) mediated by an active transcellular pathway. In contrast to the mechanisms of Mg^{2+} uptake that have been well characterized at the tissue and cell level, there are only few data regarding basolateral Mg^{2+} extrusion from the epithelial cells. This scare information points to an Na^+/Mg^{2+} exchanger. The present study was performed to show its existence in ruminal epithelial cells (REC) and to characterize the transport protein at the functional and molecular level.

Experiments were carried out with cultured REC (prepared as described previously). Na^+-dependent Mg^{2+} uptake was determined by use of the ion sensitive fluorescent probes mag-fura 2 and SBFI to measure the $[Mg^{2+}]_i$ and $[Na^+]_i$, respectively. Mg^{2+} efflux from Mg^{2+}-loaded (6 mM) REC into a completely Mg^{2+}-free solution have also been measured. The Mg^{2+} extrusion rate was calculated from the increase of the $[Mg^{2+}]$ in the incubation medium determined with the aid of the fluorescent probe mag-fura 2 (K^+-salt). The Na^+ dependency of the Na^+/Mg exchanger activity, effects of putative inhibitors (quinidin, imipramine) and regulation of Mg^{2+} efflux mechanisms have been focused primarily. The presence of a Na^+/Mg exchanger was investigated by Western blotting. In this study monoclonal antibodies (mabs) that have been raised against the porcine erythrocyte Na^+/Mg^{2+}-exchanger have been used.

An increase of $[Mg^{2+}]_i$ was observed after reversing the transmembrane Na^+ gradient. This rise in $[Mg^{2+}]_i$ was potential- and pH-independent, dependent on $[Mg^{2+}]_e$, imipramine-sensitive and accompanied by a decrease of $[Na^+]_i$. Furthermore, under more physiological conditions (extracellular Na^+ concentration, $[Na^+]_e > [Na^+]_i$) Mg^{2+}-loaded REC show a Mg^{2+} efflux which was predominantly mediated by the Na^+/Mg^{2+} exchanger (97 percent of total efflux). The Mg^{2+} extrusion rate increased from 0.018 ± 0.009 in a Na^+-free medium to 0.73 ± 0.3 mM/l cells/min in a 145 mM-Na^+ medium, respectively, and relates to $[Na^+]_e$ according to a typical saturation kinetic (K_m-value for $[Na^+]_e = 24$ mM; $V_{max} = 11$ mM/l cells/min). Imipramine, a known inhibitor of the Na^+/Mg^{2+} exchanger, decreased the Mg^{2+} efflux by 48%. A rise of $[cAMP]_i$ or stimulation of REC with PGE_2 increased the Mg^{2+} efflux rate by 55% or 17% compared to the control values. A $[cAMP]_i$ increase was accompanied by a measurable decrease of $[Mg^{2+}]_i$ by 300 μM. These effects are completely abolished in Na^+-free media. By use of the anti-Na^+/Mg^{2+} exchanger mabs a specific 70-kDa immunoreactive band was detected in protein lysates of REC. In parallel functional studies the mabs also inhibited Na^+-dependent Mg^{2+} influx by 50%.

These data clearly demonstrate that a Na^+/Mg^{2+} exchanger is existent in the cell membrane of REC. Under physiological conditions, this transport protein is the main pathway for Mg^{2+} efflux and its activity can be modulated by hormones or mediators via the cAMP-PKA pathway. The ruminal Na^+/Mg^{2+} can be assumed to play a considerable role in the process of Mg^{2+} resorption as well as the maintenance of the cellular Mg^{2+} homeodynamics.

Acknowledgements

This study was supported by the Deutsche Forschungsgemeinschaft (M. Schweigel) and by the H. Wilhelm Schaumann Stiftung (H.-S. Park).

High potassium diet, sodium and magnesium in ruminants: the story is not over

F. Stumpff, I. Brinkmann, M. Schweigel and H. Martens
Department of Veterinary Physiology, Free University of Berlin, Germany

High dietary intake of potassium leads to a marked increase in sodium absorption[1]. Simultaneously, the rate of magnesium uptake into ruminal epithelial cells[2] and across the ruminal epithelium[3] is reduced, which can lead to grass tetany if high potassium intake continues. Previous experiments[4] suggest that the rumen epithelium expresses a non-selective cation channel that is regulated by external divalent cations. In this study, we were interested in investigating whether cytosolic Ca^{++} and Mg^{++} concentrations alter sodium conductance in ruminal cells.

Experiments were carried out with primary cultures of sheep ruminal epithelial cells using the patch-clamp technique. Both choline chloride and potassium gluconate pipette solutions were used. The effect of divalent cations on channel conductance was examined by varying the concentration of Ca^{++} and Mg^{++} in the external and internal solutions.

Removal of divalent cations from the external solution induced a highly significant increase in inward current at negative potential levels in the presence of external sodium and potassium, but not choline. This inward current could be blocked by 5 mM barium, but not by 5 mM TEACl. Removal of external divalents changed reversal potential from -30 ± 3 mV to -20 ± 3 mV (potassium gluconate $+Ca^{++} +Mg^{++}$, n = 14, P=0.001), from -23 ± 4 mV to -16 ± 2 mV (potassium gluconate $\varnothing Ca^{++} \varnothing Mg^{++}$, n = 14, P=0.04), from 10 ± 2 mV to 14 ± 2 mV (choline chloride $+Ca^{++} +Mg^{++}$, n = 21, P=0.0001) and from 5 ± 2 to 6 ± 1 mV (choline chloride $+Ca^{++} \varnothing Mg^{++}$, n = 13, P=0.01). Removal of external Mg^{++} alone had a smaller effect. Relative permeability P_K/P_{Na} of the conductance varied between potassium gluconate and choline chloride pipette solution (2.1 versus 1.4). *Removal* of internal Mg^{++} from the choline chloride pipette solution resulted in a highly significant rise in *inward current* level in divalent containing external NaCl solution and significant rises in both inward and outward current in divalent free solution, corresponding to a drop in channel rectification. The component of current that was sensitive to external divalent cations could be calculated by subtracting the current in external divalent–containing NaCl solution from that in divalent-free NaCl solution. The chloride concentration of the pipette solution had no impact on the divalent-sensitive current, while the conductance level and the degree of inward rectification changed with the concentration of Ca^{++} and Mg^{++} in the pipette solution.

Ruminal epithelial cells express a channel that conducts both sodium and potassium and opens when either cytosolic or external levels of Ca^{++} and Mg^{++} are reduced. We propose that high ruminal potassium leads to an enhancement of sodium absorption through this channel mediated by a drop in cytosolic magnesium. This mechanism should be useful in ensuring osmoregulation after an occasional high-potassium meal. Side effects due to the reduction in magnesium uptake should only become apparent if ruminal potassium concentrations remain high for extended periods of time, a condition unknown in the physiological habitate of the ruminant before the introduction of artificial fertilizing techniques.

References

[1] Sellers, A.F. and A. Dobson, 1960. Studies on reticulo-rumen sodium and potassium concentration and electrical potentials in sheep. Res. Vet. Sci. 1, 95.

[2] Schweigel, M., I. Lang and H. Martens, 1999. Mg(2+) transport in sheep rumen epithelium: evidence for an electrodiffusive uptake mechanism. Am. J. Physiol. 277, G976-982.

[3] Leonhard-Marek, S. and H. Martens, 1996. Effects of potassium on magnesium transport across rumen epithelium. Am. J. Physiol. 271, G1034-1038.

[4] Lang, I. and H. Martens, 1999. Na transport in sheep rumen is modulated by voltage-dependent cation conductance in apical membrane. Am. J. Physiol. 277, G609-618.

Functional characterization of the time course of rumen epithelium adaptation to a high energy diet

B. Etschmann, A. Suplie and H. Martens
Department of Veterinary-Physiology, Free University Berlin, D-14163 Berlin, Germany

The adaptation of the rumen epithelium to different forms of diet is of utmost importance to domesticated ruminant animals. Insufficient adaptation in dairy cows during the transition period results in various disorders, including emaciation, ketosis and rumen acidosis. The morphological adaptation of the rumen epithelium has been extensively characterized[1]. In these experiments, it was found that morphological adaptation to a high energy diet shows a sigmoid time course and takes approximately six weeks to reach peak levels[3]. The functional adaptation of the rumen epithelium has also been repeatedly examined[2,4]. In most experiments, a concentrate diet was administered over a period of six weeks before adaptation was monitored. However, no clear data regarding the time course of functional adaptation has previously been established. Therefore, in this study, a time course of the adaptation of the rumen epithelium to a high energy diet was examined.

A total of 18 sheep of the same age, breed and origin was divided into six groups of three animals each. In preparation to the experiment, all animals were fed a hay diet over a period of eight weeks in order to ensure adaptation to a low energy, roughage diet. Following this period, the groups of animals were fed a mixed concentrate and hay diet over a varying period of time ranging from 0 to 12 weeks. Following the period of hay / concentrate diet, the animals were slaughtered and rumen epithelium was examined in ussing chamber experiments. For each animal, a total of four Na^+-absorption (J_{net} Na^+) values were recorded under short circuit conditions. During the experiments, values for I_{SC} and G_t were continuously recorded.

The results of the ussing chamber experiments are summarized in Table 1.

The net absorption of sodium increases significantly following a change from a roughage diet to a concentrate diet. The largest part of this increase is seen within the first two weeks after changing the diet. As functional adaptation appears to out-perform morphological adaptation markedly during the first weeks of adaptation, we hypothesize that functional adaptation during the first weeks after dietary change is due to changes at the cellular level. Further investigations are necessary to substantiate this hypothesis.

Table 1.

	I_{sc}	G_t	J_{ms} Na^+	J_{sm} Na^+	J_{net} Na^+
Group 1	1.13 ± 0.07	2.30 ± 0.14	3.57 ± 0.25	1.41 ± 0.26	2.15 ± 0.43
Group 2	1.67 ± 0.06	3.01 ± 0.14	5.37** ± 0.73	1.65 ± 0.25	3.73** ± 1.00
Group 3	1.32 ± 0.07	2.97 ± 0.18	5.88** ± 0.71	1.98* ± 0.13	3.90** ± 0.94
Group 4	1.43 ± 0.09	3.20 ± 0.20	6.62** ± 1.58	2.01* ± 0.47	4.55** ± 0.50
Group 5	0.68 ± 0.11	3.03 ± 0.29	6.27** ± 0.96	1.71 ± 0.24	4.50** ± 0.69
Group 6	0.08 ± 0.06	3.55 ± 0.18	5.32** ± 0.60	1.48 ± 0.17	3.92** ± 0.36

Acknowledgements

This work is part of the project "Animal protection, performance and health", which is supported by the Margarete-Markus charity.

References

[1] Dirksen, G., H.G. Liebich, *et al.*, 1984. [Morphology of the rumen mucosa and fatty acid absorption in cattle-important factors for health and production]. Zentralbl. Veterinarmed. A 31, 414-430.

[2] Gäbel, G., M. Bestmann and H. Martens, 1991. Influences of diet, short-chain fatty acids, lactate and chloride on bicarbonate movement across the reticulo-rumen wall of sheep. Zentralbl. Veterinarmed. A 38, 523-529.

[3] Liebich, H.G., G. Dirksen, A. Arbel, S. Dori and E. Mayer, 1987. [Feed-dependent changes in the rumen mucosa of high-producing cows from the dry period to eight weeks post partum]. Zentralbl. Veterinärmed. A 34, 661-672.

[4] Shen Z., H.M. Seyfert, B. Lohrke, F. Schneider, R. Zitnan, *et al.*, 2004 An energy-rich diet causes rumen papillae proliferation associated with more IGF type 1 receptors and increased plasma IGF-1 concentrations in young goats. J. Nutr. 134, 11-17.

Characterization of an ovine vacuolar H$^+$-ATPase as a new mechanism for the energization of ruminal transport processes

M. Schweigel[1,2], B. Etschmann[2], E. Froschauer[3], S. Heipertz[2] and H. Martens[2]
[1]Research Institute for the Biology of Farm Animals, Nutritional Physiology Research Unit;
[2]Department of Veterinary-Physiology, Free University Berlin, D-14163 Berlin; [3]Department of Microbiology and Genetics, Vienna Biocenter, A-1030 Vienna, Austria

Recently we found functional evidence for a vacuolar-type H$^+$-adenosine triphosphatase (vH$^+$-ATPase) to be expressed in ruminal epithelial cells, REC[1]. Such electrogenic H$^+$-ATPases have been shown to energize the apical membrane of several epithelia by generating an inside negative membrane potential[2]. The existence of an active H$^+$-extruding mechanism in REC has been explored in the present study.

The presence of vH$^+$-ATPase was investigated by RT-PCR, Western blotting and immunofluorescence microscopy. Using homologous cDNA consensus sequences sets from other species, degenerated, specific primers were generated for RT-PCR. The primers were used in RT-PCR experiments to detect vH$^+$-ATPase mRNA molecules in rumen epithelium cells. PCR-products of the expected size were cloned into bacterial vectors and subjected to cycle sequencing. The resulting sequence data was compared to homologous cDNA from *Bos taurus*. H$^+$ secretion was estimated from the rate of intracellular pH (pH$_i$) recovery after acid-loading in cultured REC. Acid-loading was done by exposure of REC to butyrate and/or CO$_2$/HCO$_3^-$ and pH$_i$ was measured using the fluorescence probe BCECF. Different types of H$^+$-extruding systems were identified by using Na$^+$-containing or Na$^+$-free media and by pharmacological means.

Sequence comparisons between PCR products and bovine sequences showed a 96% (433/448) and 99% (277/278) homology for vH$^+$-ATPase B and vH$^+$-ATPase E, respectively. Furthermore, a 60-kDa immunoreactive band contributing to vH$^+$-ATPase subunit B was detected in protein lysates of REC by immunoblotting. Using immunocytochemistry, REC stained positive with antibodies directed against the B (60-kDa)-subunit of the vH$^+$-ATPase.

Exposure of REC to butyrate and/or CO$_2$/HCO$_3^-$ caused a prompt acidification of the cytosol, followed by a pH$_i$ recovery of 0.21 ± 0.01 pH-units over 10 minutes. About 60% of this pH$_i$ increase results from the Na$^+$/H$^+$ exchanger (NHE) activity. Application of the vH$^+$-ATPase inhibitors decreased the rate of pH$_i$ recovery by 0.05 pH-units showing that its relative contribution to this process amounts to 23%. Depolarisation of the REC membrane potential by incubation in high-K$^+$ medium activates and temperature reduction decreases the vH$^+$-ATPase activity. Inhibition of the vH$^+$-ATPase reduced Mg^{2+} uptake into REC by at least 36%.

These findings clearly confirm that a vH$^+$-ATPase is present in REC. As has been shown for Mg^{2+}-absorption, the proton-motive force generated by such electrogenic H$^+$-ATPases in the cell membrane can be utilized as a driving force for numerous transport processes. The activity of ruminal vH$^+$-ATPase seems to be modulated by bacterial fermentation products, explaining their effects on the transport of several solutes (*e.g.* Na$^+$, Cl$^-$, Mg^{2+}) across the rumen epithelium.

References

[1] Schweigel, M. and H. Martens, 2003. Anion-dependent Mg2+ influx and a role for a vacuolar H+-ATPase in sheep ruminal epithelial cells. Am. J. Physiol. Gastrointest. Liver Physiol. 285, G45-G53.

[2] Wieczorek, H., D. Brown, S. Grinstein, J. Ehrenfeld and W.R. Harvey, 1999. Animal plasma membrane energization by proton-motive V-ATPases. BioEssays 21, 637-648.

The absorptive capacity of sheep omasum is modulated by the diet

O. Ali[1], H. Martens[2] and C. Wegeler[2]
[1]Department of Physiology, University of Khartoum, Sudan; [2]Department of Veterinary Physiology, Free University of Berlin, Germany

The omasum is an important site of water and electrolyte absorption in the proximal part of the digestive tract[1,2,3,4]. Feeding regimes influence mineral and electrolyte transport across the rumen epithelium. All transport mechanisms of electrolytes – so far studied – are significantly higher in epithelia from concentrate fed sheep. Corresponding studies with tissues from the omasum are not known despite its important absorptive capacity. The aim of the present study has therefore been to examine the effect of diet on transport of Na^+, SCFA and of HCO_3^- across the isolated omasal epithelium of hay-fed and concentrate-fed sheep.

Animals and diet: Sheep with a body weight of 30 - 50 kg were fed two different dietary regimens: (a) hay *ad libitum* or (b) 800 g concentrate diet in equal portions at 7:00 a.m. and 3:00 p.m., plus hay *ad libitum*. The animals were fed on these diets at least three weeks before being slaughtered. Water and lick-stones were available *ad libitum*. Isolation of epithelia and *in vitro* incubation.

Na transport: The omasal epithelium of hay-fed animals exhibits an net transport of Na in mucosal-serosal direction. Jms Na (7.99 ± 2.71 meq×cm^{-2}×h^{-1}) exceded Jsm (3.42 ± 1.15 meq×cm^{-2}×h^{-1}) more than twofold, which leads to the well known Jnet of 4.57 ± 1.99 meq×cm^{-2}×h^{-1}. Feeding the concentrate diet increased all flux rates significantly: Jms 12.46 ± 3.62, Jsm 5.24 ± 1.85 and Jnet 7.22 ± 2.42 meq×cm^{-2}×h^{-1}.

Transport of SCFA: The unidirectional flux rates of acetate were relatively low in epithelia of hay-fed animals ($1 - 2$ meq×cm^{-2}×h^{-1}) and Jms and Jsm were not significantly different, which indicates that the transport of acetate occurs by simple diffusion. The transport rates of SCFA were almost doubled in epithelia from concentrate-fed sheep.

Transport of HCO_3^-: We used the pH-stat method for the determination of HCO_3^- flux rates in the mucosal-serosal direction and did not observe differences of HCO_3^- transport in epithelia form hay-fed or concentrate-fed sheep ($1.5 - 2.0$ meq×cm^{-2}×h^{-1}).

The omasal epithelia of sheep exhibit a net absorption of Na, SCFA and of HCO_3^-. The transport of Na and SCFA are significantly influenced by the feeding regime. Because SCFA are buffers (pK 4.80), an enhanced absorption of SCFA from the omasum facilitates the acidification of the ingesta in the abomasum. Surprisingly, transport rates of HCO_3^- did not respond to a change of diet.

References

[1] Edrise, B.M., R.H. Smith and D. Hewitt, 1986. Exchanges of water and certain water soluble minerals during passage of digesta through the stomach compartments of young ruminating bovines. Br. J. Nutr. 55, 157-167.

[2] Engelhardt, W. and R. Hauffe, 1975. Role of the omasum in absorption and secretion of water and electrolytes in sheep and goats. In: I.W. McDonald and A.C.I. Warner (editors). Digestive Physiology and Metabolism in ruminants. Armidale, University of New England Publishing Unit, pp. 216-230.

[3] Harrison, F.A., R.D. Keynes and L. Zurich, 1970. Ion transport across isolated omasal epithelium of the sheep. J. Physiol. 207, 24-25.

[4] Martens, H. and G. Gäbel, 1988. Transport of Na and Cl across the epithelium of ruminant forestomachs: Rumen and Omasum. A Review. Comp. Biochem. Physiol. 90A, 569-575.

Section M.
Application of genomics to production diseases

Microarray analysis of bovine neutrophils around parturition: implications for mammary gland and reproductive tract health

Jeanne L. Burton[1,2], Sally A. Madsen[1], Ling-Chu. Chang[1], Kelly R. Buckham[1], Laura E. Neuder[1] and Patty S.D. Weber[1]
[1]*Immunogenetics Laboratory;* [2]*Center for Animal Functional Genomics, Department of Animal Science, Michigan State University, East Lansing, Michigan 48824, USA, burtonj@msu.edu*

Abstract

Neutrophils play a pivotal role in innate immune defense against opportunistic bacteria, but can also cause significant tissue pathology if their potent oxidative bactericidal activities are not tightly regulated. Accordingly, the cells begin a program of apoptotic cell death as soon as they terminally differentiate in bone marrow and normally live in the circulation for only 6-12 hours. If, however, circulating neutrophils encounter inflamed blood vessels in sites of tissue infection, life span is extended for 24-48 h to enable the cells to migrate into the tissue and clear pathogens by phagocytosis. In cattle, parturition appears to change the normal trafficking pattern of blood neutrophils, reflected in the fact that neutrophilia is pronounced at this time. In addition, oxidative burst activity of the cells is depressed. These "dysfunctions" are commonly used to explain why parturient cows have elevated susceptibility to mastitis. The objective of this study was to understand these phenomena at a molecular level. RNA isolated from blood neutrophils of 4 Holstein cows (collected on days –7, 0, 0.25, and 1 relative to parturition on day 0) was used to interrogate cDNA microarrays and profile expression changes for 1056 spotted genes. Based on results, serum steroid analyses and *in vitro* apoptosis phenotyping were performed to identify hormonal factors responsible for apparent pro-survival reprogramming of neutrophils during parturition. For example, we observed that parturition induced acute expression of A1 and BAFF genes, main anti-apoptotic molecules in blood neutrophils. Parturition also increased gene expression for interleukin-8 (IL-8), which is anti-apoptotic in neutrophils. In contrast, expression of normally constitutive pro-apoptotic Bak was markedly reduced during parturition, as were several other pro-apoptotic death receptor-associated genes (Fas, FasL, FADD, RIP, FLASH, Daxx). Expression of genes encoding the Fc-gamma receptor (FcR; antibody-mediated phagocytosis) and several genes that regulate redox homeostasis (*e.g.*, PSST) were modestly to severely depressed during parturition, while expression of matrix metalloproteinase genes (MMP-8 and MMP-9) were increased and their tissue inhibitors (TIMP-2 and TIMP-3) decreased. This prompted us to study factors in parturient serum that may have contributed to these gene expression signatures. We discovered strong correlations between serum cortisol concentrations and many of the gene expression changes observed at parturition, and found that parturient serum dramatically extended *in vitro* survival of blood neutrophils due to its high content of this glucocorticoid hormone (GC). We also showed that GC-activation of glucocorticoid receptors was responsible for the numerous apoptosis gene expression changes and pro-survival phenotypes that we observed in parturient serum-treated and GC-treated neutrophils. Our results thus highlight GC-mediated changes in expression of key apoptosis, immune function, and tissue remodeling genes as likely events that may alter functional priorities of bovine neutrophils around parturition. We are currently determining if GC-induced survival associates with altered functional capacities of neutrophils, the hypothesis being (based on our gene expression and *in vitro* phenotyping results) that the GC surge at parturition extends the life span of blood neutrophils to prioritize long-lived tissue remodeling activities while down-regulating shorter lived anti-bacterial activities of the cells. If true, results of our studies could point to GC-induced reprogramming of neutrophils as critical for the reproductive tract remodeling that must occur at parturition and help explain why this time of the cow's production cycle is associated with increased susceptibility to opportunistic infectious diseases, such as mastitis.

Traditional bactericidal role of neutrophils

Neutrophils are best known for their roles in inflammation and innate immune defense against pathogens that infect dairy cattle, especially mastitis-causing bacteria (Burton and Erskine, 2003). These normally short-lived phagocytic leukocytes develop from pluripotent hematopoietic stem cells in bone marrow, after which they are released into the circulation so they can carry out first line immune functions (Figure 1). Circulating blood neutrophils use an adhesion molecule called CD62L (L-selectin) to marginate on blood vessel endothelia in their search for signs of infection and inflammation in underlying tissues (Weber *et al.*, 2004). If no such trauma is detected, marginating neutrophils gradually down regulate CD62L expression, dislodge from the blood vessel wall, and die within 6-12 h by a complex genetic program of cell death (apoptosis). Circulating apoptotic neutrophils are rapidly cleared as intact cells by the body's phagocytic network (macrophages and fibroblasts) so they do not linger in blood vessels where they could cause significant inflammatory damage to endothelial cells and surrounding tissues if they died by necrosis as they aged.

Marginating neutrophils that do detect signs of infection from tissue adjacent to blood vessels are induced to rapidly (within seconds) shed CD62L and up regulate β2-integrin molecules (especially CD11b/CD18), facilitating tight adhesion and migration through the vessel into the infected tissue (Figure 1; Burton *et al.*, 1995). Neutrophil migration into the focus of infection demands that the cells release granule stores of enzymes capable of degrading extracellular matrix proteins in the blood vessel wall and tissue, including matrix metalloproteinases such as MMP-8 and MMP-9. Migrating neutrophils also use gradients of proinflammatory molecules released by damaged tissue to direct them quickly to the precise point of infection. Of particular importance in this regard is the chemokine, IL-8, which acts both to extend the life-span of migrating neutrophils and to activate the cells for enhanced phagocytic clearance of pathogens once they reach the infection focus (Burton *et al.*, 2005).

Phagocytosis contains microorganisms in neutrophil vesicles called phagosomes (Figure 1), keeping them safely away from other host cells. This is accomplished primarily via surface receptors that can be clustered into two main categories (Kehrli and Harp, 2001; Burton and Erskine, 2003). One category of receptors binds to pathogen-specific pattern recognition molecules on the surface of infecting organisms (*e.g.*, scavenger receptors and Toll-like receptors). The other receptors bind with high affinity to the cow's own soluble immune defense molecules (such as antibodies, complement components, and acute phase proteins) that attach to pathogens and provide molecular bridges for subsequent neutrophil attachment and phagocytosis. Both receptor-mediated phagocytosis events stimulate neutrophils to consume large amounts of oxygen, enabling the cells to unleash a battery of toxic reactive oxygen species (ROS) onto the internalized pathogens to initiate their oxidative destruction (Figure 1). This respiratory burst is a critical step in initiating pathogen killing inside the phagosome. Ultimately, enzyme-filled granules (lysosomes) already present inside the neutrophil participate in the killing of pathogens

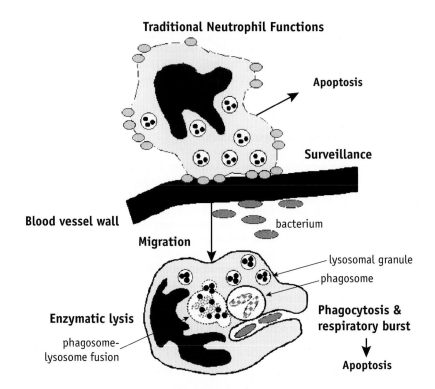

Traditional Neutrophil Functions

Apoptosis

Surveillance

Blood vessel wall

bacterium

Migration

lysosomal granule

phagosome

Enzymatic lysis

Phagocytosis &
respiratory burst

phagosome-
lysosome fusion

Apoptosis

Figure 1. Traditional bactericidal functions of neutrophils. Blood neutrophils marginate in blood vessel endothelia to survey peripheral tissues for signs of bacterial infection. If no such signs are encountered, circulating neutrophils die by apoptosis within 6-12 h of their release from bone marrow and are cleared by the body's phagocytic network. If marginating neutrophils do encounter infection signals, the cells migrate out of blood into the infection focus and phagocytose the invading bacteria. Phagocytosis causes the cells to mount a respiratory burst, generating reactive oxygen species (ROS) that begin oxidative killing of the internalized bacteria. Ultimately, lysosomal granules already present in the neutrophils fuse with the phagosome to complete bacterial killing by enzymatic lysis. The act of phagocytosis and the presence of ROS inside and outside the neutrophils send signals to the cells that they must die by apoptosis. Tissue macrophages phagocytose the dying neutrophils, thus preventing further inflammatory damage. In parturient dairy cows, margination and respiratory burst activities are depressed, which is thought to be a cause of the increased susceptibility to mastitis at this time.

by fusing with the phagosome and dumping their contents onto the dying organisms (Figure 1). These processes, in combination with the dwindling cytokine milieu of the inflammatory exudate as pathogens are cleared by neutrophils, send signals into the cells that their job is done and they should now die by apoptosis (Figure 1). This is important because apoptotic neutrophils have a considerable disarmament of their bactericidal activities and thus cannot release their granule contents into the surrounding tissues to cause more damage there. Apoptotic neutrophils are phagocytosed by neighboring macrophages, preventing unnecessary tissue damage. Thus, when the inflammatory response works properly, tissue neutrophils typically live for 24-48 h, after which they are rapidly cleared as they undergo apoptosis.

Parturition causes deficiencies in margination and bactericidal activities of neutrophils

Neutrophils of dairy cows do not function properly around calving (Kehrli and Harp, 2001). In fact, blood neutrophils of parturient cows lose their ability to marginate on blood vessels resulting in an accumulation of circulating neutrophils (neutrophilia) that have reduced ability to survey for infections present in tissues underlying the blood vessels (Hill *et al.*, 1979; Burton and Kehrli, 1995; Weber *et al.*, 2001). This means that peripheral tissue infections can proliferate unchecked by this first line of immune defense, enhancing susceptibility to disease. While some aspects of neutrophil phagocytosis appear to be normal or even increased around calving, energy metabolism and respiratory burst activity are often reduced (Nagahata *et al.*, 1988; Cai *et al.*, 1994; Kehrli and Harp, 2001; Madsen *et al.*, 2002). Thus, even if neutrophils already present in infected tissue of parturient cows could phagocytose pathogens, they may not be able to kill the internalized microorganisms.

Given that these key margination/migration and bactericidal dysfunctions occur in neutrophils around parturition, it is not surprising that parturient dairy cows suffer from increased rates of production diseases caused by opportunistic pathogens, such as intramammary coliforms that cause mastitis (Burvenich *et al.*, 2003). We have asked the question, "what molecular changes occur in blood neutrophils of parturient cows that may explain the apparent immune dysfunctions of these cells?" To begin to address this question, we embarked on a molecular "fishing mission" using our group's recently developed functional genomics tools, to characterize changes in neutrophil expressed genes as dairy cows transition from the relatively disease resistant dry period to the disease susceptible parturient period. Key highlights of our methodological approaches and preliminary results are presented next and are detailed more fully in recent papers (Burton *et al.*, 2005; Madsen *et al.*, 2004).

Microarray analysis of gene expression in blood neutrophils from parturient cows

Our Center for Animal Functional Genomics (CAFG) at Michigan State University has created specialized cDNA libraries and microarrays for studies using bovine leukocytes, mammary cells, and ovarian cells, and general microarrays for the study of most other organ systems in cattle. The tens of thousands of genes (cDNAs) included on these microarrays can be viewed on our CAFG web site at http://nbfgc.msu.edu, and some are described in recent publications (Burton *et al.*, 2001; Yao *et al.*, 2001; Suchyta *et al.*, 2003, 2004). In our neutrophil study, we used our specialized bovine immunobiology microarray called the BOTL array (for Bovine Total Leukocyte; Madsen *et al.*, 2004; Weber *et al.*, 2004). This microarray was spotted with cDNAs representing 1,056 test genes that are expressed in blood leukocytes at different times during the life and immune functioning of the cells. The BOTL microarrays are set up as series of 48 patches per array, each patch containing a 9 x 9 series of cDNA spots with each unique gene spotted in triplicate within patch. A wide variety of control genes (*e.g.*, β-actin, GAPDH, *etc.*) and blanks is also included in each patch to assist in quality control checks for the gene expression data that are obtained.

Our general approach to microarray experimentation is outlined in Figure 2. Briefly, microarrays are hybridized using dual–labeled cDNA samples that are obtained following reverse transcription of test sample mRNA. As detailed in Madsen *et al.* (2004), neutrophils (> 93% pure) for our

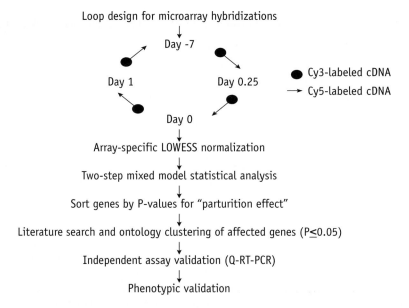

Figure 2. Our microarray experimental approach. Neutrophil mRNA was converted to cDNA and differentially labeled with Cy3 or Cy5 fluorescent dyes for hybridization to BOTL microarrays in a loop design. Resulting Cy3 and Cy5 fluorescence intensity values for each cDNA spot on the microarrays were $\log_2$-transformed and then normalized using LOWESS (SAS). Following normalization, the data were subjected to a two-step mixed model analysis to determine which genes were significantly ($P \leq 0.05$) affected in expression due to parturition. Functions of the protein products of the significantly affected genes were determined through a literature search, and the genes were clustered by ontology. Expression changes for genes of interest within these ontological clusters were validated by quantitative real time RT-PCR (Q-RT-PCR), and neutrophil phenotypes were checked to determine if the gene expression changes equated to phenotypic changes in the cells.

microarray experiment were isolated on Percoll density gradients from blood of four cows, which was collected on days –7, 0, 0.25, and 1 relative to parturition (on day 0).

Total RNA was isolated from each neutrophil sample using the TRIzol Reagent method, assessed for quantity and quality using spectrophotometry and agarose checking gels, and converted to cDNA using reverse transcription (RT). Resulting cDNAs from the various samples were then pooled within day across cows and labeled with either Cy5 (red) or Cy3 (green) fluorescent dyes. The dual-labeled samples were mixed in pairs, one Cy3 and one Cy5 labeled, and the mixed two-color cDNA sample allowed to hybridize to complementary cDNAs spotted on the BOTL microarrays using a loop design (Figure 2). Resulting two-color total fluorescence intensity data from each spot on the arrays were then $\log_2$-transformed to obtain normal distributions and normalized within array (to eliminate gross dye biases) using a local regression and spline curve approach called LOWESS (available in SAS).

Our next step was to statistically analyze the $\log_2$ and LOWESS normalized intensity data to look for differential gene expression in neutrophils over day relative to parturition. This was done using a two-step mixed model approach (in SAS) detailed in Madsen *et al.* (2004). We rejected the null hypothesis (that parturition had no effect on expression of individual genes) when $P \leq 0.05$ for the main effect of parturition. Once the analysis was complete, results for all genes were loaded into an Excel spreadsheet, sorted by *P*-value, and a thorough search of the literature done to determine the main function of protein products of genes that were significantly affected by parturition. We did this so that we could sort groups of affected genes into key ontological clusters (Figure 3). While many (~ 36%) of the identified genes still have no known function, most (~ 64%) had clear functions that enabled such ontology clustering.

When doing microarray experiments, it is important to validate gene expression changes using independent assay methods (Figure 2) and starting RNA samples from additional animals because of the inherently high animal variation and false positive rates that can occur (Brazma *et al.*, 2001;

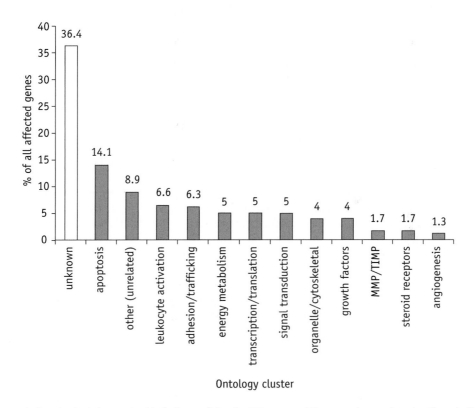

Figure 3. Ontological clusters (as % of all genes) for the 302 neutrophils genes detected as significantly (P ≤ 0.05) affected by parturition in our microarray experiment. The cluster labeled "Unknown" are genes for which DNA sequence is available in GenBank and TIGR but the function of the gene is not known. The cluster labeled "Other" represents a large number of genes with known, but unrelated, functions. Genes were assigned to all other clusters based on the most commonly documented function of their protein products. Some genes had overlapping functions and could have been placed in multiple clusters. The clusters shown were used simply as a means of selecting interesting genes for subsequent Q-RT-PCR validation of their expression changes (see Figure 2). Data are adapted from Madsen et al., 2004.

also see http://www.mged.org/miame). However, such validation can become cost prohibitive when large numbers of genes are identified through the microarray experiment, as was the case in our study. We thus decided to concentrate first on validation of approximately half of the genes that occurred in our largest ontological cluster, which were apoptosis regulatory genes (Figure 3). Neutrophils from 3 additional peripartum cows were collected on days –7, 0, 0.25, and 1 relative to parturition and their total RNA obtained for gene expression change validation using quantitative real time RT-PCR (Q-RT-PCR; described in Madsen *et al.*, 2004). Based on these Q-RT-PCR results, we next performed neutrophil phenotyping assays (Figure 2) and serum steroid analyses to establish if the gene expression changes we validated were meaningful to the cells and what blood factors might have been involved (details in Burton *et al.*, 2005; Madsen *et al.*, 2004). After this, we selected a variety of interesting genes from other ontological clusters and subjected these to additional Q-RT-PCR validation and correlation analyses with our steroid data sets. The goal here was to gain a preliminary holistic picture about how neutrophils respond to parturition and if the fluctuations in blood steroids during this period might be related to the gene expression changes and phenotypes that we confirmed.

Results of our microarray experiment were quite extraordinary because 302 neutrophil genes were identified as either induced or repressed due to parturition ($P \leq 0.05$). Genes in the largest ontological cluster of known genes affected by parturition are well known to be involved in the regulation of neutrophil apoptosis.

As depicted in Figure 4, neutrophil apoptosis can occur via two main pathways: one initiated from the outside of the cell via Fas receptor signaling and the other initiated from the inside of the cell via mitochondrial signaling. Fas is a member of the TNF receptor family of molecules. At the plasma membrane, ligation of Fas by soluble or cell-associated Fas ligand (FasL) recruits a variety of adaptor proteins [the death initiating signaling complex (DISC) molecules: FADD, FLASH, RIP, and Daxx] to Fas' cytoplasmic tails, which then recruit and sequentially activate caspases 8 and 3 to affect cell death. The potent anti-apoptotic molecule, BAFF, which is a newly discovered member of the family of TNF ligands, inhibits such death signaling by binding to its plasma membrane receptor on the same cell (Figure 4). At the mitochondrial membrane, pro-apoptotic Bcl-2 family members Bak and Bax induce release of cytochrome c (cyt-c), which associates with Apaf-1 and caspase 9 to form an apoptosome (not shown in Figure 4). This activates caspase 9, which can then activate caspase 3 directly or through actions on caspase 8 to affect cell death. Cytochrome c release is normally uninhibited in circulating neutrophils because the main anti-apoptotic molecules A1 (Bcl-2 homologue in neutrophils) and IL-8 (chemokine) are only lowly expressed in the cells. However, factors that induce A1 and IL-8 expression in neutrophils extend the cells' life span by blocking the actions of Bak and (or) Bax on mitochondrial membranes (Figure 4).

Confirming our microarray results, Q-RT-PCR showed that expression of the Fas-related genes, Fas, FasL, RIP, FLASH, FADD, and Daxx, were all down regulated in blood neutrophils at parturition and into the first day of lactation (Figure 5). This suggested that parturition reprogrammed the cells for extended survival with regard to the Fas pathway of apoptosis. Survival gene expression reprogramming in the neutrophils was also supported by the fact that anti-apoptotic A1, BAFF, and IL-8 gene expression was induced by ~ 2-5 fold in the cells (Figure 6), while pro-apoptotic Bak gene expression was dramatically repressed (Figure 7). Thus, bovine parturition induced a predominantly survival gene expression signature in circulating neutrophils. We confirmed that the gene expression signature was meaningful by showing that

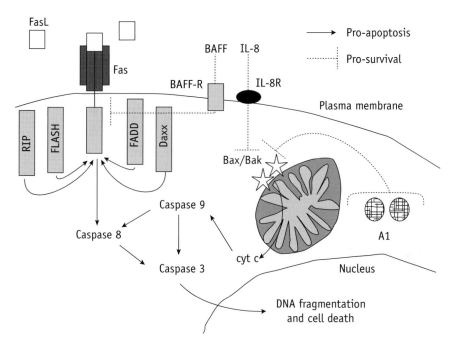

Figure 4. Neutrophils use two main pathways for apoptosis signaling, one stemming from the plasma membrane (Fas-FasL interaction) and the other from the mitochondria (Bax/Bak attack of the mitochondrial membrane). Blood neutrophils normally have high expression of the pro-apoptotic molecules shown (Fas, FasL, RIP, FLASH, FADD, Daxx, Bax, Bak) and low or no expression of the anti-apoptotic molecules shown (IL-8, A1, BAFF). This is why the normal life span of circulating neutrophils is a short 6-12 hours. Figure is adapted from Burton et al., 2005.

neutrophils from young steers lived significantly ($P \leq 0.05$) longer *in vitro* when cultured for 12 or 24 hours in 20% serum obtained from blood of our trial cows on day 0 versus that obtained on days –7, 0.25, or 1 (Madsen *et al.*, 2004). In additional apoptosis phenotyping experiments, we manipulated effective levels of cortisol, progesterone, and estradiol using charcoal extraction, steroid receptor inhibition, and steroid supplementation of the parturient serum prior to its use in culture to support steer neutrophils.

We found that the high level of cortisol contained in sera from day 0 was responsible for neutrophil survival induction *in vitro*. Progesterone and estradiol were without effect. Furthermore, serum cortisol profiles of the trial cows correlated ($P \leq 0.05$) with expression profiles for most of the apoptosis genes we observed in neutrophils subjected to the steroid during parturition *in vivo* (Burton *et al.*, 2005). Since then, we have also shown that a synthetic form of cortisol (dexamethasone) potently induces *in vitro* survival of bovine blood neutrophils (Burton *et al.*, 2005), causes down regulation of Fas and Bak gene and protein expression and up regulation of A1 expression *in vivo* (Chang *et al.*, 2004; and our unpublished data), and inhibits FasL-induced caspase 8 activation *in vitro* (Chang *et al.*, 2004). These observations are well supported by biomedical literature, which shows that glucocorticoids (cortisol and dexamethasone) induce

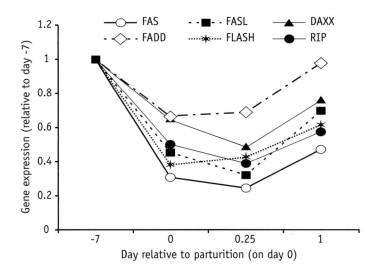

Figure 5. Q-RT-PCR validation of six Fas-related genes. These genes are normally expressed at high levels in short-lived blood neutrophils but were down regulated at parturition and into the first day of lactation. This gene expression signature suggests that bovine blood neutrophils are reprogrammed for extended survival at parturition. Data are adapted from Madsen et al., 2004 and Burton et al., 2005.

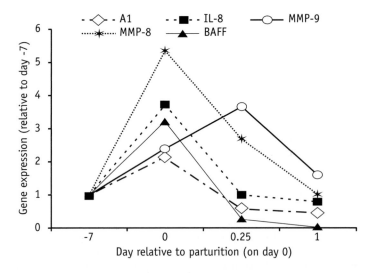

Figure 6. Q-RT-PCR validation of three apoptosis regulatory genes (A1, BAFF, and IL-8) and two extracellular matrix remodeling genes (MMP-8 and MMP-9) in neutrophils from parturient cows. These genes are normally expressed at low levels in short-lived blood neutrophils but were up regulated at parturition and into the first day of lactation. This gene expression signature suggests that bovine blood neutrophils are reprogrammed for extended survival and enhanced tissue remodeling activity at parturition. Data are adapted from Madsen et al., 2004 and Burton et al., 2005.

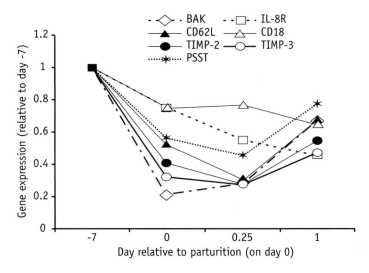

Figure 7. Q-RT-PCR validation of one apoptosis regulatory gene (Bak), three margination/migration genes (CD62L, CD18, IL-8R), one respiratory metabolism gene (PSST), and two genes that inhibit MMP-8 and MMP-9 (and thus prevent extracellular matrix remodeling; TIMP-2 and TIMP-3) in neutrophils from parturient cows. These genes are normally expressed at moderate to high levels in short-lived blood neutrophils but were down regulated at parturition and into the first day of lactation. This gene expression signature suggests that bovine blood neutrophils are reprogrammed for extended survival, decreased oxidative metabolism, and enhanced tissue remodeling capacity at parturition. Data are adapted from Madsen et al., *2004 and Burton* et al., *2005.*

extended survival in human and rodent neutrophils (Sendo *et al.*, 1996; Fanning *et al.*, 1999; Maianski *et al.*, 2004). We thus concluded from our combined gene expression and phenotyping work that the parturient surge in blood cortisol causes a reprogramming of apoptosis pathways in circulating neutrophils that favors extended longevity of the cells. This, combined with the known margination dysfunction of blood neutrophils at parturition, may help explain the pronounced neutrophilia that occurs during calving (Preisler *et al.*, 2000; Weber *et al.*, 2001; Madsen *et al.*, 2004).

While interesting, it was not intuitive why parturition and its associated rise in blood cortisol would reprogram neutrophils for extended survival, especially since the cells' ability to marginate and mount respiratory burst are also depressed at this time. In order to get a better sense of the neutrophils' overall functional state, we used Q-RT-PCR to characterize gene expression profiles for additional genes in other ontological clusters and correlated these with the blood cortisol profiles. What we found was an intriguing scenario in which genes involved in margination (CD62L), migration (CD18, IL-8R), and ROS generation (PSST) were all down regulated around parturition (Figure 7), while expression of key extracellular matrix remodeling genes were changed in favor of enhanced collagen degrading activity [increased MMP-8 and MMP-9 (Figure 6) and decreased TIMP-2 and TIMP-3 (Figure 7)]. Furthermore, these gene expression profiles correlated well ($P \leq 0.09$) with the serum cortisol profile of the animals. Our combined gene expression and apoptosis phenotyping data suggested to us that the parturient surge in blood cortisol acts to redirect blood neutrophils away from their traditional short-lived innate immune defense function (Figure 1) towards "non-traditional" tissue remodeling activities that may, of necessity, require cells with extended longevity (Figure 8).

The reproductive tract has a high demand for blood neutrophils at parturition

A review of available human and bovine literature shows that neutrophils have a predominate role in the induction of parturition (for example, see Junqueira *et al.*, 1980; Klucinski *et al.*, 1990; Hussain and Daniel, 1991, 1992; Eiler and Hopkins, 1992; Hoedemaker *et al.*, 1992b; Osmers *et al.*, 1992; Kelly *et al.*, 1994; Smith, 1994; Kelly, 1996; Maj and Kankofer, 1997; Thomson *et al.*, 1999; Winkler *et al.*, 1999; Luo *et al.*, 2000; Maymon *et al.*, 2000; Bowen *et al.*, 2002; Fortunato and Menon, 2002; Kimura *et al.*, 2002;; Mateus *et al.*, 2002; Osman *et al.*, 2003; Owen *et al.*, 2004).

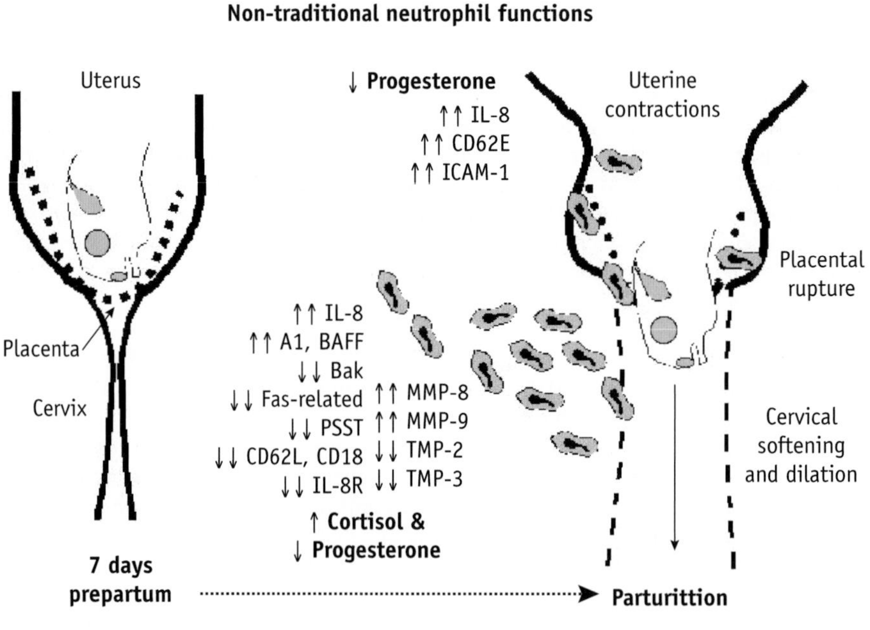

Figure 8. Blood-derived neutrophils are key initiators of parturition in humans and cattle. When fetal-derived cortisol removes the progesterone block on pregnancy at term, the cervix, lower uterine tract and placenta actively secrete IL-8 and up regulate vascular expression of CD62E and ICAM-I to recruit blood neutrophils to these tissue sites. Our gene expression, phenotyping, and correlation results suggest that fetal cortisol also works on the dam's blood neutrophils, causing a genetic reprogramming of the cells that extends their life span, prevents them from normal trafficking into peripheral tissues, reduces their traditional innate immune activities, and enhances their capacity for tissue remodeling. Thus, the reproductive tract may have a more urgent demand for long lived neutrophils with elevated tissue remodeling capacity and decreased innate immune activity during parturition than peripheral tissues have of short lived neutrophils endowed with potent antimicrobial defense capacity. If true, parturition would occur at the expense of innate immune defense of peripheral tissues, possibly explaining the heightened susceptibility to mastitis at this time.

The lower uterine tract, cervix, and placenta actively recruit large numbers of blood neutrophils during parturition because of a new ability to secrete massive amounts of IL-8 and up regulate blood vessel expression of CD62E and ICAM-I adhesion molecules as the progesterone block on parturition is released under the influence of cortisol (Figure 8). The cortisol derives from the fetus: as the fetus perceives its lack of space in the uterus, it mounts a classical HPA-axis response resulting in cortisol secretion. We propose that the dam's (and fetus') neutrophils are enabled to traffic preferentially to the reproductive tract tissues because of cortisol-induced loss of CD62L expression, which prevents circulating neutrophils from marginating on blood vessels in peripheral tissues. However, we also propose that the pronounced CD62L down regulation and modest down regulation of CD18 and IL-8R on circulating neutrophils may be overcome by the heightened expression of CD62E and ICAM-I on endothelial cells in reproductive tract tissues, and by up regulated IL-8 expression by these tissues and the recruited neutrophils, because neutrophils readily accumulate in the cervix, lower uterine tract, and placenta during normal parturition (Figure 8). Under the influence of IL-8, the recruited neutrophils become activated to release their granule contents of MMP-8 and MMP-9, which rapidly degrade collagen and enable separation of maternal-fetal membranes, cervical softening and dilation, and rupture of the fetal membranes so the calf is born alive (Figure 8). In fact, inadequate levels/activity of these pro-inflammatory mediators at parturition is linked to retained placentas in cattle (Maj and Kankofer, 1997; Kimura *et al.*, 2002). In addition, the bovine placenta secretes an unknown factor that helps shut down oxidative burst activity in recruited neutrophils (Hoedemaker *et al.*, 1992a, b; Kimura *et al.*, 2002), possibly to avoid inflammatory damage to host endometrial cells. Supporting this notion, we also observed that expression of genes encoding granzyme B and TGF-β were profoundly down regulated in the blood neutrophils of our parturient cows (Burton *et al.*, 2005). Thus, the reproductive tract may have a more urgent demand for long lived neutrophils with elevated extracellular remodeling capacity and decreased cell-damaging activity during parturition than peripheral tissues have of short lived neutrophils endowed with potent antimicrobial immune defense capacity. If true, parturition would occur at the expense of innate immune defense of peripheral tissues, possibly explaining the heightened susceptibility to mastitis at this time. Certainly, our gene expression and apoptosis phenotyping results, as well as past observations of so many mastitis researchers support such a paradigm. While we have yet to demonstrate that our gene expression profiles (Figures 6 and 7) equate to heightened extracellular matrix remodeling activity of blood-derived neutrophils during the cortisol surge at parturition, it would make sense that this was the case because of the critical role these cells play in collagenolysis of the reproductive tract. Our future studies are geared towards testing these possibilities, with the hope of gaining a better understanding of how we might manipulate the bovine neutrophil system to better fight intramammary infections while still enabling parturition to occur normally.

Acknowledgements

We would like to thank ABS Global, Inc. for sponsoring Session XII: Application of Genomics to Production Diseases at the 12th International Conference on Production Diseases (held July 19-22, 2004; Michigan State University, East Lansing, MI, USA). We also thank our colleagues at the Center for Animal Functional Genomics, Drs. Paul Coussens, Steve Suchyta, Guilherme Rosa, Robert Tempelman, and Ms. Sue Sipkovsky, without whom we would not have embarked on this fantastic journey into the world of neutrophil functional genomics. The research reported in this presentation was funded in part by the Michigan Agricultural Experiment Station (project numbers MICL01691 and MICL01836) and USDA's NRI and IFAFS competitive grants programs (project numbers 2001-35204-10798 and 2001-52100-11211, respectively).

References

Bowen, J.M., L. Chamley, J.A. Keelan and M.D. Mitchell, 2002. Cytokines of the placenta and extra-placental membranes: roles and regulation during human pregnancy and parturition. Placenta 23, 257-273.

Brazma, A., P. Hingamp, J. Quackenbush, G. Sherlock, P. Spellman, C. Stoeckert, J. Aach, W. Ansorge, C.A. Ball, H.C. Causton, T. Gaasterland, P. Glenisson, F.C. Holstege, I.F. Kim, V. Markowitz, J.C. Matese, H. Parkinson, A. Robinson, U. Sarkans, S. Schulze-Kremer, J. Stewart, R. Taylor, J. Vilo and M. Vingron, 2001. Minimum information about a microarray experiment (MIAME)-toward standards for microarray data. Nat. Genet. 29, 365-371.

Burton, J.L. and M.E. Kehrli, Jr., 1995. Regulation of neutrophil adhesion molecules, and shedding of *Staphylococcus aureus* in milk of cortisol- and dexamethasone-treated cows. Am. J. Vet. Res. 56, 997-1006.

Burton, J.L. and R.J. Erskine, 2003. Mastitis and immunity: some new ideas for an old disease. Vet. Clinics N. Am. Food Anim. Pract. 19, 1-45.

Burton, J.L., M.E., Kehrli, Jr., S. Kapil and R.L. Horst, 1995. Regulation of L-selectin and CD18 on bovine neutrophils by glucocorticoids: effects of cortisol and dexamethasone. J. Leuk. Biol. 57, 317-325.

Burton, J.L., S.A. Madsen, J. Yao, S.S. Sipkovsky and P.M. Coussens, 2001. An immunogenomics approach to understanding periparturient immunosuppression and mastitis susceptibility in dairy cows. Acta Vet Scand. 42, 407-24.

Burton. J.L., S.A. Madsen, L-C. Chang, P.S.D. Weber, K.R. Buckham, R. van Dorp, M-C. Hickey and B. Earley, 2005. Gene Expression Signatures in Neutrophils Exposed to Glucocorticoids: a New Paradigm to Explain "Neutrophil Dysfunction" in Parturient Dairy Cows. Vet. Immunol. Immunopathol. 105, 197-219.

Burvenich, C., Van Merris, V., Mehrzad, J., Diez-Fraile, A. and L. Duchateau, 2003. Severity of *E. coli* mastitis is mainly determined by cow factors. Vet. Res. 34, 521-564.

Cai, T.Q., Weston, P.G., Lund, L.A., Brodie, B., D.J. McKenna and W.C. Wagner, 1994. Association between neutrophil functions and periparturient disorders in cows. Am. J. Vet. Res. 55, 934-943.

Chang, L-C., S.A. Madsen, T. Toelboell, P.S.D. Weber and J.L. Burton, 2004. Effects of glucocorticoids on Fas gene expression in bovine blood neutrophils. J. Endocrinol. 183, 569-83.

Eiler, H. and F.M. Hopkins, 1992. Bovine retained placenta: effects of collagenase and hyaluromidase on detachment of placenta. Biol. Reprod. 46, 580-585.

Fanning, N.F., H.P. Redmond and D. Bouchier-Hayes, 1999. Neutrophils and apoptosis. *In*: The Neutrophils: New Outlook for Old Cells. D.I. Gabrilovich (ed). Imperial College Press, London, UK. Ch. 6, pp 231-242.

Fortunato, S.J. and R. Menon, 2002. Screening of novel matrix metalloproteinases (MMPs) in human fetal membranes. J. Assisted Reprod. Genet. 19, 483-486.

Hill, A.W., A.L. Shears and K.G. Hibbitt, 1979. The pathogenesis of experimental *Escherichia coli* mastitis in newly calved dairy cows. Res. Vet. Sci. 26, 97-101.

Hoedemaker, M., L.A. Lund, P.G. Weston and W.C. Wagner, 1992a. Influence of conditioned media from bovine cotyledon tissue cultures on function of bovine neutrophils. Am. J. Vet. Res. 53, 1530-1533.

Hoedemaker, M., L.A. Lund and W.C. Wagner, 1992b. Function of neutrophils and chemoattractant properties of fetal placental tissue during the last month of pregnancy in cows. Am. J. Vet. Res. 53, 1524-1529.

Hussain, A.M. and R.C.W. Daniel, 1991. Bovine endometritis: current and future alternative therapy. J. Vet. Med. Ass. 38, 641-651.

Hussain, A.M. and R.C.W. Daniel, 1992. Phagocytosis by uterine fluid and blood neutrophils and haematological changes in postpartum cows following normal and abnormal parturition. Theriogenol. 37, 1253-1267.

Junqueira, L.C.U., M. Zugaib, G.S. Montes, O.M.S. Toledo, R.M. Krisztán and K.M. Shigihara, 1980. Morpholic and histochemical evidence for the occurrence of collagenolysis and for the role of neutrophilic polymorphonuclear leukocytes during cervical dilation. Am. J. Obstet. Gynecol. 138, 273-281.

Kehrli, M.E. Jr. and J.A. Harp, 2001. Immunity in the mammary gland. Vet. Clinics N. Am. Food Anim. Pract. 17, 495-516.

Kelly, R.W., 1996. Inflammatory mediators and parturition. Rev. Reprod. 1, 89-96.

Kelly, R.W., P. Illingworth, G. Baldie, R. Leask, S. Brouwer and A.A. Calder, 1994. Progesterone control of interleukin-8 production in endometrium and chorio-decidual cells underlies the role of the neutrophil in menstruation and parturition. Human Reprod. 9, 253-258.

Kimura, K., J.P. Goff, M.E. Kehrli, Jr. and T.A. Reinhardt, 2002. Decreased neutrophil function as a cause of retained placenta in dairy cattle. J. Dairy Sci. 85, 544-550.

Klucinski, W., S.P. Targowski, E. Miernick-Degórska and A. Winnicka, 1990. The phagocytic activity of polymorphonuclear leukocytes isolated from normal uterus and that with experimentally induced inflammation in the cow. J. Vet. Med. Ass. 37, 506-512.

Luo, L., T. Ibaragi, M. Nozawa, T. Kasahara, M. Sakai, Y. Sasaki, K. Tanebe and S. Sait, 2000. Interleukin-8 levels and granulocyte counts in cervical mucus during pregnancy. Am. J. Repro. Immunol. 43, 78-84.

Madsen, S.A., L-C. Chang, M-C. Hickey, G.J. M. Rosa, P.M. Coussens and J.L. Burton, 2004. Microarray analysis of gene expression in blood neutrophils of parturient cows. Physiol. Genomics 16, 212-221.

Madsen, S.A., P.S.D. Weber and J.L. Burton, 2002. Altered blood neutrophil gene expression in periparturient dairy cows. Vet. Immunol. Immunopathol. 86, 159-175.

Maianski, N.A., A.N. Maianski, T.W. Kuijpers and D. Roos, 2004. Apoptosis of neutrophils. Acta Hematologica 111, 56-66.

Maj, J.G. and M. Kankofer, 1997. Activity of 72-kDa and 92-kDa matrix metalloproteinases in placental tissues of cows with and without retained fetal membranes. Placenta 18, 683-687.

Mateus, L., L. Lopes da Costa, H. Carvalho, P. Serra and R. Silva, 2002. Blood and intrauterine leukocyte profile and function in dairy cows that spontaneously recovered from postpartum endometritis. Reprod. Dom. Anim. 37, 176-180.

Maymon, E., R. Romero, P. Pacora, R. Gomez, N. Athayde, S. Edwin and B.H. Yoon, 2000. Human neutrophil collagenase (matrix metalloproteinase 8) in parturition, premature rupture of the membranes, and intrauterine infection. Am. J. Obstet. Gynecol. 183, 94-99.

Nagahata, H., S. Makino, S. Takeda, H. Takahashi and H. Noda, 1988. Assessment of neutrophil function in the dairy cow during the perinatal period. J. Vet. Med. 35, 747-751.

Preisler, M.T., P.S.D. Weber, R.J. Tempelman, R.J. Erskine, H. Hunt and J.L. Burton, 2000. Glucocorticoid receptor down-regulation in neutrophils of periparturient cows. Am. J. Vet. Res. 61, 14-19.

Osman, I., A. Young, M.A. Ledingham, A.J. Thomson, F. Jordan, I.A. Greer and J.E. Norman, 2003. Leukocyte density and pro-inflammatory cytokine expression in human fetal membranes, decidua, cervix, and myometrium before and during labour at term. Mol. Human Reprod. 9, 41-45.

Osmers, R., W. Rath, B.C. Adelmann-Grill, C. Fittkow, M. Kuloczik, M. Szeverényi, H. Tschesche and W. Kuhn, 1992. Origin of cervical collagenase during parturition. Am. J. Obstet. Gynecol. 166, 1455-1460.

Owen, C.A., Z. Hu, C. Lopez-Otin and S.D. Shapiro, 2004. Membrane-bound matrix metalloproteinase-8 on activated polymorphonuclear cells is a potent, tissue inhibitor of metalloproteinase-resistant collagenase and serpinase. J. Immunol. 172, 7791-7803.

Sendo, F., H. Tsuchida, Y. Takeda, S. Gon, H. Takei, T. Kato, O. Hachiya and H. Watanabe, 1996. Regulation of neutrophil apoptosis – its biological significance in inflammation and the immune response. Human Cell 9, 215-222.

Smith, J.A., 1994. Neutrophils, host defense, and inflammation: a double-edged sword. J. Leuk. Biol. 56, 672-686.

Suchyta, S.P., S. Sipkovsky, R.G. Halgren, R. Kruska, M. Elftman, M. Weber-Nielsen, M.J. Vandehaar, R.J. Tempelman and P.M. Coussens, 2004. Bovine mammary gene expression profiling using a cDNA microarray enhanced for mammary specific transcripts. Physiol. Genomics 16, 8-18.

Suchyta, S.P., S. Sipkovsky, R. Kruska, A. Jeffers, A. McNulty, M.J. Coussens, R.J. Tempelman, R.G. Halgren, P.M. Saama, D.E. Bauman, Y.R. Boisclair, J.L. Burton, R.J. Collier, E.J. DePeters, T.A. Ferris, M.C. Lucy, M.A. McGuire, J.F. Medrano, T.R. Overton, T.P. Smith, G.W. Smith, T.S. Sonstegard, J.N. Spain, D.E. Spiers, J. Yao and P.M. Coussens, 2003. Development and testing of a high-density cDNA microarray resource for cattle. Physiol. Genomics 15, 158-164.

Thomson, A.J. J.T. Telfer, A. Young, S. Campbell, C.J.R. Stewart, I.T. Cameron, I.A. Greer and J.E. Norman, 1999. Leukocytes infiltrate the myometrium during human parturition: further evidence that labour is an inflammatory process. Human Reprod. 14, 229-236.

Weber, P.S.D., S.A. Madsen, G.W. Smith, J.J. Ireland and J.L. Burton, 2001. Pre-translational regulation of neutrophil CD62L in glucocorticoid-challenged cattle. Vet. Immunol. Immunopathol. 83, 213-240.

Weber, P.S.D., T. Toelboell, L-C. Chang, J. Durrett Tirrell, P.M. Saama, G.W. Smith and J.L. Burton, 2004. Mechanisms of glucocorticoid-induced down regulation of neutrophil L-selectin: Evidence for effects at the gene expression level and primarily on blood neutrophils. J. Leuk. Biol. 75, 815-827.

Winkler, M., Fischer, D-C., Ruck, P., Marx, T., Kaiserling, E., Oberpichler, A., Tschesche, H. and W. Rath, 1999. Parturition at term: parallel increases in interleukin-8 and proteinase concentrations and neutrophil count in the lower uterine segment. Human Reprod.14, 1096-1100.

Yao, J., Burton, J.L., Saama, P., Sipkovsky, S. and P.M. Coussens, 2001. Generation of EST and cDNA microarray resources for the study of bovine immunobiology. Acta Vet. Scand. 42, 391-405.

Functional genomics analysis of bovine viral diarrhea virus-infected cells: unraveling an enigma

John D. Neill and Julia F. Ridpath
National Animal Disease Center, USDA, Agricultural Research Service, Ames IA, USA

Bovine viral diarrhea viruses (BVDV), members of the *Flaviviridae*, are ubiquitous pathogens worldwide. BVDV cause respiratory, gastrointestinal, immune and reproductive problems in infected cattle. These viruses, particularly genotype 2 viruses, can cause severe acute disease characterized by very high fever (>1058 F) and severe lymphopenia and thrombocytopenia (>50% decline each). In the US, it is estimated that acute disease and reproductive problems caused by BVDV account for $25 to $35/calf born in annual losses to the beef and dairy industries. Recent evidence suggests that loss of productivity in BVDV-infected animals can persist for a significant period of time following recovery. To further complicate the matter, BVDV can establish a persistent infection in fetuses during the first 150 days of pregnancy, resulting in lifelong viral replication and immunotolerance to the infecting virus. These calves are often normal in appearance and act as Trojan horses, spreading the virus to other animals and herds. This is one of the major means of dissemination of the virus. Recent advances in functional genomics have provided innovative means to examine old questions in new ways. It is now possible to globally examine gene expression changes that occur in cells following viral infection to determine how the cell responds. This type of information can provide insight into the mechanism(s) behind pathologies observed in infected animals.

We have chosen serial analysis of gene expression (SAGE) for the global analysis of gene expression in BVDV2-infected cells. This sequence-based technique can both detect and quantify virtually every transcript in the cell type studied. SAGE is based on sequencing of 14 base 'tags' derived from the 3' end of each transcript in the cell type being examined. These tags are used to identify the transcript from which it was derived as well as provide a sequence anchor to clone the transcript if no other sequence data is available. We have constructed and sequenced SAGE libraries from noninfected BL3 cells (a bovine B cell lymphosarcoma cell line) and BL3 cells acutely infected with BVDV2 strains 1373 (noncytopathic high virulence), 28508 (noncytopathic low virulence), and 296c (cytopathic), as well as BL3 cells persistently infected with noncytopathic strain 28508.

The expression profiling of the non-infected BL3 cells revealed the normal levels of gene expression. Many of the genes that are characteristic of this cell type and integral to their function were identified. Comparison of the SAGE databases from noninfected and BVDV2-infected BL3 cells revealed genes that were differentially expressed. Additionally, altered levels of gene expression were observed that were specific to acutely infected cells, specific to persistent infections and specific to infection by cytopathic viruses.

These results are leading to the formation of hypotheses concerning the mechanisms behind BVDV2-induced immunosuppression, non-apoptotic cell death observed in infected animals and the establishment of persistent infections. Furthermore, these results may lead to drug discovery trials for the suppression of observed disease symptoms with a subsequent increase in animal recovery and production rates.

Early postpartum ketosis in dairy cows and hepatic gene expression profiles using a bovine cDNA microarray

J.J. Loor, H.M. Dann, D.E. Morin, R.E. Everts, S.L. Rodriguez-Zas, H.A. Lewin and J.K. Drackley
University of Illinois, Urbana IL, USA

Primary ketosis (PK) decreases milk yield and hinders reproductive performance, resulting in economic losses for dairy farmers. We used a simple model for induction of PK in Holstein cows during the early postpartum period to examine gene expression profiles using a bovine cDNA microarray.

Cows had *ad libitum* (150% of NRC requirements) or restricted (80% of requirements) feed intake during the dry period. All cows were fed a common lactation diet after parturition. At 4 days in milk (DIM), 7 cows classified as healthy after a physical examination were fed at 50% of intake at day 4 from day 5 to signs of PK (anorexia, ataxia, or abnormal behavior) or until 14 DIM. Another group of 8 healthy cows served as controls (NK). Liver was biopsied at 10-14 DIM (PK) or 14 DIM (NK). A microarray consisting of 7,872 cDNA inserts was used for transcript profiling. Annotation was based on similarity searches using BLASTN and TBLASTX against human and mouse UniGene databases and the human genome. Cy3- and Cy5-labelled cDNA from liver and a reference standard (derived from a mixture of cattle tissues) were used for hybridizations (30 microarrays). Loess-normalized log-transformed ratios (liver/standard) were used to detect differential gene expression.

Analysis of variance using Benjamini and Hochberg's False Discovery Rate (P=0.05) indicated that 2,726 transcripts were differentially expressed between PK and NK. Hierarchical clustering of differentially expressed genes showed that their relative expression differed by 42% between PK and NK. Among differentially expressed genes, 454 were decreased 1.5-fold and 69 were decreased 2-fold or greater in cows with PK compared with NK. In contrast, there were 458 genes with expression greater than 1.5-fold and 63 with expression greater than 2-fold in PK compared with NK. Expression of some genes with key functions in oxidative stress responses, inflammation, and complement activation were decreased 2-fold or greater in cows with PK compared with NK. In cows with PK, there was a 5-fold increase in expression of a gene involved in intracellular activation and transport of long-chain fatty acids. Other upregulated genes (1.5-fold or greater) with PK were associated with intracellular binding of fatty acids, the cytochrome P450 system, steps of peroxisomal and mitochondrial oxidation, and gluconeogenesis. Liver from PK cows showed a 2-fold reduction in expression of a crucial gene in mitochondrial beta-oxidation, and also genes associated with cholesterol synthesis and fatty acid desaturation.

Our data show that primary ketosis during the early postpartum period is associated with diminished hepatic inflammatory and oxidative stress responses, enhanced intracellular transport of fatty acids, and altered fatty acid oxidation.

Acknowledgements

Supported by award 2001-35206-10946 from NRI Competitive Grants Program/CSREES/ USDA.

Genetic improvement of dairy cattle health

J.B. Cole[1], P.D. Miller[2], and H.D. Norman[1]
[1]Animal Improvement Programs Laboratory, USDA-ARS, Beltsville, Maryland; [2]Department of Dairy Science, University of Wisconsin-Madison, Madison, WI, USA

Genetic response has been well documented when cows are selected based upon breeding values from national genetic evaluations of field-recorded traits. This response is steady and the gains accumulate over time. There is increasing interest in breeding dairy cows that will be healthier and remain in the herd longer. Direct and indirect costs associated with disease represent a major expense to the producer, and selection for improved health may reduce these costs significantly. Considerable genetic variability exists for several health traits in dairy cattle, including mastitis, ketosis, displaced abomasum and lameness. This genetic variation is not currently being efficiently utilized for genetic improvement. Somatic cell score has been used to select for improved resistance to mastitis in the United States. Significant genetic gains in productive life are being achieved by selecting sires whose daughters produce more, conceive faster, calve easier, and live longer than their contemporaries.

Four Scandinavian countries currently provide genetic evaluations for clinical mastitis and place more emphasis on health and fitness traits in their national selection programs. The statistical tools and computing resources needed for routine health genetic evaluations are available, but the necessary data are not. A standard set of health and fitness event codes is not used by participants in the Dairy Herd Improvement program and on-farm management software.

The USDA Animal Improvement Programs Laboratory is currently developing a data exchange format for the routine collection of health and fitness data from dairy cattle throughout the United States that includes a set of standard event codes. Events have been included in the format based on frequency of occurrence in field data, economic impact, and consultation with veterinarians. A health event record includes detailed cow identification, a health event code, an event date, and an optional detail field.

This format can provide the data necessary for research and the implementation of routine genetic evaluations for economically-important health traits. Uniform management codes that can also be used for genetic evaluations for health traits of economic significance will provide dairy producers with a valuable tool for improving the efficiency and profitability of their cows.

Parturition-induced gene expression signatures in bovine peripheral blood mononuclear cells

L-C. Chang[1], R. van Dorp[2], P.S.D. Weber[1], K.R. Buckham[1], and J.L. Burton[1,2]
[1]Immunogenetics Laboratory; [2]Center for Animal Functional Genomics, Department of Animal Science, Michigan State University, East Lansing, MI 48824 USA

Bovine peripheral blood mononuclear cells (PBMC) collected around parturition do not perform as well as cells collected before parturition in assays of mitogen- and antigen-induced proliferation and antibody secretion. In addition, expression of some pro-inflammatory cytokines by PBMC collected from blood or milk of parturient cows can be depressed (*e.g.*, IFN-gamma) or exaggerated (*e.g.*, TNF-alpha), suggesting that parturition induces an imbalance in PBMC sensitivity and responsiveness to noxious challenges[1]. If true, this could contribute to the heightened occurrence and severity of infectious and metabolic diseases as cows transition from the dry period through parturition into early lactation. In order to develop new management tools and strategies that help protect the animals from such diseases, a more sophisticated understanding of how parturition affects PBMC is required. The objective of our study was to elucidate PBMC genes affected by parturition that might explain the cells' immune imbalance and provide logical targets for future research.

PBMC were isolated from blood of 3 Holstein cows, on d -8, 0.5, 2, 7, and 14 relative to parturition (on d 0). Total RNA was obtained and used in a loop design[2] to interrogate our group's bovine immunobiology cDNA microarrays[3] spotted with cDNA from 1056 genes. Resulting gene expression data were $\log_2$-transformed and LOWESS normalized prior to statistical analysis using a 2-step mixed models approach[2]. Daily means for genes significantly ($P<0.05$) affected by parturition were imported into GeneSpring® and a tree built that clustered them based on similarity of expression pattern changes. In silico approaches were used to establish probable ontology of affected genes and to help guide future validation experiments.

Our statistical analysis revealed 243 genes that were significantly influenced by parturition. The estimated false discovery rate was 9%. The gene tree showed that while ~ 50% of these genes were induced ~2-5 fold immediately following parturition (d 0.5), the majority (> 90%) were dramatically repressed on days 7 and (or) 14 of lactation. While 43.2% of the affected genes have functions that are yet unknown, the rest encode proteins with critical roles in processes such as antigen recognition, cell proliferation and adhesion, cytokine production and signaling, and apoptosis regulation. We have designed PCR primers for 40 of these genes and are currently validating their expression changes using quantitative real time RT-PCR. Thus, our study has identified greater than 200 parturition-sensitive gene targets in bovine PBMC that could be pursued in future immunology, nutrition, genetics, and drug studies aimed at understanding and manipulating PBMC gene expression for improved transition cow health.

References

[1] Burton, J.L., and R.J. Erskine, 2003. Immunity and mastitis: some new ideas for an old disease. Vet. Clin. North Am. Food. Anim. Pract. 19, 1-45.

[2] Madsen, S.A., L.-C. Chang, M.-C. Hickey, *et al.*, 2004. Microarray analysis of gene expression in blood neutrophils of parturient cows. Physiol. Genom. 16, 212-221.

[3] Burton, J.L., S.A. Madsen, J. Yao, *et al.*, 2001. An immunogenomics approach to understanding periparturient immunosuppression and mastitis susceptibility in dairy cows. Acta Vet. Scand. 42, 407-424.

Mechanisms of glucocorticoid-induced L-selectin (CD62L) down-regulation in bovine blood neutrophils

Patty S.D. Weber, Ling-Chu. Chang, Trine Toelboell and Jeanne L. Burton
Immunogenetics Laboratory, Department of Animal Science, Michigan State University, East Lansing, MI 48824, USA

Neutrophil CD62L is vital to successful inflammatory responses that result in pathogen clearance from sites of infection. CD62L is a leukocyte adhesion molecule that mediates initial contact of circulating neutrophils with vascular endothelia, which allows the fast flowing cells to tether and roll along the blood vessel wall surveying for signs of infection. Glucocorticoids (GC) are well known for their potent anti-inflammatory actions, an important component of which is their ability to modulate expression of a variety of adhesion molecules on leukocytes, including CD62L. In particular, CD62L mRNA and protein are profoundly down regulated in neutrophils in GC challenged cattle[1,2]. Currently the mode of action by which this hormone elicits its effects is unclear. We hypothesize that GC induced down regulation of CD62L in neutrophils is mediated directly at the gene transcription level, via hormone-activated GC receptors (GR).

The objective of this study was to monitor CD62L gene transcription, mRNA abundance, and protein expression in isolated neutrophils treated with the GC, dexamethasone (Dex), both in the presence and absence of the GR antagonist, RU486.

Neutrophils were isolated from 4 steers and incubated with 10^{-7} M Dex $\pm$ RU486 for 4h prior to purification of nuclei, RNA, and protein. CD62L gene transcription was assessed by nuclear run-on assay, mRNA abundance by quantitative real-time RT-PCR, and CD62L protein expression by flow cytometry and Western blot analyses.

Dex dramatically (P<0.01) reduced CD62L mRNA abundance *in vitro*, an effect that was inhibited by RU486. Nuclear run-on data showed that the rate of CD62L gene transcription was moderately reduced by Dex, but this did not fully account for the massive decreases in CD62L mRNA abundance. Also, Dex had variable effects on surface and intracellular levels of CD62L protein expression in cultured neutrophils, showing clear down-regulating effects in cells from two animals and no effects in cells from the other two animals.

We conclude that Dex has a direct, GR-mediated inhibitory effect on CD62L gene expression that is in part due to the receptor's ability to interfere with CD62L gene transcription. However, our results cannot rule out that increases in CD62L mRNA turnover are also influenced by Dex-activated GR. The failure of Dex to consistently down regulate CD62L protein *in vitro*, despite its clear effects on CD62L mRNA abundance, is contrary to effects of the steroid *in vivo*. Indeed, Dex is well known to repress CD62L protein in the cytoplasm and surface of neutrophils when administered into cattle. Thus, removal of neutrophils from the circulation for *in vitro* culture appears to uncouple their CD62L mRNA and protein expression systems, and may not be the best way to study CD62L at the protein level. Nonetheless, our results clearly implicate steroid-activated GR as a key negative regulator of CD62L gene expression in bovine blood neutrophils, partly explaining heightened risk for infectious diseases in stressed animals.

References

[1] Weber P.S., S.A. Madsen, G.W. Smith, *et al.*, 2001. Pre-translational regulation of neutrophil L-selectin in glucocorticoid-challenged cattle. Vet. Immunol. Immunopathol. 83, 213-240.
[2] Burton, J.L., M.E. Kehrli, Jr., S. Kapil, *et al.*, 1995. Regulation of L-selectin and CD18 on bovine neutrophils by glucocorticoids: effects of cortisol and dexamethasone. J. Leukoc. Biol. 57, 317-325.

T-cell receptor Vb gene repertoire analysis of the mammary gland T-cells on the *Staphylococcus aureus* causing bovine mastitis

T. Hayashi[1], H. Ohtsuka[2], M. Kohiruimaki[2], K. Katsuda[3], Y. Yokomizo[3], S. Kawamura[2] and R. Abe[1]
[1]Division of Immunobiology, Research Institute for Biological Sciences, Tokyo University of Science, Noda, Chiba 278 0022, Japan; [2]Veterinary Internal Medicine, School of Veterinary Medicine and Animal Sciences, Kitasato University, Towada, Aomori 034 8628, Japan; [3]Department of Immunology, National Institute of Animal Health, Tsukuba, Ibaraki 305 0856, Japan

We analyzed the T-cell receptor (TCR) Vb gene repertoire in peripheral blood T-cells (PBTC) and mammary gland T-cells (MGTC) from 14 *Staphylococcus aureus* (*S. aureus*) strains causing bovine mastitis by using newly developed semi quantitative RT-PCR. We compared the TCR repertoire in PBTC with MGTC, and in MGTC of affected quarter with that of unaffected quarter in same individuals, respectability.

It was found that 5 out of 14 mastitic cows; BoVb45, BoVb90 and BoVb93 gene segments were overrepresented in the MGTC of affected quarter compared to the PBTC and MGTC of unaffected quarter counterpart. In order to examine whether this alternation was caused by the TCR specific stimulation of enterotoxin produced by *S. aureus* as bacterial superantigens (SEs), we compared the Vb expression patterns of MGTC in affected quarter with those of cultured bovine PBTC stimulated with SEA, SEB, SEC, SED, SEE and TSST1. It was found that the Vb usage pattern in the MGTC from affected quarters extremely resembled the Vb restricted T-cell expansion pattern against SEC stimulation *in vitro*. Consistent with these results, we detected SEC in milk from these affected quarters by ELISA.

We believe that these observations provide strong evidence that bacterial superantigen produced by *S. aureus* is involved in the development of *S. aureus* causing bovine mastitis.

A new sensitive microarray for studying metabolic diseases in cattle

B.E. Etchebarne, W. Nobis, M.S. Allen, and M.J. VandeHaar
Department of Animal Science, Michigan State University, East Lansing, MI 48824, USA

First generation microarrays employing extensive cDNA libraries have allowed high numbers of both known and unidentified genes to be surveyed. Many of these arrays have only one spot per gene, giving no information on within-plate variance. The Human Genome Project has provided extensively annotated databases, such as LocusLink, the Kyoto Encyclopedia of Genes and Genomes (KEGG), Gene Ontology (GO), The Institute for Genomics Research (TIGR) and BioCarta. These publicly available resources, paired with recent price reductions in oligonucleotide synthesis, allow researchers to feasibly design and produce microarrays with gene sets tailored to specific research areas.

Using these databases, we identified approximately 2000 bovine genes representing enzymes of metabolic pathways, metabolic regulators and receptors, transport and binding proteins, intracellular signaling cascades, and cell cycle and apoptotic pathways. Three individual 70mer oligonucleotide probes per gene were designed for triplicate spotting onto glass slides, giving nine spots per gene. Each oligonucleotide was designed within specific parameters to standardize hybridization behavior.

Use of multiple oligonucleotides per gene improves representation of the expressed fraction of each gene, including splice variants. Spot replication improves within-array quality control and increases the statistical power of accurately detecting small changes in expression at a lower cost than slide replication. Reduction in technical error to increase statistical power is especially important for metabolic research, in which changes in gene expression are often subtle. In addition, our focus on only those genes that are relevant to metabolism improves downstream bioinformatics and data analysis for integration of metabolic gene networks. Because all genes included in this design are annotated with corresponding human homologs, the design can be applied to other species to promote our understanding of comparative metabolism.

In conclusion, our design of a focused oligonucleotide microarray with multiple spots per gene will facilitate research in the metabolic genomics of cattle and can be easily applied to other species and disciplines.

Ontogenetic development of mRNA levels and binding sites of hepatic beta-adrenergic receptors in cattle

J. Carron, H.M. Hammon, C. Morel and J.W. Blum
Division of Nutrition and Physiology, Vetsuisse Faculty, University of Berne, Berne, Switzerland

Catecholamines affect hepatic glucose production mainly through β_2-adrenergic receptors (AR). We have investigated hepatic mRNA levels and receptor binding of β-AR of neonatal calves born before term (preterm), born at term (full term) and after feeding colostrum and milk replacer for 4 d. In addition, we have studied hepatic β-AR of veal calves (VC) to investigate whether gene expression and binding sites of β-AR are influenced by age.

Calves of P0 (n=7) were born on d 277 of gestation within 42 ± 3 h, after cows were injected with 500 µg prostaglandin $F_{2\alpha}$ and 5 mg Flumethason. Calves of M0 (n=7) were spontaneously born after normal length of pregnancy (290 ± 2 d). Calves of P0 and M0 were euthanized immediately after birth. Calves of M5 (n=7) were spontaneously born after normal length of pregnancy, were fed colostrum up to d 3 and milk replacer on d 4, and were euthanized on d 5. Calves of VC (n=7) received colostrum for 3 d and were raised with milk and milk replacer until weighing 200 kg. Liver samples were taken and mRNA concentrations of β_1-, β_2- and β_3-AR were measured by real-time PCR[1]. Relative quantification was done using glyceraldehyde-3-phosphate dehydrogenase as reference gene transcript. Binding of βR was evaluated by radioreceptor binding assays[2] using ^{3}H-CGP-12177 (β-antagonist) as a ligand. Competitive binding studies were performed to evaluate βR subtypes with alprenolol, propranolol (both β_1- and β_2-antagonists), ICI-188,551 (β_2-antagonsit), atenolol (β_1-antagonist), and adrenaline as competitors. For quantification of receptor sites saturation binding assays were performed[2].

Abundance of hepatic mRNA for β_2-AR was highest ($P<0.01$) and for β_1-AR was higher ($P<0.01$) than for β_3-AR, which could not be detected in VC. Abundance of mRNA for β_2-AR was higher ($P < 0.01$) in VC than in all groups of neonatal calves. Within neonatal calves mRNA abundance of β_3-receptors were higher ($P<0.05$) in M5 than in P0. Competitive binding studies revealed highest affinities for alprenolol, propranolol and ICI-188,551, which did not significantly differ from each other. Atenolol in concentrations up to 10^{-5} M did not displace ^{3}H-CGP-12177 from specific binding. Binding curve for adrenaline was best fitted by a two-receptor model. Numbers of βR were higher ($P<0.01$) in VC than in all groups of neonatal calves. Within neonatal calves numbers of β-adrenergic binding sites were higher ($P<0.05$) in M5 than in P0 and tended to be higher ($P<0.1$) in M0 than in P0. Binding affinity of ^{3}H-CGP-12177 was not different among groups. Binding sites correlated positively with mRNA levels of β_2-AR (r = 0.67; $P<0.01$), but not with mRNA levels of β_1- or β_3-AR.

In conclusion, mRNA of all 3 β-AR subtypes was found in liver of neonatal calves, but the β_2-AR was the dominant subtype. Binding studies indicated most binding of ^{3}H-CGP-12177 to β_2-AR. β_2-AR increased with age and were mainly regulated at the transcriptional level.

References

[1] Inderwies, T., M.W. Pfaffl, H.H.D. Meyer, *et al.*, 2003. Detection and quantification of mRNA expression of alpha- and beta-adrenergic receptor subtypes in the mammary gland of dairy cows. Domest. Anim. Endocrinol. 24, 123-135.

[2] Hammon, H.M., R.M. Bruckmaier, U.E. Honegger, *et al.*, 1994. Distribution and density of alpha and beta-adrenergic receptor binding sites in the bovine mammary gland. J. Dairy Res. 61, 47-57.

Bovine lactase messenger RNA levels determined by real time PCR

E.C. Ontsouka[1], B. Korczack[2], H.M. Hammon[1] and J.W. Blum[1]
[1]Div. of Nutrition and Physiology, Inst. of Animal Genetics Nutrition and Housing; [2]Inst. of Bacteriology, Vetsuisse Faculty, University of Berne, Berne, Switzerland

Lactase is an important disaccharidase expressed in the intestinal brush border membrane of the small intestine (SI) of milk-fed animals. We aimed (i) to develop and validate a real-time PCR protocol for the amplification of bovine lactase, (ii) to study the distribution of lactase mRNA along the gastrointestinal tract and within the SI wall of calves, and (iii) to investigate the ontogeny of lactase mRNA expression in calves.

Primers were designed in highly conserved regions with 100% identity based on multiple species alignment (rat, rabbit, human). The specificity of primers and of PCR products was validated by testing the bovine lactase mRNA in pooled reverse transcribed RNA from full-tickness walls of esophagus, rumen, fundus, pylorus, duodenum, jejunum, ileum and colon. Furthermore, mRNA levels of lactase were measured in fractionized layers of SI mucosa (villus tips and crypt fractions) and submucosa (fraction containing mainly Peyer's patches, PP). The ontogeny of lactase expression was assessed by measuring mRNA levels in SI mucosa (duodenum, jejunum and ileum) of calves at birth (preterm born and full-term born; born at 277 and 290 days of pregnancy), and (in full-term born calves) on day 5 and 159 of life. Lactase mRNA was expressed relative to mean levels of glyceraldehyde phosphate hydrogenase, β-actin, ubiquitin and 18S that were used as housekeeping genes, using real-time PCR.

The 1.5% agarose gel electrophoresis and melting curve analysis have evidenced lactase mRNA in the total tissue (mucosa plus submucosa) of SI, but not of other sites of the gastrointestinal tract. The amplified partial bovine lactase sequence showed 82 to 87% similarity with human, rabbit or rat sequences. The lactase mRNA levels were significantly (jejunum) and numerically (ileum) higher in villus than in crypt fractions. No lactase mRNA could be detected in the duodenal and jejunal submucosal fraction containing mainly PP. At all investigated age stages, lactase mRNA levels in the mucosa of the duodenum and jejunum were significantly higher than in the mucosa of the ileum. In fullterm calves lactase mRNA levels were similar during the first 5 days of life in the SI mucosa at all sites, then significantly increased up to day 159 in jejunum and ileum. In conclusion, the developed protocol allowed quantitative measurements of lactase mRNA abundance in the intestine of calves. Lactase mRNA was present primarily in villi of the SI, but was not detectable in the submucosa. Abundance of lactase mRNA in the SI mucosa was different (jejunum ≥ duodenum > ileum). Lactase mRNA was absent in esophagus, rumen, fundus, pylorus and colon. Lactase mRNA in SI mucosa was present already in preterm calves. The expression of mRNA in duodenal and jejunal mucosa in veal calves increased with age.

Design and application of a bovine metabolism long oligonucleotide microarray

B.E. Etchebarne[1], W. Nobis[2], K.J. Harvatine[3], M.S. Allen[1], P.M. Coussens[1] and M.J. VandeHaar[1]
[1]Department of Animal Science, Michigan State University, East Lansing, Michigan 48824, USA;
[2]Vanderbilt University School of Medicine, Vanderbilt University, Nashville, Tennessee 37232,
USA; [3]Department of Animal Science, Cornell University, Ithaca, New York 14853, USA

Abstract

The Human Genome Project provided a catalyst for the development of extensive publicly available databases housing genome sequence information for many species of research interest, including production animals. When coupled with advancements in oligonucleotide synthesis, these bioinformatics resources provide a powerful tool for in-house generation of custom oligonucleotide microarrays designed to test specific hypotheses. Nutritional genomics is the integration of genomics into the nutritional sciences that has been used to elucidate the functions and interactions of genes and nutrients which can then be used to design diet-based therapeutic interventions to avoid disease manifestation. We have designed a bovine metabolism-focused microarray containing known genes using publicly available genomic internet database resources provided by NCBI, TIGR, KEGG, Swiss-Prot and BioCarta. 70mer oligonucleotides for each gene were designed and spotted on glass slides at sufficient replication to facilitate accurate detection of changes in gene expression. A microarray containing a robust collection of genes known to be involved in metabolism and regulation of metabolism will provide a better understanding of the etiology of common metabolic diseases in cattle.

Keywords: metabolism, gene expression, microarray, disease

Nutrition and functional genomics

Nutritional genomics is a new tool to study the interactions between dietary nutrients, environmental factors and cellular and genetic processes (Kaput and Rodriguez, 2004; Muller and Kersten, 2003; Stover, 2004). Advances in a number of scientific fields have allowed for parallel assessment of expression of thousands of messenger RNAs (mRNA) to be performed in a single experiment. Oligonucleotide production has improved for both yield and quality of gene probes. Robotics technology now allows the arraying of hundreds or thousands of PCR amplified complementary DNA (cDNA) clones or genes at high density on derivatized glass slides or chip substrates. Genetics research has provided identification of a large fraction of the total gene products present in mammalian systems. Mathematical aspects of computational biology have made analysis of data from large scale genomics research increasingly possible. These advances will accelerate discovery of genes responsible for various nutritionally-related diseases and syndromes (Barsh and Schwartz, 2002; German *et al.*, 2003).

A complete understanding of nutrition requires research in many facets of biology. Chemical analysis techniques must be performed to isolate and characterize the structures of essential nutrients. Biochemical and physiological experiments must be performed to determine nutrient metabolic and signaling pathways responsible for homeostasis in the organism of interest. In addition, research on inborn errors of metabolism or other genetic studies is used to identify gene-nutrient interactions. Examples of this include the human recessive defects galactosemia, in which defective lactose-1-phosphate uridyltransferase (GALT) causes galactose to accumulate in the blood, causing health problems including mental retardation, and phenylketonuria (PKU), a trait that results in accumulation of phenylalanine in the blood that can cause neurological damage (Kaput and Rodriguez, 2004). One limitation of this approach is that one is unable to sufficiently predict and quantify interactions among dietary components and polymorphic alleles. Microarray technology provides a high-throughput functional genomics approach to begin to provide a greater understanding of the complex and reciprocal interactions within the genome at the molecular level (Stover, 2004). Both genetic and dietary variables can be manipulated to design an effective nutritional functional genomics study to examine gene interactions affecting normal and diseased physiological states.

Developing a microarray

Advances in robotics and biotechnology in the past decade have made possible the fabrication of microarrays for expression screening of tens-of-thousands of genes in a single experiment (Harkin, 2000; Lipshutz *et al.*, 1999; Schena *et al.*, 1998; Schulze and Downward, 2001; Southern *et al.*, 1999; Watson *et al.*, 1998). Microarrays are thousands of DNA fragments (probes) precisely positioned at a high density on a solid support where they can act as molecular detectors to measure relative mRNA abundance between biological samples (Figure 1). Microarrays vary in the type of solid support used (glass slides or filters), the type of probe used (cDNA or oligonucleotides) and the manner in which the probe is deposited on the array (synthesized *in situ* or spotted).

Microarray overview: spotted DNA arrays

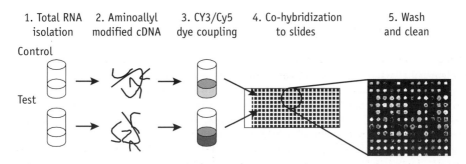

Figure 1. Spotted DNA Arrays. Plasmid clones are propagated in bacteria, and the cloned inserts are amplified by polymerase chain reaction (PCR) and then purified. These purified PCR products are robotically printed onto solid supports. Alternatively, oligonucleotides can be used instead of PCR products. Synthesized oligonucleotides are printed directly onto the glass support. Fluorescently labeled control and test samples are fluorescently labeled with Cy3 or Cy5 dyes and hybridized simultaneously onto the glass slide array. Following washes, hybridization is detected by phosphor imaging or measurement of the excitation of the two fluors at the appropriate wavelengths by a laser scanner.

The first-generation arrays were built by spotting cDNAs amplified from clone libraries onto nylon membranes (Harrington *et al.*, 2000; Schena *et al.*, 1995). A cDNA microarray is generated by amplifying gene specific sequences using polymerase chain reaction (PCR). This approach requires development of a clone library and thus is limited by the presence or absence of genes within the clone library. A clone library is generated by extracting DNA from a tissue or organism of interest, fragmenting the DNA using restriction enzymes, and then "cloning" the DNA fragments into plasmid vectors. Once inserted into the plasmid vector the sequence can be amplified by growing it in competent bacteria which are able to take up the plasmid. These competent bacteria also contain DNA coding for antibiotic resistance so that clone containing colonies can be selected in bacterial culture. Antibiotic resistant colonies now each contain different plasmids (and thus gene sequence fragments) and can be collected and sequenced to create a gene library. One limitation of cDNA microarrays is that cDNA clones and PCR amplicons must be tracked, which can lead to misidentification of up to 10-30% of clones (Wang *et al.*, 2003; Watson *et al.*, 1998). Nylon membrane arrays are limited in their spot density but development of spotting procedures for modified glass slides increased the maximal spot density and allowed addition of more genes probes to an array.

Oligonucleotide arrays are a more recent advance that allows designing of probes to ensure unique probe sequences, optimal melting temperature, and similar hybridization efficiency between probes (Figure 1). Control of these factors standardizes hybridization conditions from gene to gene to improve expression data. This approach is not limited by the genes present in clone libraries and requires only a gene sequence from a database. Although the probes used in oligonucleotide arrays are shorter in length than cDNA probes, they experience less problems with dimerization (probes annealing to each other), secondary structural formation (probes annealing to themselves), low hybridization efficiency, and cross-hybridization. Oligonucleotide microarrays can be composed of short oligonucleotides (25 bases) synthesized directly onto a solid support using photolithographic technology (Affymetrix arrays) (Figure 2) (Lipshutz *et al.*, 1999; Lockhart *et al.*, 1996) or constructed from long oligonucleotides (55-130 bases) spotted onto glass slides (Call *et al.*, 2001; Kane *et al.*, 2000; Zhao *et al.*, 2001). Large differences in expression data among genes have been reported across microarray platforms (cDNA, oligonucleotide, and Affymetrix arrays), but this variation may be due to probe sequence and annotation (Bloom *et al.*, 2004; Kuo *et al.*, 2002; Yuen *et al.*, 2002). However, data from one study directly comparing gene expression data from a spotted long (80mer) oligonucleotide array to those on an Affymetrix 25mer chip showed that for the majority of genes (92%), no significant effect of platform on gene expression was detected (Larkin *et al.*, 2004). Here, treatment effects were stronger than platform effects. Only 8% of genes showed divergent results between the two platforms, and this divergence may represent incorrect probe identification on the array, incorrect gene identification, or alternative splicing (Larkin *et al.*, 2004). Medium length (70mer) oligonucleotide probes spotted onto glass slides at high density are an excellent and cost-effective microarray platform increasingly used by researchers in many disciplines (Wang *et al.*, 2003).

Conducting a microarray experiment consists of four steps: 1) generation of the array, 2) RNA isolation and labeling with fluorescence dye, 3) hybridization of labeled sample to the array and measurement of resulting hybridization, and 4) analysis and interpretation of the data (Clarke *et al.*, 2004; Duggan *et al.*, 1999; Eisen and Brown, 1999; Hegde *et al.*, 2000) (Figures 1, 2).

The first step in developing a microarray is to generate a set of sequence-validated probes, each with a unique sequence that has little cross-hybridization to other probes. The gene set should

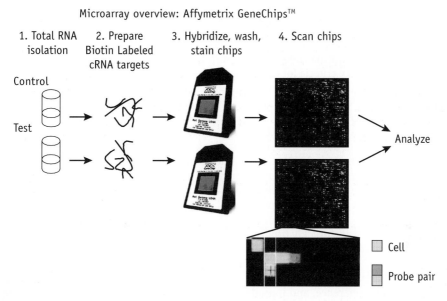

Microarray overview: Affymetrix GeneChips™

1. Total RNA isolation   2. Prepare Biotin Labeled cRNA targets   3. Hybridize, wash, stain chips   4. Scan chips

Control
Test
Analyze

Cell
Probe pair

Figure 2. Affymetrix Arrays. The GeneChip gene "probes" are synthesized in situ on the chip using photolithographic techniques. Gene probes are made up of 25mer oligonucleotides called perfect match probes (PM), and each is paired with a "mismatch" probe (MM) that contains a base pair substitution to interfere with hybridization between the probe and its "target", present in the sample. Test and control samples are labeled with biotin and hybridized to separate chips. The hybridized "target" probes are then stained with streptavidin phycoerythrin and scanned for light emitted at 570 nm, which indicates the sample's signal. Match and mismatch probe intensities can be used for expression data analysis (PM-MM), or match pairs can be compared between the control and test samples alone (PM only) (Seo et al., 2004).

contain a comprehensive representation of the expressed fraction of the genome or a collection of genes specifically tailored to an experimental system of interest (*i.e.* a set specific to a research field such as metabolism, or type of tissue such as the mammary gland). Following collection of the genes of interest, microarrays can be fabricated in a number of ways (Duggan *et al.*, 1999; Eisen and Brown, 1999; Hegde *et al.*, 2000; Lipshutz *et al.*, 1999). In general there are two main types of arrays: either the oligonucleotides are synthesized on the area *in situ* (Affymetrix arrays) or they are printed or deposited onto their solid supports.

Microarrays rely on a labeled representation of mRNA generated using reverse transcriptase to make a single-stranded cDNA. Following cDNA synthesis different colored fluorescent labels (Cy3 = red, and Cy5 = green) are attached to the aminoallyl modified cDNA targets in the experiment (control or treated) and hybridized to the array. Labeled Cy3 and Cy5 fluoresce at different wavelengths and are detected using a laser scanner. The signal presented by each fluor indicates relative mRNA abundance. To analyze expression, data spots must be correctly identified and aligned to a grid specifying spot location. Background and signal intensities are calculated based on the expected position, size and shape of each spot. To standardize hybridization across an array, normalization procedures are applied to analysis. Normalization procedures may make use of genes that are known to be unaffected by treatments and remain constant in their expression (housekeeping genes), or the average spot intensity across the entire slide, based on the assumption that the majority of genes are not modified between treatments (Wang *et al.*, 2003). A local weighted linear regression procedure (lowess) is considered the most robust normalization and is often used for data normalization (Cleveland, 1979). Fold change cutoffs (*i.e.* greater than 1.5 fold) are a rudimentary criteria for detecting differentially expressed genes still used by some researchers (Quackenbush, 2002); (Cleveland, 1979). More recently, linear and mixed models allow analysis of variance and determination of statistically significant genes (*i.e.* $P<0.05$) (Baldi and Long, 2001; Long *et al.*, 2001). In addition, many data visualization techniques have been developed for microarray data analysis including clustering and principle component analysis. These methods provide deductive insight especially in data across multiple treatment levels for time sequences (German *et al.*, 2003). Known information on gene activity and relationship to other genes should be used to analyze microarray data in a biological context (Quackenbush, 2002). Recent approaches include determination of biochemical and cell signaling pathways that are coordinately regulated between experimental conditions. Bayesian approaches also allow integration of previously described biological data. Statistical and bioinformatics applications to microarray data analysis is a quickly progressing field and the interested reader is directed to the many reviews of this topic (Baldi and Long, 2001; Churchill, 2004; Long *et al.*, 2001).

A number of different designs may be employed in microarray experiments, including direct comparison, dye-flip, reference, and loop designs (Churchill, 2002; Kerr and Churchill, 2001a, b; Woo *et al.*, 2004) (Figure 3). Often the type of comparison being made in the study as well as availability of biological replicates and funding will determine the type of design used. Although biological replicates (an increased number of individual animals included for cDNA comparisons) are preferred to technical replicates (repeated sampling of a single or pooled sample over a number of different microarrays), the cost of each sample will often determine the number of samples analyzed. Dye-flip experiments, in which two samples are compared on two different microarrays with the labeling dye reversed for each microarray, are preferred to account for preferential binding of either the Cy3 or Cy5 toward given genes. Dye-flips, however, are technical replicates, not biological replicates (Oleksiak *et al.*, 2002).

Nutritional genomics and animal production diseases

Discovery of specific genes responsible for disease phenotype will enable improved prediction of disease risk and development of novel methods to prevent and treat diseases specific to an animal's genotype. Diet has been identified as a causative agent for a number of diseases in dairy cattle including ketosis, hepatic lipidosis, ruminal acidosis, laminitis, and periparturient paresis. Each of these diseases is incompletely understood, and prevention solely through careful formulation of diets specific to the metabolic status of the animal is difficult. The precise molecular mechanisms causing diseased phenotypes are not clear. It is highly likely that some diet-regulated genes, including their normal, common splice-variants (genes with DNA sequence rearrangements), play a part in dictating the onset, incidence, progression and severity of each disease. One major factor determining disease susceptibility lies in individual genetic makeup, as many animals in the group will be phenotypically normal with others failing in health. It is possible, however, that dietary intervention based on knowledge of nutritional requirement, nutritional status and genotype can be used to prevent or mitigate these diseases by improving group diet formulation.

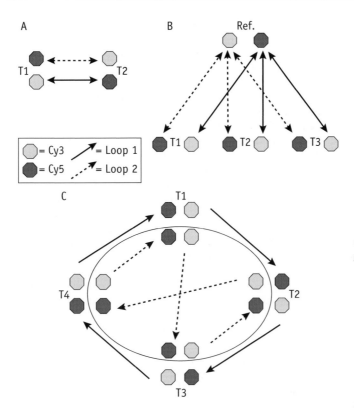

Figure 3. Microarray experimental designs for two-color arrays. A) Direct comparison with dye-flip; B) Reference design for three treatments with dye-flip; C) Augmented loop design for four treatments. All Affymetrix GeneChips are scanned individually based on a one color system and all chips can be directly compared to each other.

Dietary factors influence gene expression

Multiple mechanisms exist by which dietary factors can elicit cellular responses in biological systems (Figure 4). Nutrients may directly modify gene expression as ligands for transcription factors or stimulate signal transduction pathways (Francis *et al.*, 2003; Kaput and Rodriguez, 2004; Lu *et al.*, 2001; Muller and Kersten, 2003). Nutrients may also indirectly modify gene expression via their metabolites or through the modification of other cellular pathways (Kaput and Rodriguez, 2004).

One example of a dietary factor acting as a transcription factor ligand (Figure 4A) is illustrated in fatty acid metabolism, which is known to be regulated by the peroxisome proliferator-activated receptor (PPAR) family of transcription factors. The fatty acids palmitic, linoleic and arachidonic acid and eicosinoids have been shown to be ligands for the nuclear receptors of the PPARs, which act as fatty acid sensors (Adhami *et al.*, 2004). (Auwerx, 1992; Forman *et al.*, 1995; Kliewer *et al.*,

2001) The PPAR lipid sensors heterodimerize with the retinoic acid receptor (RXR), and RXR expression can be regulated by dietary retinol (Dauncey *et al.*, 2001). In a similar manner, the sterol regulatory element binding proteins (SREBP) are activated by protease cleavage and are regulated by low oxysterol concentrations and changes in insulin/glucose and poly unsaturated fatty acids (Edwards *et al.*, 2000), and the carbohydrate responsive element binding proteins (ChREBP) are a proteins activated in response to high glucose and regulated by phosphorylation events (Kaput and Rodriguez, 2004; Uyeda *et al.*, 2002).

An example of the metabolic conversion of dietary chemicals altering gene expression (Figure 4B) is seen in steroid hormone biosynthesis. Concentrations of steroid hormones derived from cholesterol are regulated by 10 steps in their biosynthetic pathway. The various intermediates produced along this pathway are also available to branch into other metabolic pathways[1] (Nobel *et al.*, 2001; Pegorier *et al.*, 2004). Specific combinations of alleles regulating the enzymatic steps in these assorted pathways will regulate the concentration of any given ligand branching off of steroid biosynthesis.

Dietary chemicals and nutrients have been linked to effects on numerous established signal transduction pathways (Figure 4C). These include polyphenols (PI3K/Akt pathway) (Adhami *et al.*, 2004), resveratrol (MAPK pathway) (Bachrach *et al.*, 2001), phenethyl isothiocyanate (NF-kappa B pathway) (Jeong *et al.*, 2004), genistein (NF-kappa B and PI3K/Akt pathways) (Sarkar and Li, 2003), and retinoids (PI3K/Akt pathway) (Niles, 2004).

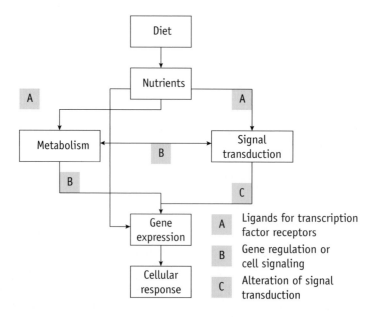

Figure 4. Nutritional genomics model. Nutrients can act directly as ligands for transcription factor receptors (A); may be metabolized by primary or secondary metabolic pathways to alter concentrations of substrates or intermediates involved in gene regulation or cell signaling (B); or might alter signal transduction pathways and signaling (C).

Nutrition and genome interactions

Selection pressures exerted on the bovine species in the processes of gene mutation, purifying selection and random drift as well as human selection have given rise to the primary sequence of the bovine genome as well as the genetic variation that exists today within the species (Duret and Mouchiroud, 2000). One of the most persistent and variable environmental pressures working to shape an organism's genome is nutrition, which can contribute greatly to genetic variation (Fenech, 2002; Stover, 2004) Nutrients and environmental factors can heavily influence the development and viability of a fetus and a growing portion of literature indicate that gene imprinting is an important factor in the penetrance of deleterious genetic flaws. The ability of dietary components to affect postnatal growth and gene imprinting shows that nutrition is an important *in utero* selection pressure that may contribute to fixation of altered phenotypes and genetic defects in populations (Boland *et al.*, 2001; Khosla *et al.*, 2001) Genomic responses to nutritional insults can work at the level of DNA transcription, mRNA translation, and protein and mRNA stability to maintain homeostasis (Kaput and Rodriguez, 2004). Maternal nutritional status can alter the epigenetic state of the fetal genome and imprint gene expression levels with lifelong consequences (Waterland and Garza, 1999). These epigenetic alterations of the fetal genome do not alter the primary DNA sequence, but affect gene expression. This could explain the subtle differences observed in monozygotic twins (Dennis, 2003). Methylation of DNA has been shown to be a major factor in regulating gene expression. These methylation patterns are established early in life and remain stable throughout life (Ho *et al.*, 1995; Khosla *et al.*, 2001). Cells can use thus genomic adaptation to regulate nutrient transport and nutrient status, alter nutrient storage capacity, alter the flux of intermediates through metabolic branch points, and restructure the cellular transcriptome and proteome to initiate cellular differentiation, proliferation or apoptosis.

Cattle metabolic disease and nutritional genomics

Similar limitations exist for assessing the regulation of individual or multiple genes by dietary factors in cattle. One must first separate cause from effect for each gene to evaluate which genes are responsible for a given phenotype. Next, one must determine if expression patterns for one strain or genotype are particular to that genotype. Different breeds of cattle may indeed have unique patterns of gene expression based on diet and genotype, based on inbred mouse studies (Frankel, 1995; Kaput and Rodriguez, 2004; Linder, 2001). This could also be used to help explain differences in disease susceptibility seen in different breeds of cattle. Increased susceptibility to copper toxicity and periparturient paresis by the Jersey breed, relative to the Holstein would be an example of this difference (Du *et al.*, 1996; Mangelsdorf *et al.*, 1995).

Directed dietary intervention to treat or prevent diseases can be a great challenge. It is difficult to determine the precise way to confront a disease state when multiple genes and environmental variables interact to cause the phenotype. Identifying the genes with the greatest contribution to the disease state must first be performed and then linked to their regulation by dietary variables. To use the example of hepatic lipidosis, a disorder which can arise in overconditioned lactating dairy cows in the immediate postpartum period, much is known about the etiology of the disease, but we do not have a true understanding of the genes responsible. A bovine specific microarray will facilitate the discovery of genes regulating the diseased condition in dairy cattle and foregoes the need to use model organism microarrays (*i.e.* mouse) to study these problems.

Development of the bovine metabolism (BMET) microarray

First generation microarrays employing extensive cDNA libraries have allowed high numbers of both known and unidentified genes to be surveyed. Many of these arrays have no or limited spot replication, impeding observation of within-plate variance and increasing the technical variance of expression observations (Black and Doerge, 2002). The Human Genome Project has provided extensively annotated databases, such as Entrez Gene, the Kyoto Encyclopedia of Genes and Genomes (KEGG), Gene Ontology (GO), The Institute for Genomics Research (TIGR), and BioCarta. These publicly available resources, paired with recent price reductions in oligonucleotide synthesis, allow researchers to feasibly design and produce microarrays with gene sets tailored to specific research areas.

To build a comprehensive metabolism directed gene database, a directed search intended to collect metabolic enzymes and signal transduction related genes was performed using internet database resources and nucleotide sequences gathered from United States Department of Agriculture sequencing projects. Genes were collected from several functional categories: metabolic enzymes, mitogen and mitogen binding proteins, growth factors and cytokines, intracellular signaling proteins, transcriptional control regulatory proteins, apoptotic regulators, and cell-cycle regulatory proteins. A gene list of human DNA sequences pertaining to metabolic genes and genes specific to pathways of interest were extracted from the Swiss-Prot Metabolic Pathway website[2], the Kyoto Encyclopedia of Genes and Genomes (KEGG)[3], and BioCarta[4] websites. The resulting genes from this search were annotated using NCBI Gene[5], Swiss-Prot[2], and Online Mendelian Inheritance in Man (OMIM)[6], along with their KEGG[3], BioCarta[4] and Gene Ontology[7] classifications. The human sequences from metabolism related genes were searched against bovine sequences in the GenBank database using the Basic Local Alignment Search Tool (BLAST). Those genes with an NCBI Reference Sequence (RefSeq), complete sequence, or 3'-end sequence were saved for oligonucleotide design. Next, we used keywords corresponding to categories of interest to search the Gene Ontology, UniGene and Swiss-Prot databases to gather human sequences of known genes and collected their bovine sequences in the same manner. This process generated a large list of genes corresponding to a UniGene cluster unique ID number with a specific accession number as an identifier (Figure 5). A cluster is a compilation of sequences of overlapping DNA that represent one gene[8]. Only those bovine genes with a RefSeq link, complete sequence, 3'-end sequencing, or strong TIGR[9]cluster match were used to facilitate oligonucleotide probe design. All Perl scripts written to generate this data are available upon request under an open source license.

From these databases 4,010 human RefSeq genes were identified. From the nucleotide sequences of these 4,010 genes 2,371 matching bovine sequence homologues with an expectation value (e-value) of $1e10^{-35}$ or smaller were found. The expectation value is a measurement of sequence similarity, and values below 0.01 tending to show strong sequence homology between two sequences being compared. A very low e-value score was chosen to ensure correct gene identification. This gene set included 513 GenBank RefSeq genes, 109 completely sequenced genes, and 901 3'-end United States Department of Agriculture sequenced genes and all genes had a TIGR cluster match. Of the 1207 KEGG metabolism genes extracted, greater than 1,100 are represented in this dataset.

Using the Oligopicker program (Wang and Seed, 2003) 70mer oligonucleotide probes were successfully designed from 2,349 of the genes and splice variants. Each oligonucleotide was

Bovine sequence extraction

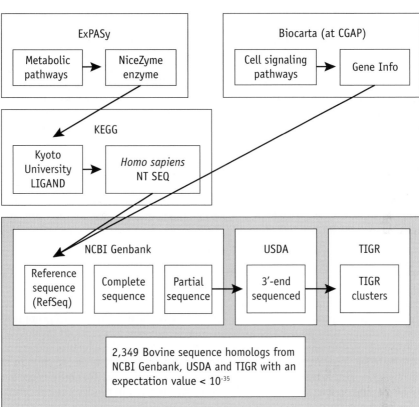

Figure 5. Metabolic gene database assembly. Using database resources such as ExPASy[2], the Kyoto Encyclopedia of Genes and Genomes[3], BioCarta[4], and NCBI GenBank[5] 2320 human genomic sequences were identified. Following a BLAST search against bovine sequences provided by NCBI, the United States Department of Agriculture, and The Institute for Genomic Research[9], 2349 bovine sequence homologues were detected at an expectation value better than 1e10[-35].

designed within specific parameters to standardize hybridization behavior. These 70mers were then synthesized at the Massachusetts General Hospital Oligonucleotide Synthesis Facility[10] and spotted four times per gene on microarray slides at the Massachusetts General Hospital Microarray Core Facility[11] on Poly-L-lysine coated glass slides. Replicate oligonucleotide spots were printed on each array in a global fashion (spot replicates are distributed across the slide to account for spatial variation in hybridization conditions) to ensure adequate spot replication for downstream data analysis. The BMET array contains 9,582 spots including housekeeping (including GAPDH and Beta-actin) and negative (sequences not found in the bovine genome) control genes (10 *Arabidopsis thaliana* metabolic genes) (Wang *et al.*, 2003).

As part of our effort to build a strong bovine metabolic genome resource, we are currently developing a web-based analytical tool to help with downstream analysis of the BMET array. Known human metabolic and cell regulatory pathways represented within KEGG and BioCarta will be adapted to this bovine-specific gene set using the GenMAPP[12] tool and a GeneLink resource (Figure 5). This will be housed at our home websites[13, 14].

Spot replication improves within-array quality control and increases the statistical power of accurately detecting small changes in expression at a lower cost than slide replication. Reduction in technical error to increase statistical power is especially important for metabolic research, in which changes in gene expression are often subtle and the cost per experimental unit is very high. In addition, our focus on only those genes that are relevant to metabolism increases the efficiency of downstream bioinformatics and data analysis for integration of metabolic gene networks. Because all genes included in this design are annotated with corresponding human homologs, the design can be applied to other species to promote our understanding of comparative metabolism. Our design of a focused oligonucleotide microarray with multiple spots per gene will facilitate research in the study of cattle metabolic disease genomics of cattle and can be easily applied to other species and disciplines.

Conclusions

The merging of nutritional and genomics approaches to the understanding of interactions between diet and genotype is revealing a system of great complexity. Nutritional genomics studies have given researchers the opportunity to target specific dietary interventions to avoid and treat metabolic diseases in humans and cattle. Mechanisms underlying the effects of dietary nutrients on genome stability, imprinting, expression and signaling may become more clear in the near future through the employment of targeted microarrays and experimental design to answer questions remaining on specific metabolic diseases encountered in modern animal production.

The design of a microarray containing a large fraction of the genome with known links to metabolism and signal transduction will increase the quality and utility of gene expression data that can be acquired from microarray technology and lead to advancements in the field of nutritionally related production animal diseases.

Acknowledgments

We would like to thank our collaborators Randy Baldwin, Rick Baumann, Kyle McLeod, and Dale Bauman for their help in funding the production of the BMET array, and the staff of the Michigan State University Center for Animal Functional Genomics, especially Peter Saama and Steven Suchyta, for their help in assembling the gene database.

Website References

[1] http://www.genome.jp/dbget-bin/show_pathway?hsa00100+3156; Kyoto Encyclopedia of Genes and Genomes, Steroid Biosynthesis Pathway.

[2] http://us.expasy.org/; The ExPASy (Expert Protein Analysis System) proteomics server of the Swiss Institute of Bioinformatics.

[3] http://www.genome.jp/kegg/pathway.html; Kyoto Encyclopedia of Genes and Genomes pathways website

[4] http://cgap.nci.nih.gov/Pathways/BioCarta; Pathways obtained from BioCarta via the Cancer Genome Anatomy Project.

[5] http://www.ncbi.nih.gov/entrez/query.fcgi?db=gene; Entrez Gene.

[6] http://www.ncbi.nlm.nih.gov/entrez/query.fcgi?db=OMIM; The Online Mendelian Inheritance in Man database, a catalog of human genes and genetic disorders.

[7] http://www.geneontology.org/; The Gene Ontology consortium website.

[8] http://www.ncbi.nlm.nih.gov/entrez/query.fcgi?db=unigene; NCBI's UniGene link. This system automatically partitions GenBank sequences into a non-redundant set of gene-oriented clusters.

[9] http://www.tigr.org/tigr-scripts/tgi/T_index.cgi?species=cattle; The Institute for Genome Research's bovine gene index.

[10] https://dnacore.mgh.harvard.edu/synthesis/index.shtml; Massachusetts General Hospital Oligonucleotide Synthesis Facility.

[11] https://dnacore.mgh.harvard.edu/microarray/index.shtml; Massachusetts General Hospital Microarray Core Facility.

[12] http://www.genmapp.org; GenMAPP is a free computer application designed to visualize gene expression data on maps representing biological pathways and groupings of genes.

[13] http://www.nutri-genomics.org/; Home of the BMET microarray database.

[14] http://nbfgc.msu.edu/; Michigan State University Center for Animal Functional Genomics database.

References

Adhami, V.M., I.A. Siddiqui, N. Ahmad, S. Gupta and H. Mukhtar, 2004. Oral consumption of green tea polyphenols inhibits insulin-like growth factor-I-induced signaling in an autochthonous mouse model of prostate cancer. Cancer Res 64, 8715-22.

Auwerx, J., 1992. Regulation of gene expression by fatty acids and fibric acid derivatives: an integrative role for peroxisome proliferator activated receptors. The Belgian Endocrine Society Lecture 1992. Horm Res 38, 269-77.

Bachrach, U., Y.C. Wang and A. Tabib, 2001. Polyamines: new cues in cellular signal transduction. News Physiol Sci 16, 106-9.

Baldi, P. and A.D. Long, 2001. A Bayesian framework for the analysis of microarray expression data: regularized t-test and statistical inferences of gene changes. Bioinformatics 17, 509-19.

Barsh, G.S. and M.W. Schwartz, 2002. Genetic approaches to studying energy balance: perception and integration. Nat Rev Genet 3, 589-600.

Black, M.A. and R.W. Doerge, 2002. Calculation of the minimum number of replicate spots required for detection of significant gene expression fold change in microarray experiments. Bioinformatics 18, 1609-16.

Bloom, G., I.V. Yang, D. Boulware, K.Y. Kwong, D. Coppola, S. Eschrich, J. Quackenbush and T.J. Yeatman, 2004. Multi-platform, multi-site, microarray-based human tumor classification. Am J Pathol 164, 9-16.

Boland, M. P., P. Lonergan, and D. O'Callaghan, 2001. Effect of nutrition on endocrine parameters, ovarian physiology, and oocyte and embryo development. Theriogenology 55, 1323-40.

Call, D.R., D.P. Chandler and F. Brockman, 2001. Fabrication of DNA microarrays using unmodified oligonucleotide probes. Biotechniques 30, 368-72, 374, 376 passim.

Churchill, G.A., 2002. Fundamentals of experimental design for cDNA microarrays. Nat Genet 32 Suppl, 490-5.

Churchill, G.A., 2004. Using ANOVA to analyze microarray data. Biotechniques 37, 173-5, 177.

Clarke, P.A., R. te Poele and P. Workman, 2004. Gene expression microarray technologies in the development of new therapeutic agents. Eur J Cancer 40, 2560-91.

Cleveland, W.S., 1979. Robust locally weighted regression and smoothing scatterplots. J Am Stat Assoc 74, 829-836.

Dauncey, M.J., P. White, K.A. Burton and M. Katsumata, 2001. Nutrition-hormone receptor-gene interactions: implications for development and disease. Proc Nutr Soc 60, 63-72.

Dennis, C., 2003. Epigenetics and disease: Altered states. Nature 421, 686-8.

Du, Z., R.W. Hemken and R.J. Harmon, 1996. Copper metabolism of Holstein and jersey cows and heifers fed diets high in cupric sulfate or copper proteinate. J Dairy Sci 79, 1873-80.

Duggan, D.J., M. Bittner, Y. Chen, P. Meltzer and J.M. Trent, 1999. Expression profiling using cDNA microarrays. Nat Genet 21, 10-4.

Duret, L. and D. Mouchiroud, 2000. Determinants of substitution rates in mammalian genes: expression pattern affects selection intensity but not mutation rate. Mol Biol Evol 17, 68-74.

Edwards, P.A., D. Tabor, H.R. Kast and A. Venkateswaran 2000. Regulation of gene expression by SREBP and SCAP. Biochim Biophys Acta 1529, 103-13.

Eisen, M.B. and P.O. Brown, 1999. DNA arrays for analysis of gene expression. Methods Enzymol 303, 179-205.

Fenech, M., 2002. Micronutrients and genomic stability: a new paradigm for recommended dietary allowances (RDAs). Food Chem Toxicol 40, 1113-7.

Forman, B.M., P. Tontonoz, J. Chen, R.P. Brun, B.M. Spiegelman, and R.M. Evans, 1995. 15-Deoxy-delta 12, 14-prostaglandin J2 is a ligand for the adipocyte determination factor PPAR gamma. Cell 83, 803-12.

Francis, G.A., E. Fayard, F. Picard and J. Auwerx, 2003. Nuclear receptors and the control of metabolism. Annu Rev Physiol 65, 261-311.

Frankel, W.N., 1995. Taking stock of complex trait genetics in mice. Trends Genet 11, 471-7.

German, J.B., M.A. Roberts and S.M. Watkins, 2003. Genomics and metabolomics as markers for the interaction of diet and health: lessons from lipids. J Nutr 133, 2078S-2083S.

Harkin, D.P., 2000. Uncovering functionally relevant signaling pathways using microarray-based expression profiling. Oncologist 5, 501-7.

Harrington, C.A., C. Rosenow and J. Retief, 2000. Monitoring gene expression using DNA microarrays. Curr Opin Microbiol 3, 285-91.

Hegde, P., R. Qi, K. Abernathy, C. Gay, S. Dharap, R. Gaspard, J.E. Hughes, E. Snesrud, N. Lee and J. Quackenbush, 2000. A concise guide to cDNA microarray analysis. Biotechniques 29, 548-50, 552-4, 556 passim.

Ho, Y., K. Wigglesworth, J.J. Eppig and R.M. Schultz, 1995. Preimplantation development of mouse embryos in KSOM: augmentation by amino acids and analysis of gene expression. Mol Reprod Dev 41, 232-8.

Jeong, W.S., I.W. Kim, R. Hu and A.N. Kong, 2004. Modulatory properties of various natural chemopreventive agents on the activation of NF-kappaB signaling pathway. Pharm Res 21, 661-70.

Kane, M.D., T.A. Jatkoe, C.R. Stumpf, J. Lu, J.D. Thomas and S.J. Madore, 2000. Assessment of the sensitivity and specificity of oligonucleotide (50mer) microarrays. Nucleic Acids Res 28, 4552-7.

Kaput, J. and R.L. Rodriguez, 2004. Nutritional genomics: the next frontier in the postgenomic era. Physiol Genomics 16, 166-77.

Kerr, M.K. and G.A. Churchill, 2001b. Experimental design for gene expression microarrays. Biostatistics 2, 183-201.

Kerr, M.K. and G.A. Churchill 2001a. Statistical design and the analysis of gene expression microarray data. Genet Res 77, 123-8.

Khosla, S., W. Dean, W. Reik and R. Feil, 2001. Culture of preimplantation embryos and its long-term effects on gene expression and phenotype. Hum Reprod Update 7, 419-27.

Kliewer, S.A., H.E. Xu, M.H. Lambert and T.M. Willson, 2001 Peroxisome proliferator-activated receptors: from genes to physiology. Recent Prog Horm Res 56, 239-63.

Kuo, W.P., T.K. Jenssen, A.J. Butte, L. Ohno-Machado and I.S. Kohane 2002. Analysis of matched mRNA measurements from two different microarray technologies. Bioinformatics 18, 405-12.

Larkin, J., H. Gavras and J. Quackenbush, 2004. Cardiac transcriptional response to acute and chronic Angiotensin II treatments. Advances in Genome Technology and Bioinformatics Workshop Presentation.

Linder, C.C., 2001. The influence of genetic background on spontaneous and genetically engineered mouse models of complex diseases. Lab Anim (NY) 30, 34-9.

Lipshutz, R.J., S.P. Fodor, T.R. Gingeras and D.J. Lockhart, 1999. High density synthetic oligonucleotide arrays. Nat Genet 21, 20-4.

Lockhart, D.J., H. Dong, M.C. Byrne, M.T. Follettie, M.V. Gallo, M.S. Chee, M. Mittmann, C. Wang, M. Kobayashi, H. Horton and E.L. Brown, 1996. Expression monitoring by hybridization to high-density oligonucleotide arrays. Nat Biotechnol 14, 1675-80.

Long, A.D., H.J. Mangalam, B.Y. Chan, L. Tolleri, G.W. Hatfield and P. Baldi, 2001. Improved statistical inference from DNA microarray data using analysis of variance and a Bayesian statistical framework. Analysis of global gene expression in Escherichia coli K12. J Biol Chem 276, 19937-44.

Lu, T.T., J.J. Repa and D.J. Mangelsdorf, 2001. Orphan nuclear receptors as eLiXiRs and FiXeRs of sterol metabolism. J Biol Chem 276, 37735-8.

Mangelsdorf, D.J., C. Thummel, M. Beato, P. Herrlich, G. Schutz, K. Umesono, B. Blumberg, P. Kastner, M. Mark, P. Chambon, and R.M. Evans, 1995. The nuclear receptor superfamily: the second decade. Cell 83, 835-9.

Muller, M. and S. Kersten, 2003. Nutrigenomics: goals and strategies. Nat Rev Genet 4, 315-22.

Niles, R.M., 2004. Signaling pathways in retinoid chemoprevention and treatment of cancer. Mutat Res 555, 81-96.

Nobel, S., L. Abrahmsen and U. Oppermann, 2001. Metabolic conversion as a pre-receptor control mechanism for lipophilic hormones. Eur J Biochem 268, 4113-25.

Oleksiak, M.F., G.A. Churchill and D.L. Crawford, 2002. Variation in gene expression within and among natural populations. Nat Genet 32, 261-6.

Pegorier, J.P., C. Le May and J. Girard, 2004. Control of gene expression by fatty acids. J Nutr 134, 2444S-2449S.

Quackenbush, J., 2002. Microarray data normalization and transformation. Nat Genet 32 Suppl, 496-501.

Sarkar, F.H. and Y. Li, 2003. Soy isoflavones and cancer prevention. Cancer Invest 21, 744-57.

Schena, M., R.A. Heller, T.P. Theriault, K. Konrad, E. Lachenmeier and R.W. Davis, 1998. Microarrays: biotechnology's discovery platform for functional genomics. Trends Biotechnol 16, 301-6.

Schena, M., D. Shalon, R.W. Davis and P.O. Brown, 1995. Quantitative monitoring of gene expression patterns with a complementary DNA microarray. Science 270, 467-70.

Schulze, A. and J. Downward, 2001. Navigating gene expression using microarrays--a technology review. Nat Cell Biol 3, E190-5.

Seo, J., M. Bakay, Y.W. Chen, S. Hilmer, B. Shneiderman and E.P. Hoffman, 2004. Interactively optimizing signal-to-noise ratios in expression profiling: project-specific algorithm selection and detection p-value weighting in Affymetrix microarrays. Bioinformatics 20, 2534-44.

Southern, E., K. Mir and M. Shchepinov, 1999. Molecular interactions on microarrays. Nat Genet 21, 5-9.

Stover, P.J., 2004. Nutritional genomics. Physiol Genomics 16, 161-5.

Uyeda, K., H. Yamashita and T. Kawaguchi, 2002. Carbohydrate responsive element-binding protein (ChREBP): a key regulator of glucose metabolism and fat storage. Biochem Pharmacol 63, 2075-80.

Section M.

Wang, H.Y., R.L. Malek, A.E. Kwitek, A.S. Greene, T.V. Luu, B. Behbahani, B. Frank, J. Quackenbush and N.H. Lee, 2003. Assessing unmodified 70-mer oligonucleotide probe performance on glass-slide microarrays. Genome Biol 4, R5.

Wang, X. and B. Seed, 2003. Selection of oligonucleotide probes for protein coding sequences. Bioinformatics 19, 796-802.

Waterland, R.A. and C. Garza, 1999. Potential mechanisms of metabolic imprinting that lead to chronic disease. Am J Clin Nutr 69, 179-97.

Watson, A., A. Mazumder, M. Stewart and S. Balasubramanian, 1998. Technology for microarray analysis of gene expression. Curr Opin Biotechnol 9, 609-14.

Woo, Y., J. Affourtit, S. Daigle, A. Viale, K. Johnson, J. Naggert and G. Churchill, 2004. A comparison of cDNA, oligonucleotide, and Affymetrix GeneChip gene expression microarray platforms. J Biomol Tech 15, 276-84.

Yuen, T., E. Wurmbach, R.L. Pfeffer, B.J. Ebersole and S.C. Sealfon, 2002. Accuracy and calibration of commercial oligonucleotide and custom cDNA microarrays. Nucleic Acids Res 30, e48.

Zhao, X., S. Nampalli, A.J. Serino and S. Kumar, 2001. Immobilization of oligodeoxyribonucleotides with multiple anchors to microchips. Nucleic Acids Res 29, 955-9.

Meeting summaries

Meeting summary and synthesis: new scientific directions for production medicine research

J.K. Drackley
Department of Animal Sciences, University of Illinois, Urbana, IL 61801, USA

Introduction

My assignment for the closing address of the 12th International Conference on Production Diseases (ICPD) consisted of the following: 1) Recap the high points of the conference. 2) Point out where advances have been made. 3) Indicate what you feel are the most potentially fruitful areas for further investigation. While this certainly constituted a daunting task, my approach was to group presentations into several theme areas, and to summarize and synthesize general comments, rather than to go into details on individual presentations. By adopting this approach, I did not comment, and will not here, on every paper presented. Indeed, in these written comments I will not reference individual presentations made at the 12th ICPD. Failure to mention a specific paper or group of papers presented at this conference does not imply that I did not think it important. Interested readers are referred to the proceedings volume containing the abstracts presented at the meeting.

As an overall assessment of the 12th ICPD, I conclude that there is both good news and bad news. The bad news first: the problem of production diseases in livestock has not been solved. But now for the good news: we are justified in holding a 13th ICPD in another three years!

Theme 1: Development of improved diagnostic techniques

Several presentations described panels of tests to perform "metabolic profiles" or diagnostics, and their analysis and interpretation. While not a new topic, re-visiting metabolic profiling with proper attention to stage of lactation (far-off dry, close-up dry, early lactation) shows promise in providing useful information to clinicians and scientists in both research and field settings. Given the history of efforts in profiling and the inherent variability in the diagnostic measurements contained in them, metabolic profiles should be viewed as another tool to combine with herd level information and observations rather than providing the first line of attack in diagnosing problems or defining approaches to fix problems.

Based on information presented at the meeting, it seems that introducing some of the amino acids (of the indispensable group as well as glutamine and alanine) and components of the acute phase response (such as serum amyloid A, ceruloplasmin, zinc, *etc.*) or other "stress" or "immune" indicators might have potential to improve usefulness of profiling. Addition of such measurements may provide new tools to address the interactions of animal productivity and the animal's environment that likely form the basis for production diseases. New measurements would of course need to be evaluated in the context of field conditions across a variety of nutritional, management, and disease status situations.

Theme 2: "Something's going on before calving…"

Several papers provided evidence of prepartum changes in cows that would go on to develop production diseases after calving, such as displaced abomasum. Such data confirm other existing datasets as well as considerable speculation, and should help to develop better hypotheses on the etiology and causes of production disease as well as improve diagnostic tests. A major unresolved issue in practical management of production diseases is that even if we can identify cows at a high likelihood to develop conditions such as displaced abomasum, what do we do about it? Along with the uncertainty of etiology and cause is the uncertainty about how to prevent many of the production diseases, even if we know that they are imminent. Additional investigations in which predictive or predisposing factors are identified in large populations will be useful to advance the diagnoses, mechanisms, and causes of disease. If predictive criteria can be established, then experiments aimed to define protocols to manage the disease can and must be conducted.

Several of the papers presented here were consistent with the idea that psychological, physiological, and immune stressors interact to cause, trigger, or exacerbate production diseases. Papers dealt with factors in the environment (*e.g.*, stall comfort, electromagnetic radiation, immune challenge) that may predispose cows to production disease. Identification of such factors is useful both from a practical management and prevention standpoint, but also will be helpful in designing appropriate experimental models to address etiology, prevention, and treatment.

A common theme among many of the papers that I have grouped here is the potential involvement of cytokines and acute phase responses in the development of production disease. As argued in my opening address, this is a fertile area for investigation relative to production disease, and provides a plausible biological mechanism for Payne's thesis that the interaction of productivity demands with the animal's environment is the root cause of production disease.

Theme 3: The importance of "subclinical…"

Solid evidence was presented at this meeting for effects of several subclinical or subacute conditions (ketosis, hypocalcemia, metritis, and acidosis) in increasing associated disease. While some data in support were available previously, the current presentations strengthen the linkages and provide a unifying theme for primary factors that increase the likelihood of secondary production disease. Consequently, continued management and research efforts to minimize the occurrence of subclinical disease situations seem strongly justified.

Theme 4: Reproductive problems in dairy cows are linked to peripartal disease

While it has been assumed or speculated for some time that subsequent reproductive success is dependent on a smooth and "non-turbulent" transition period, data describing the specific mechanistic linkages have been sparse. During this meeting, data were presented that provide solid evidence for the detrimental effects of both clinical and subclinical metritis on subsequent reproduction. Moreover, subclinical metritis was shown to adversely affect milk production. Thus, it is likely that metritis can result from other production disease, but also might cause production disease. In either case, data indicate that even a subclinical degree of metritis will have pronounced negative impacts on animal well-being and producer profits.

Linkages to cytokine function and immune system dysfunction were strongly suggested by variables measured in these experiments. This seems a highly fertile (pun intended!) area for further investigation, and one that builds in the general area of interactions among the immune system and stress responses with subsequent production disease.

Theme 5: Descriptive physiology and biochemistry

Significant progress has been made in describing the basic biology underlying development of production diseases. At this meeting, fundamental processes of calcium and phosphorus metabolism, fatty liver development, and ruminal epithelial metabolism were reviewed and new data presented. Fundamental research into basic mechanisms of tissue and organ function, and their regulation in health and disease, continues to be an important component of the search to prevent or minimize production disease. Strategies then need to be developed to make use of the findings to decrease the impact of production disease on animal well-being and producer profitability. Here, information was presented on application of anionic salts to decrease incidence of hypocalcemia and its sequelae, use of propylene glycol to improve glucose status, and the potential use of glucagon to prevent or treat fatty liver. Development of improved therapeutic or preventative strategies to ameliorate production disease is of course one of the major desired outcomes of fundamental research on production disease, and it is encouraging to see at least some successes that have arisen from past research.

Theme 6: Nutrition, immunity, and disease

Another theme of the current meeting was on various aspects of nutritional impacts on immunity (and *vice versa*), as well as their interaction in development of disease. Discussion of these topics considered both theory and application. Research also was presented on relationships of specific long-chain fatty acids supplied postruminally, on amino acids such as tryptophan that may impact the immune system, and possible impacts of several feed additives.

It is important to continue to explore mechanistic relationships between nutrients and aspects of the immune system, and how they may relate to occurrence of production diseases. Nutrition remains critically important in its relationship with production disease; whether this relationship is causal or casual remains uncertain, but regardless it seems clear that nutrition is a critically important aspect in promoting a strong immune system that can withstand pathogenic challenges.

Theme 7: Genomics and production disease

A theme that was new at the 12th ICPD was the application of the "-omics" technologies to the study of production disease. Research presented here demonstrated that the promise of genomic and genetic technologies is beginning to be demonstrated in practice. For example, using DNA microarray techniques, researchers at Michigan State University have identified areas of defect or impaired function in neutrophils, and have postulated on the basis of microarray data that there may be novel functions or strategies of neutrophils during the peripartal period never before realized. These data offer new perspective to the well described suppression of immune system function during the periparturient period. As another example, our research group presented data demonstrating that patterns of gene expression in the liver are markedly altered during an induced ketosis during the early postpartum period, again pointing to interrelationships that have not been able to be examined by other methodologies. Use of these genomic techniques to determine mechanisms of pathogen infectivity was another area highlighted at this meeting, which appears to have enormous prospects to move the field of infectious disease forward.

While high-throughput 'omics technologies often have been criticized as glorified "fishing expeditions", we need to remember that the origins of all science, and all fields of science, is in observational studies. Functional genomics, proteomics, and metabolomics may in some cases be simply observational, but given the power to look into corners of biology never before accessible, or at least not so quickly and economically, the value of this effort should not be underestimated in its likelihood of leading to new testable hypotheses. The power of microarrays and other techniques in hypothesis-driven experiments represents another dimension in potential progress in combating production disease.

Challenges for the future and closing comments

In my closing comments, I return to the challenges laid down in my opening address. First, there is the need to continue to work for integrative disciplines, so that we "blur the lines" of traditional disciplines both within and between investigators. The nature of several themes of this meeting identified above makes the justification for this interdisciplinary approach obvious.

Second, scientists must continue to search for and develop appropriate models and measurements for environmental influences on animals that interact with productivity demands to cause production diseases. Appropriate models still are lacking to study displaced abomasum, retained placenta, metritis, and some of the emerging problems such as hemorrhagic bowel disease (dead gut syndrome), among others. Perhaps the fact that we cannot reliably produce such syndromes experimentally should not surprise us if we cannot prevent them in the field either.

Third, while the best new technologies that are developed in other areas of biological science need to be quickly embraced and applied to the study of production diseases wherever they can be of benefit, we must not neglect the old reliable techniques and disciplines that continue to provide clues into the big puzzle of production disease. This principle applies also to the need for those working in this area to have a thorough knowledge of the research literature, including that not accessible online by computer. In some cases, this may mean introducing a new generation of students to that wonderful tool called the library and its hard-bound journal volumes. One of my favorite sayings from a colleague, which I now share with my graduate students, is "Six months in the lab can save you an afternoon in the library." Too many of us continue to reinvent the wheel rather than continuing to build on past knowledge.

Finally, I strongly believe that basic biological scientists working on mechanistic aspects of production disease must forge linkages with on-farm professionals to develop credible hypotheses and relevant animal models. Likewise, epidemiological studies and large-scale survey or field sampling studies also can provide clues from which mechanistic hypotheses can be tested. Of course, the reverse is true too, in that those working at the field level can use the latest mechanistic theories and data to form hypotheses about which relationships and variables should be studied at the whole animal, whole farm, or whole industry levels.

A number of more practical challenges also are apparent for continued progress on production diseases. Among the most serious, and potentially threatening for continued progress, is the erosion of financial resources devoted to production diseases in livestock. The European continent has seen a huge reduction over the last three decades in the number of universities and research institutes conducting agricultural and animal health research. In the USA, similar trends are underway as the number of Land-Grant universities actively involved in animal research continues to dwindle. No effective counterpart to the National Institutes of Health exists to support research in animal health and well-being; the funds allocated to the United States Department of Agriculture National Research Initiative for competitive grants in the area of animal production in total are woefully small, much less the portion that might be available to those working in the area of production diseases. This trend becomes ever more dangerous as we seek to "embrace the new technologies" that may be more powerful but also are more expensive.

Perhaps as part of the shortage of resources, another problem that most of us are guilty of at some point or another is employing inadequate replication to test our hypotheses. Facilities for controlled studies, the number of animals on farms, and the cost of daily animal care may be such that too few animals are assigned to experiments. This is especially true of research with dairy cattle. Spending scarce resources on experiments that never have the chance of showing statistically meaningful data might best not be run at all.

Production disease is certainly not a new problem, and one that seems to have been refractory to most of the logic applied to date. As scientists hoping to move this field forward, it is imperative that we strive to "think out of the box" for creative new hypotheses and approaches. This may involve bringing in scientists from other disciplines and backgrounds, retraining via sabbaticals or other methods, and continued or enhanced interaction through meetings such as the ICPD. A personal credo of mine is to constantly seek to "challenge dogma" in all that we do. There are countless examples of practices or principles that are taken for granted with no understanding of how they arose. More importantly, there are a surprising number of these areas of dogma that have arisen with amazingly little, if any, solid data in support. Even where dogma is based on solid past research and experience, changing management systems in the industry and changing animal genetics should cause us to step back occasionally and examine critically why we do things the way we do.

In closing, the 12th ICPD, my first, was an enjoyable and informative conference. Those involved with organization of future conferences must work to keep it viable by recruiting new scientists and clinicians in the area, and continuing to develop cutting-edge programs. I offer my sincere thanks to the ICPD organizing committee for the invitation and chance to be involved.

Meeting summary and synthesis: practical applications and new directions for applied production medicine research

Walter M. Guterbock
Sandy Ridge Dairy, Scotts, MI, USA

In the press of everyday academic life, it is well to stop occasionally and remember what the goals of production medicine or production-oriented research are. They are to help our industries produce high-quality proteins and fibers of animal origin while assuring the well-being and good health of animals and the care of the environment. Animal production is necessary in the world economy to enable us to produce high-quality food from materials that humans cannot or will not eat, many of which are byproducts of producing human food. It is also important that we help producers prosper, since without adequate returns to capital and labor production cannot be sustained.

It is important that we be proactive in defending animal production against its critics. It is not incompatible with good management of the environment. On the contrary, it is essential for the preservation of many grassland ecosystems. Nor is it incompatible with animal welfare. In fact, one of the aims of our science is to maximize animals' health and wellbeing.

All animals face predictable times of increased stress. One of the aims of good management is to prepare for those times and minimize their ill effects on the animals. For the dairy cow, that time is the transition from the dry period to lactation. For the dairy calf, those times are birth and weaning. For the baby pig, it is weaning. For a beef calf, it is the time when he is commingled with cattle from other sources, transported, and sent to the feed yard.

Producers must deal with other stressful times that are less predictable. These might be times of inclement weather, of temporary overcrowding, or of poor forage caused by climatic conditions. Poor ventilation of indoor facilities and life on concrete are stressors that are built into many production facilities. Poorly designed or maintained facilities that reduce comfort also put stress on animals, as do human errors of omission or commission. Good husbandry consists in preventing these stressors or minimizing their effects.

I have identified several main themes underlying the papers that were given at this Congress.

The first is the interaction between the animal's immune status and its metabolism. The production of immune mediators in chicks affects their growth rates, and the production of acute phase proteins by dairy cows may affect their ability to export triglycerides from the liver. Endometritis is associated with the increase in acute phase proteins in fresh cows. Negative energy balance is associated with uterine infection in dairy cattle and metabolic acidosis can affect neutrophil function.

The second theme is the association between nutritional or metabolic state and mastitis in dairy cows.

The third theme is the use of science to find alternatives to the use of antibiotics in food producing animals. Both immunological and nutritional strategies can be used to reduce the populations of pathogens in target animals.

The fourth theme was the use of genomics to explain the mechanisms of stressor effects on metabolism and immunity.

On the whole, I could see in the conference the beginnings of a scientific integration of stress effects, nutrition, metabolism, genetics, and immunity. I think that further understanding of these interactions will lead us to some very powerful tools in improving the lives and productivity of domestic animals.

In the United States, government priorities for spending on animal science research currently center on food safety, and food security. As I mentioned in my opening remarks, this is congruent with the desire of producers to protect the markets for their products. However it also results in a misapplication of resources, in my opinion, the latest example being the scare over bovine spongiform encephalopathy, which really represents an infinitesimal risk to American consumers. There is also currently increased interest in foreign animal diseases and the prevention of their use in bioterrorism. Animal welfare research and environmental protection, specifically manure disposal, are also priority subjects. Large amounts of money are now going to be thrown at a nationwide system for animal identification. Unfortunately studies that address the efficiency of production or more mundane problems on the farm are unlikely to find funding either from government or from producer funds unless the proposals can be couched in such a way as to make a connection with the priority issues.

The main research priorities of producer groups are to protect markets, assure product safety, reduce unnecessary loss of animals, and to either forestall or influence environmental regulations and help producers develop economically feasible ways to comply with the regulations.

Researchers like to investigate the mechanisms of phenomena at the molecular and cellular level. They want to be able to use the latest techniques of molecular biology and biochemistry in their laboratories. They need to find and recruit students who may not have backgrounds in production agriculture. They need to be able to maintain programs and laboratories and publish papers in order to get promoted.

These apparently contradictory goals can be married. Increased animal health and welfare decrease death loss and culling and increase animal welfare. Maximal animal welfare is consistent with optimal production efficiency. Healthy animals carry fewer pathogens, require fewer antibiotics, and produce more wholesome food than sick ones. Reduced losses lead to a decreased cost of production and more viable livestock enterprises.

In the rich countries people want to feel good about the food they eat. They want to be sure that it is free of contamination with microbes or drug residues. They like animals and do not want them to suffer in order to produce food. They value environmental quality and do not want their land or their waters polluted with animal waste. Well-conceived research can help animal industries meet these desires. The big goals for all are the same: increased animal well-being, protection of the environment, viable livestock enterprises, maintenance of the rural landscape, and the production of abundant, affordable, high-quality food. Properly managed intensive animal agriculture is part of a sustainable system in which manure is used to fertilize crops that are fed back to the animals. It is not bad for the environment.

I urge those in the research community to maintain close ties to their producers, and to try to make the connections between fundamental and applied research and between the needs of livestock industries and the projects in the laboratory.

Author index

Abe, R.	265, 312
Ackerman, C.E.	263
Agenäs, S.	256
Ali, O.	289
Allen, M.S.	270, 313, 316
Ametaj, B.N.	154, 155
Arai, T.	106
Archambault, M.	140
Balog, J.M.	55
Beedle, P.	52
Beitz, D.C.	154, 155
Bekele, H.	156
Bennett, T.B.	64, 66
Berning, R.	255
Bertoni, G.	157
Blum, J.W.	53, 58, 110, 120, 121, 122, 267, 314, 315
Bobe, G.	154, 155
Bolin, S.	91
Boven, M.R. van	137
Brinkmann, I.	284
Buckham, K.R.	292, 310
Buist, W.	138
Burton, J.L.	292, 310, 311
Busby, W.D.	52
Buschmann, F.	282
Carron, J.	53, 314
Catana, N.	161
Cattin, I.	58
Chang, L.-C.	292, 310, 311
Choothesa, Apassara	159
Christen, S.	58
Coenen, M.	255
Cole, J.B.	309
Cook, N.B.	64, 66
Corah, L.R.	52
Corato, A.	244
Costello, C.E.	92
Coussens, P.M.	316
Cox, E.	61
Crandall, K.D.	263, 266
Cui, D.	71
Cui, Z.	71
Dann, H.M.	308
Delling, U.	156
Dhiman, T.R.	104
Dividich, J. Le	179
Doherr, M.G.	110, 121, 122, 267
Donovan, D.	57
Dopfer, D.	137, 138
Dorp, R. van	136, 310
Drackley, J. K.	24, 308, 332
Drillich, M.	56
Dubosson, C.	110
Duehlmeier, R.	255
Duffield, T.F.	43, 44, 126, 140
Dunkley, Kingsley	90
Ekelund, A.	216
Engel, B.	138
Erb, H.N.	116
Erminio, Trevisi	157
Etchebarne, B.E.	313, 316
Etschmann, B.	282, 286, 288
Everts, R.E.	308
Evjen, I. M.	104
Floc'h, Nathalie Le	103
Fontenot, J.P.	240
Forderung, D.	56
Frajblat, M.	116
Froschauer, E.	288
Fürll, M.	156, 223
Gao, G.	71
Geue, L.	138
Gilbert, R.O.	92, 116
Goddeeris, B.M.	61
Goeke, C.L.	68
Goff, J.P.	104, 182, 215, 260, 279
Gooijer, L.	44
Grace, N.D.	238
Graham, M.H.	68
Grooms, D.	91
Grummer, R. R.	141
Guard, C.L.	116
Guterbock, W.M.	35, 91, 263, 266, 337
Hakamada, R.	42
Hammon, D. S.	104
Hammon, H.M.	53, 120, 314, 315
Han, B.	71
Han, Bo	163
Han, H.R.	163
Hansen, S.A.	180
Harmon, Bud G.	178

Harvatine, K.J.	316
Hayashi, T.	265, 312
Hayes, G.R.	92
Heath, M.F.	256
Heeres-van der Tol, J.J.	69
Heipertz, S.	288
Hendrick, S.H.	140
Herdt, T.H.	19, 45
Herman, V.	161
Heutink, L.F.M.	69
Heuwieser, W.	56, 114
Hillman, D.	68
Hiss, S.	59, 107, 264
Hoeltershinken, M.	255
Hogan, J.S.	262
Höhling, A.	220
Holsten, N.	220
Holtenius, K.	216, 280
Höltershinken, M.	220
Homfeld, E.	110, 121, 122, 267
Hoops, M.	223
Horst, R.L.	215, 279
Hoshi, H.	42
Huërou-Luron, I. Le	179
Huff, G.R.	55
Huff, W.E.	55
Huguet, A.	179
Hultgren, J.	112
Hunter, K.H.	240
Inoue, A.	106
Jaakson, H.	46, 118, 119
Jäckel, L.	156
Janssens, G.P.J.	61
Jong, M.C.M. de	137
Joshi, N.P.	45
Kaart, T.	46, 118, 119
Kaldmae, H.	46
Kaneene, J.	91
Kärt, O.	46
Kaske, M.	102, 255
Katsuda, K.	265, 312
Kawamura, S.	42, 265, 312
Kehler, H.W.	102, 255
Kelton, D.F.	136, 140
Kim, D.	163
Kimura, K.	260
Kimura, N.	106
Kiupel, Matti	74
Klasing, Kirk C.	50
Knowles, S.O.	238
Koehler, K.J.	155
Kohiruimaki, M.	265, 312
Korczack, B.	315
Krause, K.M.	281
Kuhnert, P.	110
Kulkarni, G.B.	123
Larson, R.L.	51
Lawrence, B.V.	180
LeBlanc, S.	43, 44
Lee, C.W.	163
Leesmäe, A.	118, 119
Leslie, K.E.	43, 44, 140
Lewin, H.A.	308
Liesegang, A.	218
Ling, K.	46, 118, 119
Lissemore, K.	140
Loor, J.J.	308
Lopez, John	166
Love, B.C.	268
Lucas, J.J.	92
Luigi, Calamari	157
Madsen, Sally A.	292
Martens, H.	282, 284, 286, 288, 289
Massimo, Bionaz	157
Mathson, K.	68
McCluskey, B.J.	215
McGuire, M.A.	244
McWhinney, D. R.	90
Meiring, R.W.	117
Miller, B.	109
Miller, P.D.	309
Millet, S.	61
Morel, C.	53, 314
Morin, D.E.	308
Mosley, E.E.	244
Mosley, S.A.	244
Müller, U.	264
Nafikov, R.A.	154, 155
Neill, John D.	307
Neu-Zahren, A.	264
Neuder, L.	45
Neuder, Laura E.	292
Nixon, R.M.	256
Nobis, W.	313, 316
Nordlund, K.V.	64, 66
Norman, H.D.	309

Nudda, A.	244	Scholz, H.	220
Oba, M.	270	Schubert, H.-J.	102
Odensten, M.	280	Schweigel, M.	282, 284, 288
Oetzel, G.R.	281	Scott, F.	109
Ohtsuka, H.	265, 312	Sears, P.M.	263, 266
Ontsouka, E.C.	315	Sève, Bernard	103
Otsuka, M.	42	Shanklin, R.K.	240
Overend, D.	180	Shaw, S.G.	58
Paemel, M. Van	61	Shibaro, K.	42
Pandiri, A.R.	123	Shin, S.T.	116
Park, B.K.	163	Singh, B.N.	92
Park, H.S.	282	Smith, K.L.	262
Pascu, C.	161	Sommer, U.	92
Pehrson, B.G.	236, 241	Spears, Jerry W.	226
Peters, T.	109	Spörndly, R.	216
Pfeffer, Ernst	188	Stetzer, D.	68
Phillips, C.J.C.	256	Stika, J.F.	52
Pickworth, C.L.	240	Stockhofe, N.	69
Portmann, O.	255	Strohbehn, D.	52
Rajala-Schultz, P.J.	117, 262, 268	Stumpff, F.	284
Raphael, W.	91	Suplie, A.	286
Rath, N.C.	55	Swecker, W.S.	240
Reenen, C.G. van	69	Takeguchi, A.	106
Rehage, J.	102, 255	Tenhagen, B.-A.	114
Rérat, M.	120	Todd, Amy	221
Ridpath, Julia F.	307	Toelboell, Trine	311
Risley, Chad R.	166	Urabe, S.	106
Risteli, Juha	218	VandeHaar, M.J.	313, 316
Röchert, D.	156	VanHorn, H.H.	68
Rodriguez-Zas, S.L.	308	Varga, Gabriella	221
Roesch, M.	110, 121, 122, 267	Vázquez-Añón, M.	109
Roman, H.	116	Voelker, J.A.	270
Ruebesam, C.	114	Voigt, D.	56
Rukkwamsuk, T.	159	Vuuren, A.M. van	69
Rungruagn, Sunthorn	159	Waldmann, A.	118
Sako, T.	106	Waldmann, Andres	119
Sallmann, H.-P.	255	Waller, K. Person	280
Samarütel, J.	46, 118, 119	Weber, P.S.D.	292, 310, 311
Sánchez, Jorge M.	182	Wegeler, C.	289
Saner, R.	120	Wensing, Theo	159
Sat, S.	42	Werner, A.	102
Sattler, T.	223	Wilcox, C.J.	68
Sauerwein, H.	59, 107, 264	Wilkinson, J.M.	256
Saun, Robert J. Van	41, 47, 221	Wittek, Th.	156
Scaglia, G.	240	Yancey, R.J.	93
Schaeren, W.	267	Yokomizo, Y.	312
Schällibaum, M.	267	Yoo, H.S.	163
Schmitz, S.	59, 107	Yoshimura, I.	106

Young, J.W.   154, 155
Zbinden, Y.   120
Zhang, H.M.   123